MATHEMATICS
for
MECHANICAL ENGINEERS

MATHEMATICS
for
MECHANICAL ENGINEERS

William F. Ames
George Cain
Y.L. Tong
W. Glenn Steele
Hugh W. Coleman
Richard L. Kautz
Dan M. Frangopol
Paul Norton

CRC Press
Taylor & Francis Group
Boca Raton London New York

CRC Press is an imprint of the
Taylor & Francis Group, an **informa** business

CRC Press
Taylor & Francis Group
6000 Broken Sound Parkway NW, Suite 300
Boca Raton, FL 33487-2742

First issued in paperback 2019

ISBN-13: 978-0-367-39916-0

Library of Congress Cataloging-in-Publication Data

Catalog record is available from the Library of Congress

Visit the Taylor & Francis Web site at
http://www.taylorandfrancis.com

and the CRC Press Web site at
http://www.crcpress.com

Preface

The purpose of *Mathematics for Mechanical Engineers* is to provide, in a single volume, a ready reference for the practicing engineer in industry, government, and academia, to the essential problem solving mathematical tools used everyday. It covers applications employed in many different facets of mechanical engineering, from basic through advanced.

For the engineer venturing out of familiar territory, the chapters cover fundamentals such as physical constants, derivatives, integrals, Fourier transforms, Bessel functions, and Legendre functions. For the experts, it includes thorough sections on the more advanced topics of partial differential equations, approximation methods, and numerical methods, often used in applications. Statistics for analyzing data and making inferences are presented, allowing the reader to extract useful information even with the presence of randomness and uncertainty.

Contributors

William F. Ames
Georgia Institute of Technology
Atlanta, Georgia

George Cain
Georgia Institute of Technology
Atlanta, Georgia

Hugh W. Coleman
University of Alabama
Huntsville, Alabama

Dan M. Frangopol
University of Colorado
Boulder, Colorado

Richard L. Kautz
National Institutes of Science and Technology
Boulder, Colorado

Frank Kreith, P.E.
Engineering Consultant
Boulder, Colorado

Paul Norton
National Renewable Energy Laboratory
Golden, Colorado

W. Glenn Steele
Mississippi State University
State College, Mississippi

Y. L. Tong
Georgia Institute of Technology
Atlanta, Georgia

Contents

Chapter 1 Tables *William F. Ames*

Chapter 2 Linear Algebra and Matrices *George Cain*

Chapter 3 Vector Algebra and Calculus *George Cain*

Chapter 4 Differential Equations *William F. Ames*

Chapter 5 Differential Equations *William F. Ames*

Chapter 6 Integral Equations *William F. Ames*

Chapter 7 Approximation Methods *William F. Ames*

Chapter 8 Integral Transforms *William F. Ames*

Chapter 9 Calculus of Variations *William F. Ames*

Chapter 10 Optimization Methods *George Cain*

Chapter 11 Engineering Statistics *Y. L. Tong*

Chapter 12 Numerical Methods *William F. Ames*

Chapter 13 Experimental Uncertainty Analysis *W.G. Steele* and *H.W. Coleman*

Chapter 14 Chaos *R. L. Kautz*

Chapter 15 Fuzzy Sets and Fuzzy Logic *Dan M. Frangopol*

Appendices

1
Tables

William F. Ames
Georgia Institute of Technology

1.1 Greek Alphabet

Greek Letter	Greek Name	English Equivalent	Greek Letter	Greek Name	English Equivalent
Α α	Alpha	a	Ν ν	Nu	n
Β β	Beta	b	Ξ ξ	Xi	x
Γ γ	Gamma	g	Ο ο	Omicron	o
Δ δ	Delta	d	Π π	Pi	p
Ε ε	Epsilon	e	Ρ ρ	Rho	r
Ζ ζ	Zeta	z	Σ σ ς	Sigma	s
Η η	Eta	e	Τ τ	Tau	t
Θ θ ϑ	Theta	th	Υ υ	Upsilon	u
Ι ι	Iota	i	Φ ϕ φ	Phi	ph
Κ κ	Kappa	k	Χ χ	Chi	ch
Λ λ	Lambda	l	Ψ ψ	Psi	ps
Μ μ	Mu	m	Ω ω	Omega	o

1.2 International System of Units (SI)

The International System of units (SI) was adopted by the 11th General Conference on Weights and Measures (CGPM) in 1960. It is a coherent system of units built from seven *SI base units*, one for each of the seven dimensionally independent base quantities: the meter, kilogram, second, ampere, kelvin, mole, and candela, for the dimensions length, mass, time, electric current, thermodynamic temperature,

0-8493-0056-8/00/$0.00+$.50

amount of substance, and luminous intensity, respectively. The definitions of the SI base units are given below. The *SI derived units* are expressed as products of powers of the base units, analogous to the corresponding relations between physical quantities but with numerical factors equal to unity.

In the International System there is only one SI unit for each physical quantity. This is either the appropriate SI base unit itself or the appropriate SI derived unit. However, any of the approved decimal prefixes, called *SI prefixes*, may be used to construct decimal multiples or submultiples of SI units.

It is recommended that only SI units be used in science and technology (with SI prefixes where appropriate). Where there are special reasons for making an exception to this rule, it is recommended always to define the units used in terms of SI units. This section is based on information supplied by IUPAC.

Definitions of SI Base Units

Meter: The meter is the length of path traveled by light in vacuum during a time interval of 1/299 792 458 of a second (17th CGPM, 1983).

Kilogram: The kilogram is the unit of mass; it is equal to the mass of the international prototype of the kilogram (3rd CGPM, 1901).

Second: The second is the duration of 9 192 631 770 periods of the radiation corresponding to the transition between the two hyperfine levels of the ground state of the cesium-133 atom (13th CGPM, 1967).

Ampere: The ampere is that constant current which, if maintained in two straight parallel conductors of infinite length, of negligible circular cross section, and placed 1 meter apart in vacuum, would produce between these conductors a force equal to 2×10^{-7} newton per meter of length (9th CGPM, 1958).

Kelvin: The kelvin, unit of thermodynamic temperature, is the fraction 1/273.16 of the thermodynamic temperature of the triple point of water (13th CGPM, 1967).

Mole: The mole is the amount of substance of a system which contains as many elementary entities as there are atoms in 0.012 kilogram of carbon-12. When the mole is used, the elementary entities must be specified and may be atoms, molecules, ions, electrons, or other particles, or specified groups of such particles (14th CGPM, 1971). Examples of the use of the mole:

- 1 mol of H_2 contains about 6.022×10^{23} H_2 molecules, or 12.044×10^{23} H atoms.
- 1 mol of HgCl has a mass of 236.04 g.
- 1 mol of Hg_2Cl_2 has a mass of 472.08 g.
- 1 mol of Hg_2^{2+} has a mass of 401.18 g and a charge of 192.97 kC.
- 1 mol of $Fe_{0.91}S$ has a mass of 82.88 g.
- 1 mol of e^- has a mass of 548.60 μg and a charge of −96.49 kC.
- 1 mol of photons whose frequency is 10^{14} Hz has energy of about 39.90 kJ.

Candela: The candela is the luminous intensity, in a given direction, of a source that emits monochromatic radiation of frequency 540×10^{12} Hz and that has a radiant intensity in that direction of (1/683) watt per steradian (16th CGPM, 1979).

Names and Symbols for the SI Base Units

Physical Quantity	Name of SI Unit	Symbol for SI Unit
Length	meter	m
Mass	kilogram	kg
Time	second	s
Electric current	ampere	A
Thermodynamic temperature	kelvin	K
Amount of substance	mole	mol
Luminous intensity	candela	cd

SI Derived Units with Special Names and Symbols

Physical Quantity	Name of SI Unit	Symbol for SI Unit	Expression in Terms of SI Base Units
Frequency[a]	hertz	Hz	s^{-1}
Force	newton	N	$m\ kg\ s^{-2}$
Pressure, stress	pascal	Pa	$N\ m^{-2} = m^{-1}\ kg\ s^{-2}$
Energy, work, heat	joule	J	$N\ m = m^2\ kg\ s^{-2}$
Power, radiant flux	watt	W	$J\ s^{-1} = m^2\ kg\ s^{-3}$
Electric charge	coulomb	C	$A\ s$
Electric potential, electromotive force	volt	V	$J\ C^{-1} = m^2\ kg\ s^{-3}\ A^{-1}$
Electric resistance	ohm	Ω	$V\ A^{-1} = m^2\ kg\ s^{-3}\ A^{-2}$
Electric conductance	siemens	S	$\Omega^{-1} = m^{-2}\ kg^{-1}\ s^4\ A^2$
Electric capacitance	farad	F	$C\ V^{-1} = m^{-2}\ kg^{-1}\ s^4\ A^2$
Magnetic flux density	tesla	T	$V\ s\ m^{-2} = kg\ s^{-2}\ A^{-1}$
Magnetic flux	weber	Wb	$V\ s = m^2\ kg\ s^{-2}\ A^{-1}$
Inductance	henry	H	$V\ A^{-1}\ s = m^2\ kg \cdot s^{-2}\ A^{-2}$
Celsius temperature[b]	degree Celsius	°C	K
Luminous flux	lumen	lm	$cd\ sr$
Illuminance	lux	lx	$cd\ sr\ m^{-2}$
Activity (radioactive)	becquerel	Bq	s^{-1}
Absorbed dose (or radiation)	gray	Gy	$J\ kg^{-1} = m^2\ s^{-2}$
Dose equivalent (dose equivalent index)	sievert	Sv	$J\ kg^{-1} = m^2\ s^{-2}$
Plane angle	radian	rad	$1 = m\ m^{-1}$
Solid angle	steradian	sr	$1 = m^2\ m^{-2}$

[a] For radial (circular) frequency and for angular velocity the unit rad s^{-1}, or simply s^{-1}, should be used, and this may not be simplified to Hz. The unit Hz should be used only for frequency in the sense of cycles per second.

[b] The Celsius temperature θ is defined by the equation

$$q/°C = T/K = 237.15$$

The SI unit of Celsius temperature interval is the degree Celsius, °C, which is equal to the kelvin, K. °C should be treated as a single symbol, with no space between the ° sign and the letter C. (The symbol °K, and the symbol °, should no longer be used.)

Units in Use Together with the SI

These units are not part of the SI, but it is recognized that they will continue to be used in appropriate contexts. SI prefixes may be attached to some of these units, such as milliliter, ml; millibar, mbar; megaelectronvolt, MeV; and kilotonne, kt.

Physical Quantity	Name of Unit	Symbol for Unit	Value in SI Units
Time	minute	min	60 s
Time	hour	h	3600 s
Time	day	d	86 400 s
Plane angle	degree	°	(π/180) rad
Plane angle	minute	′	(π/10 800) rad
Plane angle	second	″	(π/648 000) rad
Length	angstrom[a]	Å	10^{-10} m
Area	barn	b	$10^{-28}\ m^2$
Volume	liter	l, L	$dm^3 = 10^{-3}\ m^3$
Mass	tonne	t	$Mg = 10^3$ kg
Pressure	bar[a]	bar	$10^5\ Pa = 10^5\ N\ m^{-2}$
Energy	electronvolt[b]	eV (= $e \times V$)	≈ 1.60218×10^{-19} J
Mass	unified atomic mass unit[b,c]	u (= m_a(12C)/12)	≈ 1.66054×10^{-27} kg

continued

[a] The angstrom and the bar are approved by CIPM for "temporary use with SI units," until CIPM makes a further recommendation. However, they should not be introduced where they are not used at present.

[b] The values of these units in terms of the corresponding SI units are not exact, since they depend on the values of the physical constants e (for the electronvolt) and N_A (for the unified atomic mass unit), which are determined by experiment.

[c] The unified atomic mass unit is also sometimes called the dalton, with symbol Da, although the name and symbol have not been approved by CGPM.

1.3 Conversion Constants and Multipliers

Recommended Decimal Multiples and Submultiples

Multiple or Submultiple	Prefix	Symbol	Multiple or Submultiple	Prefix	Symbol
10^{18}	exa	E	10^{-1}	deci	d
10^{15}	peta	P	10^{-2}	centi	c
10^{12}	tera	T	10^{-3}	milli	m
10^{9}	giga	G	10^{-6}	micro	μ (Greek mu)
10^{6}	mega	M	10^{-9}	nano	n
10^{3}	kilo	k	10^{-12}	pico	p
10^{2}	hecto	h	10^{-15}	femto	f
10	deca	da	10^{-18}	atto	a

Conversion Factors — Metric to English

To Obtain	Multiply	By
Inches	Centimeters	0.393 700 787 4
Feet	Meters	3.280 839 895
Yards	Meters	1.093 613 298
Miles	Kilometers	0.621 371 192 2
Ounces	Grams	$3.527\ 396\ 195 \times 10^{-2}$
Pounds	Kilograms	2.204 622 622
Gallons (U.S. liquid)	Liters	0.264 172 052 4
Fluid ounces	Milliliters (cc)	$3.381\ 402\ 270 \times 10^{-2}$
Square inches	Square centimeters	0.155 000 310 0
Square feet	Square meters	10.763 910 42
Square yards	Square meters	1.195 990 046
Cubic inches	Milliliters (cc)	$6.102\ 374\ 409 \times 10^{-2}$
Cubic feet	Cubic meters	35.314 666 72
Cubic yards	Cubic meters	1.307 950 619

Conversion Factors — English to Metric

To Obtain	Multiply	By[a]
Microns	Mils	25.4
Centimeters	Inches	2.54
Meters	Feet	0.3048
Meters	Yards	0.9144
Kilometers	Miles	1.609 344
Grams	Ounces	28.349 523 13
Kilograms	Pounds	0.453 592 37
Liters	Gallons (U.S. liquid)	3.785 411 784

To Obtain	Multiply	By[a]
Millimeters (cc)	Fluid ounces	29.573 529 56
Square centimeters	Square inches	6.451 6
Square meters	Square feet	0.092 903 04
Square meters	Square yards	0.836 127 36
Milliliters (cc)	Cubic inches	16.387 064
Cubic meters	Cubic feet	$2.831\,684\,659 \times 10^{-2}$
Cubic meters	Cubic yards	0.764 554 858

[a] Boldface numbers are exact: others are given to ten significant figures where so indicated by the multiplier factor.

Conversion Factors — General

To Obtain	Multiply	By[a]
Atmospheres	Feet of water @ 4°C	2.950×10^{-2}
Atmospheres	Inches of mercury @ 0°C	3.342×10^{-2}
Atmospheres	Pounds per square inch	6.804×10^{-2}
Btu	Foot-pounds	1.285×10^{-3}
Btu	Joules	9.480×10^{-4}
Cubic feet	Cords	**128**
Degree (angle)	Radians	57.2958
Ergs	Foot-pounds	1.356×10^{-7}
Feet	Miles	**5280**
Feet of water @ 4°C	Atmospheres	33.90
Foot-pounds	Horsepower-hours	1.98×10^{6}
Foot-pounds	Kilowatt-hours	2.655×10^{6}
Foot-pounds per minute	Horsepower	3.3×10^{4}
Horsepower	Foot-pounds per second	1.818×10^{-3}
Inches of mercury @ 0°C	Pounds per square inch	2.036
Joules	Btu	1054.8
Joules	Foot-pounds	1.355 82
Kilowatts	Btu per minute	1.758×10^{-2}
Kilowatts	Foot-pounds per minute	2.26×10^{-5}
Kilowatts	Horsepower	0.745712
Knots	Miles per hour	0.868 976 24
Miles	Feet	1.894×10^{-4}
Nautical miles	Miles	0.868 976 24
Radians	Degrees	1.745×10^{-2}
Square feet	Acres	**43 560**
Watts	Btu per minute	17.5796

[a] Boldface numbers are exact: others are given to ten significant figures where so indicated by the multiplier factor.

Temperature Factors

$$°F = 9/5(°C) + 32$$

$$\text{Fahrenheit temperature} = 1.8(\text{temperature in kelvins}) - 459.67$$

$$°C = 5/9[(°F) - 32]$$

$$\text{Celsius temperature} = \text{temperature in kelvins} - 273.15$$

$$\text{Fahrenheit temperature} = 1.8(\text{Celsius temperature}) + 32$$

Conversion of Temperatures

From	To		From	To	
Fahrenheit	Celcius	$t_c = \dfrac{t_f - 32}{1.8}$	Celsius	Fahrenheit	$t_f = (t_c \times 1.8) + 32$
				Kelvin	$T_K = t_c + 273.15$
				Rankine	$T_R = (t_c + 273.15) \times 18$
	Kelvin	$T_K = \dfrac{t_f - 32}{1.8} + 273.15$	Kelvin	Celsius	$t_c = T_K - 273.15$
				Rankine	$T_R = T_K \times 1.8$
	Rankine	$T_R = t_f + 459.67$	Rankine	Fahrenheit	$t_f = T_R - 459.67$
				Kelvin	$T_K = \dfrac{T_R}{1.8}$

1.4 Physical Constants

General

Equatorial radius of the earth = 6378.388 km = 3963.34 miles (statute)
Polar radius of the earth = 6356.912 km = 3949.99 miles (statute)
1 degree of latitude at 40° = 69 miles
1 international nautical mile = 1.150 78 miles (statute) = 1852 m = 6076.115 ft
Mean density of the earth = 5.522 g/cm^3 = 344.7 lb/ft^3
Constant of gravitation (6.673 ± 0.003) × 10^{-8} cm^3 g^{-1} s^{-2}
Acceleration due to gravity at sea level, latitude 45° = 980.6194 cm/s^2 = 32.1726 ft/s^2
Length of seconds pendulum at sea level, latitude 45° = 99.3575 cm = 39.1171 in.
1 knot (international) = 101.269 ft/min = 1.6878 ft/s = 1.1508 miles (statute)/h
1 micron = 10^{-4} cm
1 angstrom = 10^{-8} cm
Mass of hydrogen atom = (1.673 39 ± 0.0031) × 10^{-24} g
Density of mercury at 0°C = 13.5955 g/mL
Density of water at 3.98°C = 1.000 000 g/mL
Density, maximum, of water, at 3.98°C = 0 999 973 g/cm^3
Density of dry air at 0°C, 760 mm = 1.2929 g/L
Velocity of sound in dry air at 0°C = 331.36 m/s - 1087.1 ft/s
Velocity of light in vacuum = (2.997 925 ± 0.000 002) × 10^{10} cm/s
Heat of fusion of water, 0°C = 79.71 cal/g
Heat of vaporization of water, 100°C = 539.55 cal/g
Electrochemical equivalent of silver 0.001 118 g/s international amp
Absolute wavelength of red cadmium light in air at 15°C, 760 mm pressure = 6438.4696 Å
Wavelength of orange-red line of krypton 86 = 6057.802 Å

p Constants

π = 3.14159 26535 89793 23846 26433 83279 50288 41971 69399 37511
$1/\pi$ = 0.31830 98861 83790 67153 77675 26745 02872 40689 19291 48091
π^2 = 9.8690 44010 89358 61883 44909 99876 15113 53136 99407 24079
$\log_e \pi$ = 1.14472 98858 49400 17414 34273 51353 05871 16472 94812 91531
$\log_{10} \pi$ = 0.49714 98726 94133 85435 12682 88290 89887 36516 78324 38044
$\log_{10} \sqrt{2\pi}$ = 0.39908 99341 79057 52478 25035 91507 69595 02099 34102 92128

Constants Involving *e*

e = 2.71828 18284 59045 23536 02874 71352 66249 77572 47093 69996
$1/e$ = 0.36787 94411 71442 32159 55237 70161 46086 74458 11131 03177
e^2 = 7.38905 60989 30650 22723 04274 60575 00781 31803 15570 55185
$M = \log_{10} e$ = 0.43429 44819 03251 82765 11289 18916 60508 22943 97005 80367
$1/M = \log_e 10$ = 2.30258 50929 94045 68401 79914 54684 36420 76011 01488 62877
$\log_{10} M$ = 9.63778 43113 00536 78912 29674 98645 - 10

Numerical Constants

$\sqrt{2}$ = 1.41421 35623 73095 04880 16887 24209 69807 85696 71875 37695
$\sqrt[3]{2}$ = 1.25992 10498 94873 16476 72106 07278 22835 05702 51464 70151
$\log_e 2$ = 0.69314 71805 59945 30941 72321 21458 17656 80755 00134 36026
$\log_{10} 2$ = 0.30102 99956 63981 19521 37388 94724 49302 67881 89881 46211
$\sqrt{3}$ = 1.73205 08075 68877 29352 74463 41505 87236 69428 05253 81039
$\sqrt[3]{3}$ = 1.44224 95703 07408 38232 16383 10780 10958 83918 69253 49935
$\log_e 3$ = 1.09861 22886 68109 69139 52452 36922 52570 46474 90557 82275
$\log_{10} 3$ = 0.47712 12547 19662 43729 50279 03255 11530 92001 28864 19070

1.5 Symbols and Terminology for Physical and Chemical Quantities

Name	Symbol	Definition	SI Unit
		Classical Mechanics	
Mass	m		kg
Reduced mass	μ	$\mu = m_1 m_2/(m_1 + m_2)$	kg
Density, mass density	ρ	$\rho = m/V$	kg m^{-3}
Relative density	d	$d = \rho/\rho^{\circ}$	1
Surface density	ρ_A, ρ_S	$\rho_A = m/A$	kg m^{-2}
Specific volume	v	$v = V/m = 1/\rho$	m^3 kg^{-1}
Momentum	p	$p = mv$	kg m s^{-1}
Angular momentum, action	L	$L = r \times p$	J s
Moment of inertia	I, J	$I = \Sigma m_i r_i^2$	kg m^2
Force	F	$F = d p/d t = m a$	N
Torque, moment of a force	T, (M)	$T = r \times F$	N m
Energy	E		J
Potential energy	E_p, V, Φ	$E_p = \int F \cdot ds$	J
Kinetic energy	E_k, T, K	$E_k = (1/2)\, mv^2$	J
Work	W, w	$W = \int F \cdot ds$	J
Hamilton function	H	$H(q,p) = T(q,p) + V(q)$	J
Lagrange function	L	$L(q,\dot{q}) = T(q,\dot{q}) - V(q)$	J
Pressure	p, P	$p = F/A$	Pa, N m^{-2}
Surface tension	γ, σ	$\gamma = dW/dA$	N m^{-1}, J m^{-2}
Weight	G (W, P)	$G = mg$	N
Gravitational constant	G	$F = Gm_1m_2/r^2$	N m^2 kg^{-2}
Normal stress	σ	$\sigma = F/A$	Pa
Shear stress	τ	$\tau = F/A$	Pa
Linear strain, relative elongation	ε, e	$\varepsilon = \Delta l/l$	1
Modulus of elasticity, Young's modulus	E	$E = \sigma/\varepsilon$	Pa
Shear strain	γ	$\gamma = \Delta x/d$	1
Shear modulus	G	$G = \tau/\gamma$	Pa
Volume strain, bulk strain	θ	$\theta = \Delta V/V_0$	1
Bulk modulus, compression modulus	K	$K = V_0(dp/dV)$	Pa
Viscosity, dynamic viscosity	η, μ	$\tau_{x,z} = \eta(dv_x/dz)$	Pa s
Fluidity	ϕ	$\phi = 1/\eta$	m kg^{-1} s
Kinematic viscosity	ν	$\nu = \eta/\rho$	m^2 s^{-1}
Friction coefficient	μ, (f)	$F_{frict} = \mu F_{norm}$	1
Power	P	$P = dW/dt$	W
Sound energy flux	P, P_a	$P = dE/dt$	W
Acoustic factors			
Reflection factor	ρ	$\rho = P_r/P_0$	1
Acoustic absorption factor	α_a, (α)	$\alpha_a = 1 - \rho$	1
Transmission factor	τ	$\tau = P_{tr}/P_0$	1
Dissipation factor	δ	$\delta = \alpha_a - \tau$	1

Name	Symbol	Definition	SI Unit
Electricity and Magnetism			
Quantity of electricity, electric range	Q		C
Charge density	ρ	$\rho = Q/V$	$C\ m^{-3}$
Surface charge density	σ	$\sigma = Q/A$	$C\ m^{-2}$
Electric potential	V, ϕ	$V = dW/dQ$	$V, J\ C^{-1}$
Electric potential difference	$U, \Delta V, \Delta\phi$	$U = V_2 - V_1$	V
Electromotive force	E	$E = \int (F/Q)\ ds$	V
Electric field strength	E	$E = F/Q = -\text{grad } V$	$V\ m^{-1}$
Electric flux	ψ	$\psi = \int D\ dA$	C
Electric displacement	D	$D = \varepsilon E$	$C\ m^{-2}$
Capacitance	C	$C = Q/U$	$F, C\ V^{-1}$
Permittivity	ε	$D = \varepsilon E$	$F\ m^{-1}$
Permittivity of vacuum	ε_0	$\varepsilon_0 = \mu_0^{-1} c_0^{-2}$	$F\ m^{-1}$
Relative permittivity	ε_r	$\varepsilon_r = \varepsilon/\varepsilon_0$	1
Dielectric polarization (dipole moment per volume)	P	$P = D - \varepsilon_0 E$	$C\ m^{-2}$
Electric susceptibility	χ_e	$\chi_e = \varepsilon_r - 1$	1
Electric dipole moment	p, μ	$P = Qr$	C m
Electric current	I	$I = dQ/dt$	A
Electric current density	j, J	$I = \int j\ dA$	$A\ m^{-2}$
Magnetic flux density, magnetic induction	B	$F = Qv \times B$	T
Magnetic flux	Φ	$\Phi = \int B\ dA$	Wb
Magnetic field strength	H	$B = \mu H$	$A\ M^{-1}$
Permeability	μ	$B = \mu H$	$N\ A^{-2}, H\ m^{-1}$
Permeability of vacuum	μ_0		$H\ m^{-1}$
Relative permeability	μ_r	$\mu_r = \mu/\mu_0$	1
Magnetization (magnetic dipole moment per volume)	M	$M = B/\mu_0 - H$	$A\ m^{-1}$
Magnetic susceptibility	$\chi, \kappa, (\chi_m)$	$\chi = \mu_r - 1$	1
Molar magnetic susceptibility	χ_m	$\chi_m = V_m \chi$	$m^3\ mol^{-1}$
Magnetic dipole moment	m, μ	$E_p = -m\ B$	$A\ m^2, J\ T^{-1}$
Electrical resistance	R	$R = U/I$	Ω
Conductance	G	$G = 1/R$	S
Loss angle	δ	$\delta = (\pi/2) + \phi_I - \phi_U$	1, rad
Reactance	X	$X = (U/I) \sin \delta$	Ω
Impedance (complex impedance)	Z	$Z = R + iX$	Ω
Admittance (complex admittance)	Y	$Y = 1/Z$	S
Susceptance	B	$Y = G + iB$	S
Resistivity	ρ	$\rho = E/j$	Ω m
Conductivity	κ, γ, σ	$\kappa = 1/\rho$	$S\ m^{-1}$
Self-inductance	L	$E = -L(dI/dt)$	H
Mutual inductance	M, L_{12}	$E_1 = L_{12}(dI_2/dt)$	H
Magnetic vector potential	A	$B = \nabla \times A$	$Wb\ m^{-1}$
Poynting vector	S	$S = E \times H$	$W\ m^{-2}$
Electromagnetic Radiation			
Wavelength	λ		m
Speed of light			
In vacuum	c_0		$m\ s^{-1}$
In a medium	c	$c = c_0/n$	$m \cdot s^{-1}$
Wavenumber in vacuum	$\tilde{\nu}$	$\tilde{\nu} = \nu/c_0 = 1/n\lambda$	m^{-1}
Wavenumber (in a medium)	σ	$\sigma = 1/\lambda$	m^{-1}
Frequency	ν	$\nu = c/\lambda$	Hz
Circular frequency, pulsatance	ω	$\omega = 2\pi\nu$	s^{-1}, rad s^{-1}
Refractive index	n	$n = c_0/c$	1
Planck constant	h		J s
Planck constant/2π	$\hbar$	$\hbar = h/2\pi$	J s
Radiant energy	Q, W		J
Radiant energy density	ρ, w	$\rho = Q/V$	$J\ m^{-3}$

Name	Symbol	Definition	SI Unit
Spectral radiant energy density			
In terms of frequency	ρ_ν, w_ν	$\rho_\nu = d\rho/d\nu$	J m^{-3} Hz^{-1}
In terms of wavenumber	$\rho_{\tilde\nu}, w_{\tilde\nu}$	$\rho_{\tilde\nu} = d\rho/d\tilde\nu$	J m^{-2}
In terms of wavelength	ρ_λ, w_λ	$\rho_\lambda = d\rho/d\lambda$	J · m^{-4}
Einstein transition probabilities			
Spontaneous emission	A_{nm}	$dN_n/dt = -A_{nm}N_n$	s^{-2}
Stimulated emission	B_{nm}	$dN_n/dt = -\rho_{\tilde\nu}(\tilde\nu_{nm}) \times B_{nm}N_n$	s kg^{-1}
Stimulated absorption	B_{mn}	$dN_n/dt = \rho_{\tilde\nu}(\tilde\nu_{nm})\, B_{mn}N_m$	s kg^{-1}
Radiant power, radiant energy per time	Φ, P	$\Phi = dQ/dt$	W
Radiant intensity	I	$I = d\Phi/d\Omega$	W sr^{-1}
Radiant excitance (emitted radiant flux)	M	$M = d\Phi/dA_{source}$	W m^{-2}
Irradiance (radiant flux received)	$E, (I)$	$E = d\Phi/dA$	W m^{-2}
Emittance	ε	$\varepsilon = M/M_{bb}$	1
Stefan-Boltzmann constant	σ	$M_{bb} = \sigma T^4$	W m^{-2} K^{-4}
First radiation constant	c_1	$c_1 = 2\pi h c_0^2$	W m^{2}
Second radiation constant	c_2	$c_2 = hc_0/k$	K m
Transmittance, transmission factor	τ, T	$\tau = \Phi_{tr}/\Phi_0$	1
Absorptance, absorption factor	α	$\alpha = \Phi_{abs}/\Phi_0$	1
Reflectance, reflection factor	ρ	$\rho = \Phi_{refl}/\Phi_0$	1
(Decadic) absorbance	A	$A = \lg(1 - \alpha_i)$	1
Napierian absorbance	B	$B = \ln(1 - \alpha_i)$	1
Absorption coefficient			
(Linear) decadic	a, K	$a = A/l$	m^{-1}
(Linear) napierian	α	$\alpha = B/l$	m^{-1}
Molar (decadic)	ε	$\varepsilon = a/c = A/cl$	m^2 mol^{-1}
Molar napierian	κ	$\kappa = a/c = B/cl$	m^2 mol^{-1}
Absorption index	k	$k = \alpha/4\pi\tilde\nu$	1
Complex refractive index	$\hat{n}$	$\hat{n} = n + ik$	1
Molar refraction	R, R_m	$R = \dfrac{(n^2 - 1)}{(n^2 + 2)} V_m$	m^3 mol^{-1}
Angle of optical rotation	α		1, rad
	Solid State		
Lattice vector	$\mathbf{R}, \mathbf{R}_0$		m
Fundamental translation vectors for the crystal lattice	$\mathbf{a}_1, \mathbf{a}_2; \mathbf{a}_3, \mathbf{a}, \mathbf{b}; \mathbf{c}$	$\mathbf{R} = n_1\mathbf{a}_1 + n_2\mathbf{a}_2 + n_3\mathbf{a}_3$	m
(Circular) reciprocal lattice vector	$\mathbf{G}$	$\mathbf{G} \cdot \mathbf{R} = 2\pi m$	m^{-1}
(Circular) fundamental translation vectors for the reciprocal lattice	$\mathbf{b}_1, \mathbf{b}_2; \mathbf{b}_3, \mathbf{a}^*; \mathbf{b}^*; \mathbf{c}^*$	$\mathbf{a}_i \cdot \mathbf{b}_k = 2\pi\delta_{ik}$	m^{-1}
Lattice plane spacing	d		m
Bragg angle	θ	$n\lambda = 2d \sin\theta$	1, rad
Order of reflection	n		1
Order parameters			
Short range	σ		1
Long range	s		1
Burgers vector	$\mathbf{b}$		m
Particle position vectort	$\mathbf{r}, \mathbf{R}_j$		m
Equilibrium position vector of an ion	$\mathbf{R}_0$		m
Displacement vector of an ion	$\mathbf{u}$	$\mathbf{u} = \mathbf{R} - \mathbf{R}_0$	m
Debye-Waller factor	B, D		1
Debye circular wavenumber	q_D		m^{-1}
Debye circular frequency	ω_D		s^{-1}
Grüneisen parameter	γ, Γ	$\gamma = \alpha V/\kappa C_V$	1
Madelung constant	α, M	$E_{coul} = \dfrac{\alpha N_A z_+ z_- e^2}{4\pi\varepsilon_0 R_0}$	1
Density of states	N_E	$N_E = dN(E)/dE$	J^{-1} m^{-3}
(Spectral) density of vibrational modes	N_ω, g	$N_\omega = dN(\omega)/d\omega$	s m^{-3}

Name	Symbol	Definition	SI Unit
Resistivity tensor	ρ_{ik}	$E = \rho \cdot j$	$\Omega \cdot m$
Conductivity tensor	σ_{ik}	$\sigma = \rho^{-1}$	$S \cdot m^{-1}$
Thermal conductivity tensor	λ_{ik}	$J_q = -\lambda \cdot \text{grad } T$	$W \cdot m^{-1} \cdot K^{-1}$
Residual resistivity	ρ_R		$\Omega \cdot m$
Relaxation time	τ	$\tau = l/v_F$	s
Lorenz coefficient	L	$L = \lambda/\sigma T$	$V^2 \cdot K^{-2}$
Hall coefficient	A_H, R_H	$E = \rho \cdot j + R_H(B \times j)$	$m^3 \cdot C^{-1}$
Thermoelectric force	E		V
Peltier coefficient	Π		V
Thomson coefficient	$\mu, (\tau)$		$V \cdot K^{-1}$
Work function	Φ	$\Phi = E_v - E_F$	J
Number density, number concentration	$n, (p)$		m^{-3}
Gap energy	E_g		J
Donor ionization energy	E_d		J
Acceptor ionization energy	E_a		J
Fermi energy	E_F, ε_F		J
Circular wave vector, propagation vector	k, q	$k = 2\pi/\lambda$	m^{-1}
Bloch function	$u_k(r)$	$\psi(r) = u_k(r) \exp(ik \cdot r)$	$m^{-3/2}$
Charge density of electrons	ρ	$\rho(r) = -e\psi^*(r)\psi(r)$	$C \cdot m^{-3}$
Effective mass	m^*		kg
Mobility	m	$\mu = v_{drift}/E$	$m^2 \cdot V^{-1} \cdot s^{-1}$
Mobility ratio	b	$b = \mu_n/\mu_p$	1
Diffusion coefficient	D	$dN/dt = -DA(dn/dx)$	$m^2 \cdot s^{-1}$
Diffusion length	L	$L = \sqrt{D\tau}$	m
Characteristic (Weiss) temperature	ϕ, ϕ_w		K
Curie temperature	T_C		K
Neel temperature	T_N		K

1.6 Elementary Algebra and Geometry

Fundamental Properties (Real Numbers)

$a + b = b + a$	Commutative law for addition
$(a + b) + c = a + (b + c)$	Associative law for addition
$a + 0 = 0 + a$	Identity law for addition
$a + (-a) = (-a) + a = 0$	Inverse law for addition
$a(bc) = (ab)c$	Associative law for multiplication
$a\left(\frac{1}{a}\right) = \left(\frac{1}{a}\right)a = 1, a \neq 0$	Inverse law for multiplication
$(a)(1) = (1)(a) = a$	Identity law for multiplication
$ab = ba$	Commutative law for multiplication
$a(b + c) = ab + ac$	Distributive law

Division by zero is not defined.

Exponents

For integers m and n,

$$a^n a^m = a^{n+m}$$

$$a^n / a^m = a^{n-m}$$

$$\left(a^n\right)^m = a^{nm}$$

$$(ab)^m = a^m b^m$$

$$(a/b)^m = a^m/b^m$$

Fractional Exponents

$$a^{p/q} = \left(a^{1/q}\right)^p$$

where $a^{1/q}$ is the positive qth root of a if $a > 0$ and the negative qth root of a if a is negative and q is odd. Accordingly, the five rules of exponents given above (for integers) are also valid if m and n are fractions, provided a and b are positive.

Irrational Exponents

If an exponent is irrational (e.g., $\sqrt{2}$), the quantity, such as $a^{\sqrt{2}}$, is the limit of the sequence $a^{1.4}$, $a^{1.41}$, $a^{1.414}$, . . .

Operations with Zero

$$0^m = 0 \qquad a^0 = 1$$

Logarithms

If x, y, and b are positive $b \neq 1$,

$$\log_b(xy) = \log_b x + \log_b y$$

$$\log_b(x/y) = \log_b x - \log_b y$$

$$\log_b x^p = p \log_b x$$

$$\log_b(1/x) = -\log_b x$$

$$\log_b b = 1$$

$$\log_b 1 = 0 \qquad \textit{Note: } b^{\log_b x} = x$$

Change of Base $(a \neq 1)$

$$\log_b x = \log_a x \log_b a$$

Factorials

The factorial of a positive integer n is the product of all the positive integers less than or equal to the integer n and is denoted $n!$. Thus,

$$n! = 1 \cdot 2 \cdot 3 \cdot \cdots \cdot n$$

Factorial 0 is defined: $0! = 1$.

Stirling's Approximation

$$\lim_{n\to\infty}(n/e^n)^n\sqrt{2\pi n} = n!$$

Binomial Theorem

For positive integer n

$$(x+y)^n = x^n + nx^{n-1}y + \frac{n(n-1)}{2!}x^{n-2}y^2 + \frac{n(n-1)(n-2)}{3!}x^{n-3}y^3 + \cdots + nxy^{n-1} + y^n$$

Factors and Expansion

$$(a+b)^2 = a^2 + 2ab + b^2$$

$$(a-b)^2 = a^2 - 2ab + b^2$$

$$(a+b)^3 = a^3 + 3a^2b + 3ab^2 + b^3$$

$$(a-b)^3 = a^3 - 3a^2b + 3ab^2 - b^3$$

$$(a^2 - b^2) = (a-b)(a+b)$$

$$(a^3 - b^3) = (a-b)(a^2 + ab + b^2)$$

$$(a^3 + b^3) = (a+b)(a^2 - ab + b^2)$$

Progression

An *arithmetic progression* is a sequence in which the difference between any term and the preceding term is a constant (d):

$$a, a+d, a+2d, \ldots, a+(n-1)d$$

If the last term is denoted l [$= a + (n-1)d$], then the sum is

$$s = \frac{n}{2}(a+l)$$

A *geometric progression* is a sequence in which the ratio of any term to the preceding term is a constant r. Thus, for n terms,

$$a, ar, ar^2, \ldots, ar^{n-1}$$

The sum is

$$S = \frac{a - ar^n}{1-r}$$

Complex Numbers

A complex number is an ordered pair of real numbers (a, b).

Equality: $(a,b) = (c,d)$ if and only if $a = c$ and $b = d$

Addition: $(a,b) + (c,d) = (a + c, b + d)$

Multiplication: $(a,b)(c,d) = (ac - bd, ad + bc)$

The first element (a, b) is called the *real* part, the second the *imaginary* part. An alternative notation for (a, b) is $a + bi$, where $i^2 = (-1, 0)$, and $i = (0, 1)$ or $0 + 1i$ is written for this complex number as a convenience. With this understanding, i behaves as a number, that is, $(2 - 3i)(4 + i) = 8 + 2i - 12i - 3i^2 = 11 - 10i$. The conjugate of a $a + bi$ is $a - bi$, and the product of a complex number and its conjugate is $a^2 + b^2$. Thus, *quotients* are computed by multiplying numerator and denominator by the conjugate of the denominator, as illustrated below:

$$\frac{2+3i}{4+2i} = \frac{(4-2i)(2+3i)}{(4-2i)(4+2i)} = \frac{14+8i}{20} = \frac{7+4i}{10}$$

Polar Form

The complex number $x + iy$ may be represented by a plane vector with components x and y:

$$x + iy = r(\cos\theta + i\sin\theta)$$

(See Figure 1.1.). Then, given two complex numbers $z_1 = \tau_1(\cos\theta_1 + i\sin\theta_1)$ and $z_2 = \tau_2(\cos\theta_2 + i\sin\theta_2)$, the product and quotient are:

Product: $$z_1 z_2 = r_1 r_2[\cos(\theta_1 + \theta_2) + i\sin(\theta_1 + \theta_2)]$$

Quotient: $$z_1/z_2 = (r_1/r_2)[\cos(\theta_1 - \theta_2) + i\sin(\theta_1 - \theta_2)]$$

Powers: $$z^n = [r(\cos\theta + i\sin\theta)]^n = r^n[\cos n\theta + i\sin n\theta]$$

Roots: $$z^{1/n} = [r(\cos\theta + i\sin\theta)]^{1/n}$$

$$= r^{1/n}\left[\cos\frac{\theta + k\cdot 360}{n} + i\sin\frac{\theta + k\cdot 360}{n}\right]$$

$$k = 0, 1, 2, \ldots, n-1$$

Permutations

A permutation is an ordered arrangement (sequence) of all or part of a set of objects. The number of permutations of n objects taken r at a time is

$$p(n,r) = n(n-1)(n-2)\cdots(n-r+1)$$

$$= \frac{n!}{(n-r)!}$$

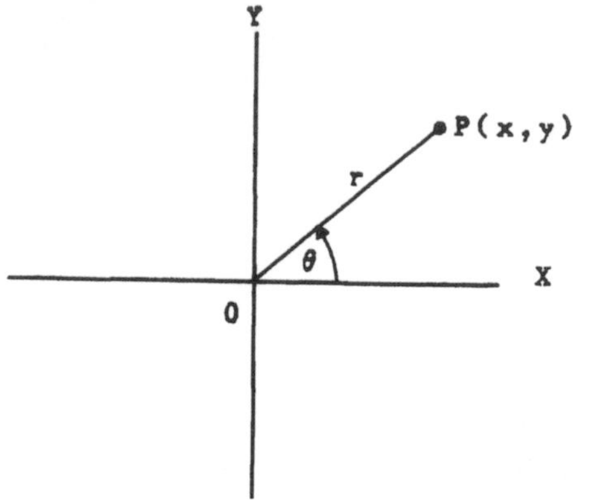

Figure 1.1 Polar form of complex number.

A permutation of positive integers is "even" or "odd" if the total number of inversions is an even integer or an odd integer, respectively. Inversions are counted relative to each integer j in the permutation by counting the number of integers that follow j and are less than j. These are summed to give the total number of inversions. For example, the permutation 4132 has four inversions: three relative to 4 and one relative to 3. This permutation is therefore even.

Combinations

A combination is a selection of one or more objects from among a set of objects regardless of order. The number of combinations of n different objects taken r at a time is

$$C(n,r) = \frac{P(n,r)}{r!} = \frac{n!}{r!(n-r)!}$$

Algebraic Equations

Quadratic

If $ax^2 + bx + c = 0$, and $a \neq 0$, then roots are

$$x = \frac{-b \pm \sqrt{b^2 - 4ac}}{2a}$$

Cubic

To solve $\mathbf{x}^3 + bx^2 + cx + d = 0$, let $x = y - b/3$. Then the *reduced cubic* is obtained:

$$y^3 + py + q = 0$$

where $p = c - (1/3)b^2$ and $q = d - (1/3)bc + (2/27)b^3$. Solutions of the original cubic are then in terms of the reduced curbic roots y_1, y_2, y_3:

$$x_1 = y_1 - (1/3)b \qquad x_2 = y_2 - (1/3)b \qquad x_3 = y_3 - (1/3)b$$

The three roots of the reduced cubic are

$$y_1 = (A)^{1/3} + (B)^{1/3}$$

$$y_2 = W(A)^{1/3} + W^2(B)^{1/3}$$

$$y_3 = W^2(A)^{1/3} + W(B)^{1/3}$$

where

$$A = -\tfrac{1}{2}q + \sqrt{(1/27)p^3 + \tfrac{1}{4}q^2}$$

$$B = -\tfrac{1}{2}q - \sqrt{(1/27)p^3 + \tfrac{1}{4}q^2}$$

$$W = \frac{-1 + i\sqrt{3}}{2}, \qquad W^2 = \frac{-1 - i\sqrt{3}}{2}$$

When $(1/27)p^3 + (1/4)q^2$ is negative, A is complex; in this case A should be expressed in trigonometric form: $A = r(\cos\theta + i\sin\theta)$ where θ is a first or second quadrant angle, as q is negative or positive. The three roots of the reduced cubic are

$$y_1 = 2(r)^{1/3}\cos(\theta/3)$$

$$y_2 = 2(r)^{1/3}\cos\left(\frac{\theta}{3} + 120°\right)$$

$$y_3 = 2(r)^{1/3}\cos\left(\frac{\theta}{3} + 240°\right)$$

Geometry

Figures 1.2 to 1.12 are a collection of common geometric figures. Area (A), volume (V), and other measurable features are indicated.

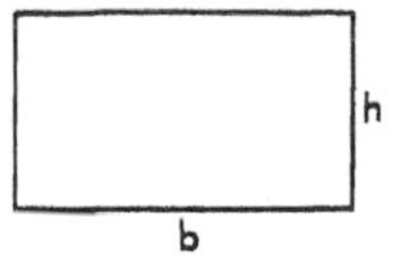

Figure 1.2 Rectangle. $A = bh$.

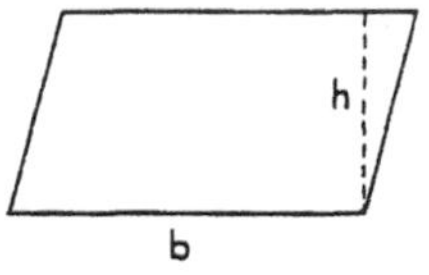

Figure 1.3 Parallelogram. $A = bh$.

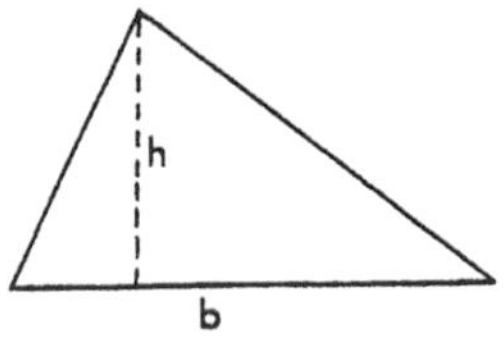

Figure 1.4 Triangle. $A = 1/2\ bh$.

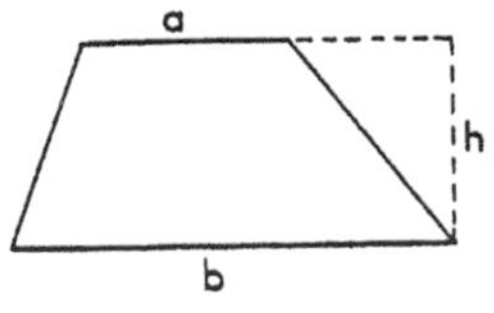

Figure 1.5 Trapezoid. $A = 1/2\ (a + b)h$.

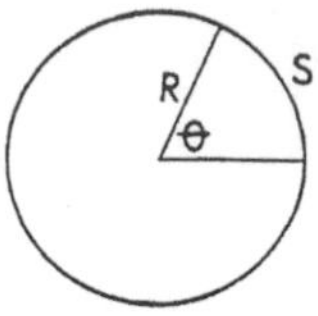

Figure 1.6 Circle. $A = \pi R^2$; circumference $= 2\pi R$, arc length $S = R\,\theta$ (θ in radians).

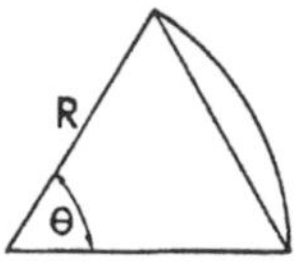

Figure 1.7 Sector of circle. $A_{sector} = 1/2R^2\theta$; $A_{segment} = 1/2R^2\,(\theta - \sin\theta)$.

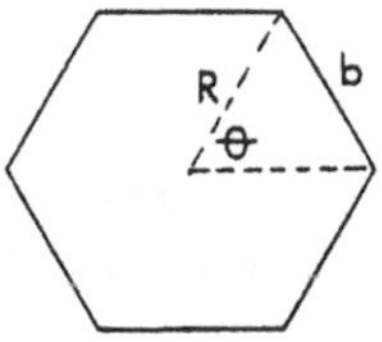

Figure 1.8 Regular polygon of n sides. $A = (n/4)b^2\,\mathrm{ctn}(\pi/n)$; $R = (b/2)\,\csc(\pi/n)$.

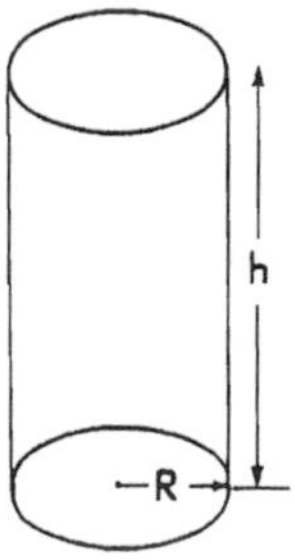

Figure 1.9 Right circular cylinder. $V = \pi R^2 h$; lateral surface area $= 2\pi Rh$

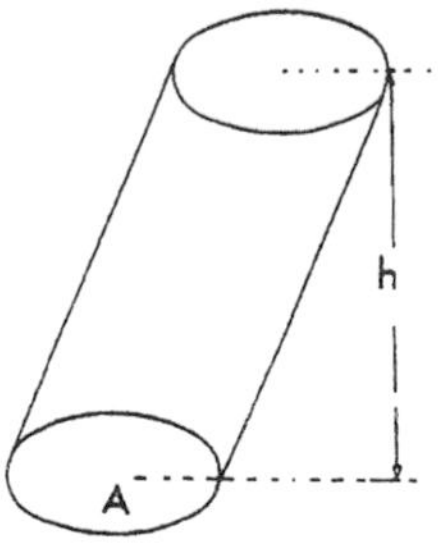

Figure 1.10 Cylinder (or prism) with parallel bases. $V = Ah$.

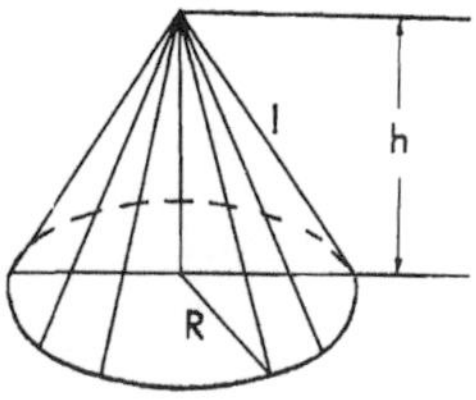

Figure 1.11 Right circular cone. $V = 1/3\;\pi R^2 h$; lateral surface area $= \pi Rl = \pi R\sqrt{R^2 + h^2}$

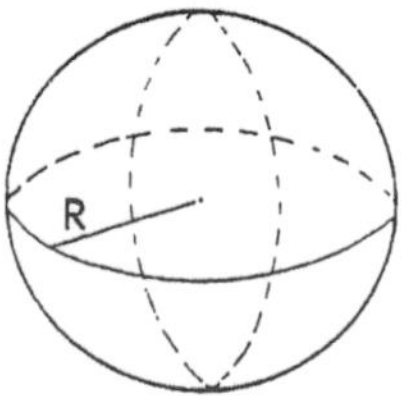

Figure 1.12 Sphere $V = 4/3\;\pi R^3$; surface area $= 4\pi R^2$

1.7 Table of Derivatives

In the following table, a and n are constants, e is the base of the natural logarithms, and u and v denote functions of x.

1. $$\frac{d}{dx}(a) = 0$$

2. $$\frac{d}{dx}(x) = 1$$

3. $$\frac{d}{dx}(au) = a\frac{du}{dx}$$

4. $$\frac{d}{dx}(u+v) = \frac{du}{dx} + \frac{dv}{dx}$$

5. $$\frac{d}{dx}(uv) = u\frac{dv}{dx} + v\frac{du}{dx}$$

6. $$\frac{d}{dx}(u/v) = \frac{v\frac{du}{dx} - u\frac{dv}{dx}}{v^2}$$

7. $$\frac{d}{dx}(u^n) = nu^{n-1}\frac{du}{dx}$$

8. $$\frac{d}{dx}e^u = e^u\frac{du}{dx}$$

9. $$\frac{d}{dx}a^u = (\log_e a)a^u\frac{du}{dx}$$

10. $$\frac{d}{dx}\log_e u = (1/u)\frac{du}{dx}$$

11. $$\frac{d}{dx}\log_a u = (\log_a e)(1/u)\frac{du}{dx}$$

12. $$\frac{d}{dx}u^v = vu^{v-1}\frac{du}{dx} + u^v(\log_e u)\frac{dv}{dx}$$

13. $$\frac{d}{dx}\sin u = \cos u\frac{du}{dx}$$

14. $$\frac{d}{dx}\cos u = -\sin u\frac{du}{dx}$$

15. $$\frac{d}{dx}\tan u = \sec^2 u\frac{du}{dx}$$

16. $$\frac{d}{dx}\operatorname{ctn} u = -\csc^2 u\frac{du}{dx}$$

17. $$\frac{d}{dx}\sec u = \sec u\tan u\frac{du}{dx}$$

18. $$\frac{d}{dx}\csc u = -\csc u\operatorname{ctn} u\frac{du}{dx}$$

19. $$\frac{d}{dx}\sin^{-1} u = \frac{1}{\sqrt{1-u^2}}\frac{du}{dx}, \quad \left(-\tfrac{1}{2}\pi \le \sin^{-1} u \le \tfrac{1}{2}\pi\right)$$

20. $$\frac{d}{dx}\cos^{-1} u = \frac{-1}{\sqrt{1-u^2}}\frac{du}{dx}, \quad \left(0 \le \cos^{-1} u \le \pi\right)$$

21. $$\frac{d}{dx}\tan^{-1} u = \frac{1}{1+u^2}\frac{du}{dx}$$

22. $$\frac{d}{dx}\operatorname{ctn}^{-1} u = \frac{-1}{1+u^2}\frac{du}{dx}$$

23. $$\frac{d}{dx}\sec^{-1} u = \frac{1}{u\sqrt{u^2-1}}\frac{du}{dx},$$
$$\left(-\pi \le \sec^{-1} u < -\tfrac{1}{2}\pi,\ 0 \le \sec^{-1} u \le \tfrac{1}{2}\pi\right)$$

24. $$\frac{d}{dx}\csc^{-1} u = \frac{-1}{u\sqrt{u^2-1}}\frac{du}{dx},$$
$$\left(-\pi < \csc^{-1} u \le -\tfrac{1}{2}\pi,\ 0 < \csc^{-1} u \le \tfrac{1}{2}\pi\right)$$

25. $$\frac{d}{dx}\sinh u = \cosh u\frac{du}{dx}$$

26. $$\frac{d}{dx}\cosh u = \sinh u\frac{du}{dx}$$

27. $$\frac{d}{dx}\tanh u = \operatorname{sech}^2 u\frac{du}{dx}$$

28. $$\frac{d}{dx}\operatorname{ctnh} u = -\operatorname{csch}^2 u\frac{du}{dx}$$

29. $$\frac{d}{dx}\operatorname{sech} u = -\operatorname{sech} u\tanh u\frac{du}{dx}$$

30. $$\frac{d}{dx}\operatorname{csch} u = -\operatorname{csch} u\operatorname{ctnh} u\frac{du}{dx}$$

31. $$\frac{d}{dx}\sinh^{-1} u = \frac{1}{\sqrt{u^2+1}}\frac{du}{dx}$$

32. $$\frac{d}{dx}\cosh^{-1} u = \frac{1}{\sqrt{u^2-1}}\frac{du}{dx}$$

33. $$\frac{d}{dx}\tanh^{-1} u = \frac{1}{1-u^2}\frac{du}{dx}$$

34. $$\frac{d}{dx}\operatorname{ctnh}^{-1} u = \frac{-1}{u^2-1}\frac{du}{dx}$$

35. $$\frac{d}{dx}\operatorname{sech}^{-1} u = \frac{-1}{u\sqrt{1-u^2}}\frac{du}{dx}$$

36. $$\frac{d}{dx}\operatorname{csch}^{-1} u = \frac{-1}{u\sqrt{u^2+1}}\frac{du}{dx}$$

Additional Relations with Derivatives

$$\frac{d}{dt}\int_a^t f(x)\,dx = f(t) \qquad \frac{d}{dt}\int_t^a f(x)\,dx = -f(t)$$

If $x = f(y)$, then $\dfrac{dy}{dx} = \dfrac{1}{dx/dy}$

If $y = f(u)$ and $u = g(x)$, then $\dfrac{dy}{dx} = \dfrac{dy}{du}\cdot\dfrac{du}{dx}$ (chain rule)

If $x = f(t)$ and $y = g(t)$, then $\dfrac{dy}{dx} = \dfrac{g'(t)}{f'(t)}$, and $\dfrac{d^2y}{dx^2} = \dfrac{f'(t)g''(t) - g'(t)f''(t)}{[f'(t)]^3}$

(Note: Exponent in denominator is 3.)

1.8 Integrals

Elementary Forms (Add an arbitrary constant to each integral)

1. $\int a\,dx = ax$

2. $\int a\,f(x)\,dx = a\int f(x)\,dx$

3. $\int \phi(y)\,dx = \int \dfrac{\phi(y)}{y'}\,dy$, where $y' = \dfrac{dy}{dx}$

4. $\int (u + v)\,dx = \int u\,dx + \int v\,dx$, where u and v are any functions of x

5. $\int u\,dv = u\int dv - \int v\,du = uv - \int v\,du$

6. $\int u\dfrac{dv}{dx}\,dx = uv - \int v\dfrac{du}{dx}\,dx$

7. $\int x^n\,dx = \dfrac{x^{n+1}}{n+1}$, except $n = -1$

8. $\int \dfrac{f'(x)\,dx}{f(x)} = \log f(x)$, $[df(x) = f'(x)\,dx]$

9. $\int \dfrac{dx}{x} = \log x$

10. $\int \dfrac{f'(x)\,dx}{2\sqrt{f(x)}} = \sqrt{f(x)}$, $[df(x) = f'(x)\,dx]$

11. $\int e^x\,dx = e^x$

12. $\int e^{ax}\,dx = e^{ax}/a$

13. $\int b^{ax}\,dx = \dfrac{b^{ax}}{a\log b}$, $(b > 0)$

14. $\int \log x\,dx = x\log x - x$

15. $\int a^x \log a\,dx = a^x$, $(a > 0)$

16. $$\int \frac{dx}{a^2+x^2} = \frac{1}{a}\tan^{-1}\frac{x}{a}$$

17. $$\int \frac{dx}{a^2-x^2} = \begin{cases} \frac{1}{a}\tan^{-1}\frac{x}{a} \\ \text{or} \\ \frac{1}{2a}\log\frac{a+x}{a-x}, \quad (a^2>x^2) \end{cases}$$

18. $$\int \frac{dx}{x^2-a^2} = \begin{cases} -\frac{1}{a}\operatorname{ctnh}^{-1}\frac{x}{a} \\ \text{or} \\ \frac{1}{2a}\log\frac{x-a}{x+a}, \quad (x^2>a^2) \end{cases}$$

19. $$\int \frac{dx}{\sqrt{a^2-x^2}} = \begin{cases} \sin^{-1}\frac{x}{|a|} \\ \text{or} \\ -\cos^{-1}\frac{x}{|a|}, \quad (a^2>x^2) \end{cases}$$

20. $$\int \frac{dx}{\sqrt{x^2\pm a^2}} = \log\left(x+\sqrt{x^2\pm a^2}\right)$$

21. $$\int \frac{dx}{x\sqrt{x^2-a^2}} = \frac{1}{|a|}\sec^{-1}\frac{x}{|a|}$$

22. $$\int \frac{dx}{x\sqrt{a^2\pm x^2}} = -\frac{1}{a}\log\left(\frac{a+\sqrt{a^2\pm x^2}}{x}\right)$$

Forms Containing $(a + bx)$

For forms containing $a + bx$, but not listed in the table, the substitution $u = (a + bx)/x$ may prove helpful.

23. $$\int (a+bx)^n\,dx = \frac{(a+bx)^{n+1}}{(n+1)b}, \quad (n \neq -1)$$

24. $$\int x(a+bx)^n\,dx = \frac{1}{b^2(n+2)}(a+bx)^{n+2} - \frac{a}{b^2(n+1)}(a+bx)^{n+1}, \quad (n \neq -1, -2)$$

25. $$\int x^2(a+bx)^n\,dx = \frac{1}{b^3}\left[\frac{(a+bx)^{n+3}}{n+3} - 2a\frac{(a+bx)^{n+2}}{n+2} + a^2\frac{(a+bx)^{n+1}}{n+1}\right]$$

26. $$\int x^m(a+bx)^n\,dx = \begin{cases} \frac{x^{m+1}(a+bx)^n}{m+n+1} + \frac{an}{m+n+1}\int x^m(a+bx)^{n-1}\,dx \\ \text{or} \\ \frac{1}{a(n+1)}\left[-x^{m+1}(a+bx)^{n+1} + (m+n+2)\int x^m(a+bx)^{n+1}\,dx\right] \\ \text{or} \\ \frac{1}{b(m+n+1)}\left[x^m(a+bx)^{n+1} - ma\int x^{m-1}(a+bx)^n\,dx\right] \end{cases}$$

27. $$\int \frac{dx}{a+bx} = \frac{1}{b}\log(a+bx)$$

28. $\int \frac{dx}{(a+bx)^2} = -\frac{1}{b(a+bx)}$

29. $\int \frac{dx}{(a+bx)^3} = -\frac{1}{2b(a+bx)^2}$

30. $\int \frac{x\,dx}{a+bx} = \begin{cases} \frac{1}{b^2}\left[a+bx-a\log(a+bx)\right] \\ \text{or} \\ \frac{x}{b}-\frac{a}{b^2}\log(a+bx) \end{cases}$

31 $\int \frac{x\,dx}{(a+bx)^2} = \frac{1}{b^2}\left[\log(a+bx)+\frac{a}{a+bx}\right]$

32. $\int \frac{x\,dx}{(a+bx)^n} = \frac{1}{b^2}\left[\frac{-1}{(n-2)(a+bx)^{n-2}}+\frac{a}{(n-1)(a+bx)^{n-1}}\right], \quad n \neq 1, 2$

33. $\int \frac{x^2\,dx}{a+bx} = \frac{1}{b^3}\left[\frac{1}{2}(a+bx)^2-2a(a+bx)+a^2\log(a+bx)\right]$

34. $\int \frac{x^2\,dx}{(a+bx)^2} = \frac{1}{b^3}\left[a+bx-2a\log(a+bx)-\frac{a^2}{a+bx}\right]$

35. $\int \frac{x^2\,dx}{(a+bx)^3} = \frac{1}{b^3}\left[\log(a+bx)+\frac{2a}{a+bx}-\frac{a^2}{2(a+bx)^2}\right]$

36. $\int \frac{x^2\,dx}{(a+bx)^n} = \frac{1}{b^3}\left[\frac{-1}{(n-3)(a+bx)^{n-3}}+\frac{2a}{(n-2)(a+bx)^{n-2}}-\frac{a}{(n-1)(a+bx)^{n-1}}\right], \quad n \neq 1, 2, 3$

37 $\int \frac{dx}{x(a+bx)} = -\frac{1}{a}\log\frac{a+bx}{x}$

38. $\int \frac{dx}{x(a+bx)^2} = \frac{1}{a(a+bx)}-\frac{1}{a^2}\log\frac{a+bx}{x}$

39. $\int \frac{dx}{x(a+bx)^3} = \frac{1}{a^3}\left[\frac{1}{2}\left(\frac{2a+bx}{a+bx}\right)^2+\log\frac{x}{a+bx}\right]$

40. $\int \frac{dx}{x^2(a+bx)} = -\frac{1}{ax}+\frac{b}{a^2}\log\frac{a+bx}{x}$

41 $\int \frac{dx}{x^3(a+bx)} = \frac{2bx-a}{2a^2x^2}+\frac{b^2}{a^3}\log\frac{x}{a+bx}$

42. $\int \frac{dx}{x^2(a+bx)^2} = -\frac{a+2bx}{a^2x(a+bx)}+\frac{2b^2}{a^3}\log\frac{a+bx}{x}$

1.9 The Fourier Transforms

For a piecewise continuous function $F(x)$ over a finite interval $0 \le x \le \pi$, the *finite Fourier cosine transform* of $F(x)$ is

$$f_c(n) = \int_0^{\pi} F(x)\cos nx\,dx \quad (n = 0, 1, 2, \ldots) \tag{1.1}$$

If x ranges over the interval $0 \le x \le L$, the substitution $x' = \pi x/L$ allows the use of this definition also. The inverse transform is written

$$\bar{F}(x) = \frac{1}{\pi} f_c(0) + \frac{2}{\pi} \sum_{n=1}^{\infty} f_c(n) \cos nx \quad (0 < x < \pi) \tag{1.2}$$

where $\bar{F}(x) = [F(x+0) + F(x-0)]/2$. We observe that $\bar{F}(x) = F(x)$ at points of continuity. The formula

$$\begin{aligned} f_c^{(2)}(n) &= \int_0^{\pi} F''(x) \cos nx \, dx \\ &= -n^2 f_c(n) - F'(0) + (-1)^n F'(\pi) \end{aligned} \tag{1.3}$$

makes the finite Fourier cosine transform useful in certain boundary value problems.

Analogously, the *finite Fourier sine transform* of $F(x)$ is

$$f_s(n) = \int_0^{\pi} F(x) \sin nx \, dx \quad (n = 1, 2, 3, \ldots) \tag{1.4}$$

and

$$\bar{F}(x) = \frac{2}{\pi} \sum_{n=1}^{\infty} f_s(n) \sin nx \quad (0 < x < \pi) \tag{1.5}$$

Corresponding to Equation (1.3), we have

$$\begin{aligned} f_s^{(2)}(n) &= \int_0^{\pi} F''(x) \sin nx \, dx \\ &= -n^2 f_s(n) - nF(0) - n(-1)^n F(\pi) \end{aligned} \tag{1.6}$$

Fourier Transforms

If $F(x)$ is defined for $x \geq 0$ and is piecewise continuous over any finite interval, and if

$$\int_0^{\infty} F(x) \, dx$$

is absolutely convergent, then

$$f_c(\alpha) = \sqrt{\frac{2}{\pi}} \int_0^{\infty} F(x) \cos(\alpha x) \, dx \tag{1.7}$$

is the *Fourier cosine transform* of $F(x)$. Furthermore,

$$\bar{F}(x) = \sqrt{\frac{2}{\pi}} \int_0^{\infty} f_c(\alpha) \cos(\alpha x) \, d\alpha \tag{1.8}$$

If $\lim_{x \to \infty} d^n F/dx^n = 0$, an important property of the Fourier cosine transform,

$$f_c^{(2r)}(\alpha) = \sqrt{\frac{2}{\pi}}\int_0^\infty \left(\frac{d^{2r}F}{dx^{2r}}\right)\cos(\alpha x)\,dx$$
$$= -\sqrt{\frac{2}{\pi}}\sum_{n=0}^{r-1}(-1)^n a_{2r-2n-1}\alpha^{2n} + (-1)^r \alpha^{2r} f_c(\alpha) \tag{1.9}$$

where $\lim_{x\to 0} d^rF/dx^r = a_r$, makes it useful in the solution of many problems.
Under the same conditions,

$$f_s(\alpha) = \sqrt{\frac{2}{\pi}}\int_0^\infty F(x)\sin(\alpha x)\,dx \tag{1.10}$$

defines the *Fourier sine transform* of $F(x)$, and

$$F(x) = \sqrt{\frac{2}{\pi}}\int_0^\infty f_s(\alpha)\sin(\alpha x)\,d\alpha \tag{1.11}$$

Corresponding to Equation (1.9) we have

$$f_s^{(2r)}(\alpha) = \sqrt{\frac{2}{\pi}}\int_0^\infty \frac{d^{2r}F}{dx^{2r}}\sin(\alpha x)\,dx$$
$$= -\sqrt{\frac{2}{\pi}}\sum_{n=1}^{r}(-1)^n \alpha^{2n-1} a_{2r-2n} + (-1)^{r-1}\alpha^{2r} f_s(\alpha) \tag{1.12}$$

Similarly, if $F(x)$ is defined for $-\infty < x < \infty$, and if $\int_{-\infty}^{\infty} F(x)\,dx$ is absolutely convergent, then

$$f(\alpha) = \frac{1}{2\pi}\int_{-\infty}^\infty F(x)e^{i\alpha x}\,dx \tag{1.13}$$

is the *Fourier transform* of $F(x)$, and

$$F(x) = \frac{1}{2\pi}\int_{-\infty}^\infty f(\alpha)e^{-i\alpha x}\,d\alpha \tag{1.14}$$

Also, if

$$\lim_{x\to\pm\infty}\left|\frac{d^nF}{dx^n}\right| = 0 \quad (n = 1, 2, \ldots, r-1)$$

then

$$f^{(r)}(\alpha) = \frac{1}{2\pi}\int_{-\infty}^{\infty} F^{(r)}(x)e^{i\alpha x}\,dx = (-i\alpha)^r f(\alpha) \tag{1.15}$$

Finite Sine Transforms

	$f_s(n)$	$F(x)$
1	$f_s(n) = \int_0^\pi F(x)\sin nx\,dx \quad (n = 1, 2, \ldots)$	$F(x)$
2.	$(-1)^{n+1} f_s(n)$	$F(\pi - x)$
3.	$\frac{1}{n}$	$\frac{\pi - x}{\pi}$
4.	$\frac{(-1)^{n+1}}{n}$	$\frac{x}{\pi}$
5.	$\frac{1-(-1)^n}{n}$	1
6.	$\frac{2}{n^2}\sin\frac{n\pi}{2}$	$\begin{cases} x & \text{when } 0 < x < \pi/2 \\ \pi - x & \text{when } \pi/2 < x < \pi \end{cases}$
7	$\frac{(-1)^{n+1}}{n^3}$	$\frac{x(\pi^2 - x^2)}{6\pi}$
8.	$\frac{1-(-1)^n}{n^3}$	$\frac{x(\pi - x)}{2}$
9.	$\frac{\pi^2(-1)^{n-1}}{n} - \frac{2[1-(-1)^n]}{n^3}$	x^2
10.	$\pi(-1)^n\left(\frac{6}{n^3} - \frac{\pi^2}{n}\right)$	x^3
11	$\frac{n}{n^2+c^2}\left[1-(-1)^n e^{c\pi}\right]$	e^{cx}
12.	$\frac{n}{n^2+c^2}$	$\frac{\sinh c(\pi - x)}{\sinh c\pi}$
13.	$\frac{n}{n^2-k^2} \quad (k \neq 0, 1, 2, \ldots)$	$\frac{\sinh k(\pi - x)}{\sinh k\pi}$
14.	$\begin{cases} \frac{\pi}{2} & \text{when } n = m \\ 0 & \text{when } n \neq m \end{cases} \quad (m = 1, 2, \ldots)$	$\sin mx$
15.	$\frac{n}{n^2-k^2}\left[1-(-1)^n\cos k\pi\right] \quad (k \neq 1, 2, \ldots)$	$\cos kx$
16.	$\begin{cases} \frac{n}{n^2-m^2}\left[1-(-1)^{n+m}\right] & \text{when } n \neq m = 1, 2, \ldots \\ 0 & \text{when } n = m \end{cases}$	$\cos mx$
17	$\frac{n}{(n^2-k^2)^2} \quad (k \neq 0, 1, 2, \ldots)$	$\frac{\pi\sin kx}{2k\sin^2 k\pi} - \frac{x\cos k(\pi - x)}{2k\sin k\pi}$
18.	$\frac{b^n}{n} \quad (\lvert b\rvert \leq 1)$	$\frac{2}{\pi}\arctan\frac{b\sin x}{1 - b\cos x}$
19.	$\frac{1-(-1)^n}{n}b^n \quad (\lvert b\rvert \leq 1)$	$\frac{2}{\pi}\arctan\frac{2b\sin x}{1 - b^2}$

Finite Cosine Transforms

	$f_c(n)$	$F(x)$
1	$f_c(n)=\int_0^\pi F(x)\cos nx\,dx \quad (n=0,1,2,\ldots)$	$F(x)$
2.	$(-1)^n f_c(n)$	$F(\pi-x)$
3.	0 when $n=1,2,\ldots;\quad f_c(0)=\pi$	1
4	$\frac{2}{n}\sin\frac{n\pi}{2},\quad f_c(0)=0$	$\begin{cases}1 & \text{when } 0<x<\pi/2\\ -1 & \text{when } \pi/2<x<\pi\end{cases}$
5.	$-\frac{1-(-1)^n}{n^2},\quad f_c(0)=\frac{\pi^2}{2}$	x
6.	$\frac{(-1)^n}{n^2},\quad f_c(0)=\frac{\pi^2}{6}$	$\frac{x^2}{2\pi}$
7	$\frac{1}{n^2},\quad f_c(0)=0$	$\frac{(\pi-x)^2}{2\pi}-\frac{\pi}{6}$
8.	$3\pi^2\frac{(-1)^n}{n^2}-6\frac{1-(-1)^n}{n^4},\quad f_c(0)=\frac{\pi^4}{4}$	x^3
9.	$\frac{(-1)^n e^{c\pi}-1}{n^2+c^2}$	$\frac{1}{c}e^{cx}$
10.	$\frac{1}{n^2+c^2}$	$\frac{\cosh c(\pi-x)}{c\sinh c\pi}$
11	$\frac{k}{n^2-k^2}\left[(-1)^n\cos\pi k-1\right] \quad (k\neq 0,1,2,\ldots)$	$\sin kx$
12.	$\frac{(-1)^{n+m}-1}{n^2-m^2},\quad f_c(m)=0 \quad (m=1,2,\ldots)$	$\frac{1}{m}\sin mx$
13	$\frac{1}{n^2-k^2} \quad (k\neq 0,1,2,\ldots)$	$-\frac{\cos k(\pi-x)}{k\sin k\pi}$
14.	0 when $n=1,2,\ldots;\quad f_c(m)=\frac{\pi}{2} \quad (m=1,2,\ldots)$	$\cos mx$

Fourier Sine Transforms

	F(x)	$f_s(\alpha)$
1	$\begin{cases}1 & (0<x<a)\\ 0 & (x>a)\end{cases}$	$\sqrt{\frac{2}{\pi}}\left[\frac{1-\cos\alpha}{\alpha}\right]$
2.	$x^{p-1} \quad (0<p<1)$	$\sqrt{\frac{2}{\pi}}\frac{\Gamma(p)}{\alpha^p}\sin\frac{p\pi}{2}$
3	$\begin{cases}\sin x & (0<x<a)\\ 0 & (x>a)\end{cases}$	$\frac{1}{\sqrt{2\pi}}\left[\frac{\sin[a(1-\alpha)]}{1-\alpha}-\frac{\sin[a(1+\alpha)]}{1+\alpha}\right]$
4.	e^{-x}	$\sqrt{\frac{2}{\pi}}\left[\frac{\alpha}{1+\alpha^2}\right]$

	F(x)	$f_s(\alpha)$
5.	$xe^{-x^2/2}$	$\alpha e^{-\alpha^2/2}$
6.	$\cos\frac{x^2}{2}$	$\sqrt{2}\left[\sin\frac{\alpha^2}{2}C\left(\frac{\alpha^2}{2}\right)-\cos\frac{\alpha^2}{2}S\left(\frac{\alpha^2}{2}\right)\right]$
7	$\sin\frac{x^2}{2}$	$\sqrt{2}\left[\cos\frac{\alpha^2}{2}C\left(\frac{\alpha^2}{2}\right)+\sin\frac{\alpha^2}{2}S\left(\frac{\alpha^2}{2}\right)\right]$

* $C(y)$ and $S(y)$ are the Fresnel integrals

$$C(y)=\frac{1}{\sqrt{2\pi}}\int_0^y \frac{1}{\sqrt{t}}\cos t\,dt$$

$$S(y)=\frac{1}{\sqrt{2\pi}}\int_0^y \frac{1}{\sqrt{t}}\sin t\,dt$$

Fourier Cosine Transforms

	$F(x)$	$f_c(\alpha)$
1	$\begin{cases}1 & (0<x<a)\\ 0 & (x>a)\end{cases}$	$\sqrt{\frac{2}{\pi}}\frac{\sin a\alpha}{\alpha}$
2	x^{p-1} $(0<p<1)$	$\sqrt{\frac{2}{\pi}}\frac{\Gamma(p)}{\alpha^p}\cos\frac{p\pi}{2}$
3.	$\begin{cases}\cos x & (0<x<a)\\ 0 & (x>a)\end{cases}$	$\frac{1}{\sqrt{2\pi}}\left[\frac{\sin[a(1-\alpha)]}{1-\alpha}+\frac{\sin[a(1+\alpha)]}{1+\alpha}\right]$
4.	e^{-x}	$\sqrt{\frac{2}{\pi}}\left(\frac{1}{1+\alpha^2}\right)$
5	$e^{-x^2/2}$	$e^{-\alpha^2/2}$
6	$\cos\frac{x^2}{2}$	$\cos\left(\frac{\alpha^2}{2}-\frac{\pi}{4}\right)$
7	$\sin\frac{x^2}{2}$	$\cos\left(\frac{\alpha^2}{2}-\frac{\pi}{4}\right)$

Fourier Transforms

	$F(x)$	$f(\alpha)$
1	$\frac{\sin ax}{x}$	$\begin{cases}\sqrt{\frac{\pi}{2}} & \lvert\alpha\rvert<a\\ 0 & \lvert\alpha\rvert>a\end{cases}$
2.	$\begin{cases}e^{wx} & (p<x<q)\\ 0 & (x<p,\ x>q)\end{cases}$	$\frac{i}{\sqrt{2\pi}}\frac{e^{q(w+i\alpha)}-e^{p(w+i\alpha)}}{(w+\alpha)}$

	$F(x)$	$f(\alpha)$
3.	$\begin{cases} e^{iwx-cx} & (x>0) \\ 0 & (x<0) \end{cases} \quad (c>0)$	$\dfrac{i}{\sqrt{2\pi}(w+\alpha+ic)}$
4.	$e^{-px^2} \quad R(p)>0$	$\dfrac{1}{\sqrt{2p}} e^{-\alpha^2/4p}$
5.	$\cos px^2$	$\dfrac{1}{\sqrt{2p}} \cos\left[\dfrac{\alpha^2}{4p} - \dfrac{\pi}{4}\right]$
6.	$\sin px^2$	$\dfrac{1}{\sqrt{2p}} \cos\left[\dfrac{\alpha^2}{4p} + \dfrac{\pi}{4}\right]$
7.	$\lvert x\rvert^{-p} \quad (0<p<1)$	$\sqrt{\dfrac{2}{\pi}} \dfrac{\Gamma(1-p)\sin\dfrac{p\pi}{2}}{\lvert\alpha\rvert^{(1-p)}}$
8.	$\dfrac{e^{-a\lvert x\rvert}}{\sqrt{\lvert x\rvert}}$	$\dfrac{\sqrt{\sqrt{(a^2+\alpha^2)}+a}}{\sqrt{a^2+\alpha^2}}$
9.	$\dfrac{\cosh ax}{\cosh \pi x} \quad (-\pi<a<\pi)$	$\sqrt{\dfrac{2}{\pi}} \dfrac{\cos\dfrac{a}{2}\cosh\dfrac{\alpha}{2}}{\cosh\alpha+\cos a}$
10.	$\dfrac{\sinh ax}{\sinh \pi x} \quad (-\pi<a<\pi)$	$\dfrac{1}{\sqrt{2\pi}} \dfrac{\sin a}{\cosh\alpha+\cos a}$
11	$\begin{cases} \dfrac{1}{\sqrt{a^2-x^2}} & (\lvert x\rvert<a) \\ 0 & (\lvert x\rvert>a) \end{cases}$	$\sqrt{\dfrac{\pi}{2}} J_0(a\alpha)$
12.	$\dfrac{\sin\left[b\sqrt{a^2+x^2}\right]}{\sqrt{a^2+x^2}}$	$\begin{cases} 0 & (\lvert\alpha\rvert>b) \\ \sqrt{\dfrac{\pi}{2}} J_0\left(\sqrt{b^2-\alpha^2}\right) & (\lvert\alpha\rvert<b) \end{cases}$
13.	$\begin{cases} P_n(x) & (\lvert x\rvert<1) \\ 0 & (\lvert x\rvert>1) \end{cases}$	$\dfrac{i^n}{\sqrt{\alpha}} J_{n+1/2}(\alpha)$
14.	$\begin{cases} \dfrac{\cos\left[b\sqrt{a^2-x^2}\right]}{\sqrt{a^2-x^2}} & (\lvert x\rvert<a) \\ 0 & (\lvert x\rvert>a) \end{cases}$	$\sqrt{\dfrac{\pi}{2}} J_0\left(a\sqrt{a^2+b^2}\right)$
15.	$\begin{cases} \dfrac{\cosh\left[b\sqrt{a^2-x^2}\right]}{\sqrt{a^2-x^2}} & (\lvert x\rvert<a) \\ 0 & (\lvert x\rvert>a) \end{cases}$	$\sqrt{\dfrac{\pi}{2}} J_0\left(a\sqrt{a^2-b^2}\right)$

The following functions appear among the entries of the tables on transforms.

Function	Definition	Name
$\mathrm{Ei}(x)$	$\int_{-\infty}^{x} \dfrac{e^v}{v}\,dv$; or sometimes defined as $-\mathrm{Ei}(-x) = \int_x^{\infty} \dfrac{e^{-v}}{v}\,dv$	Exponential integral function

Function	Definition	Name
$\mathrm{Si}(x)$	$\int_0^x \frac{\sin v}{v}\,dv$	Sine integral function
$\mathrm{Ci}(x)$	$\int_\infty^x \frac{\cos v}{v}\,dv$; or sometimes defined as negative of this integral	Cosine integral function
$\mathrm{erf}(x)$	$\frac{2}{\sqrt{\pi}}\int_0^x e^{-v^2}\,dv$	Error function
$\mathrm{erfc}(x)$	$1-\mathrm{erf}(x)=\frac{2}{\sqrt{\pi}}\int_x^\infty e^{-v^2}\,dv$	Complementary function to error function
$L_n(x)$	$\frac{e^x}{n!}\frac{d^n}{dx^n}\left(x^n e^{-x}\right),\quad n=0,1,\ldots$	Laguerre polynomial of degree n

1.10 Bessel Functions

Bessel Functions of the First Kind, $J_n(x)$ (Also Called Simply *Bessel Functions*) (Figure 1.13)

Domain: $[x > 0]$

Recurrence relation:

$$J_{n+1}(x) = \frac{2n}{x}J_n(x) - J_{n-1}(x), \quad n = 0, 1, 2, \ldots$$

Symmetry: $J_{-n}(x) = (-1)^n J_n(x)$

0. $J_0(20x)$ 3. $J_3(20x)$
1. $J_1(20x)$ 4. $J_4(20x)$
2. $J_2(20x)$ 5. $J_5(20x)$

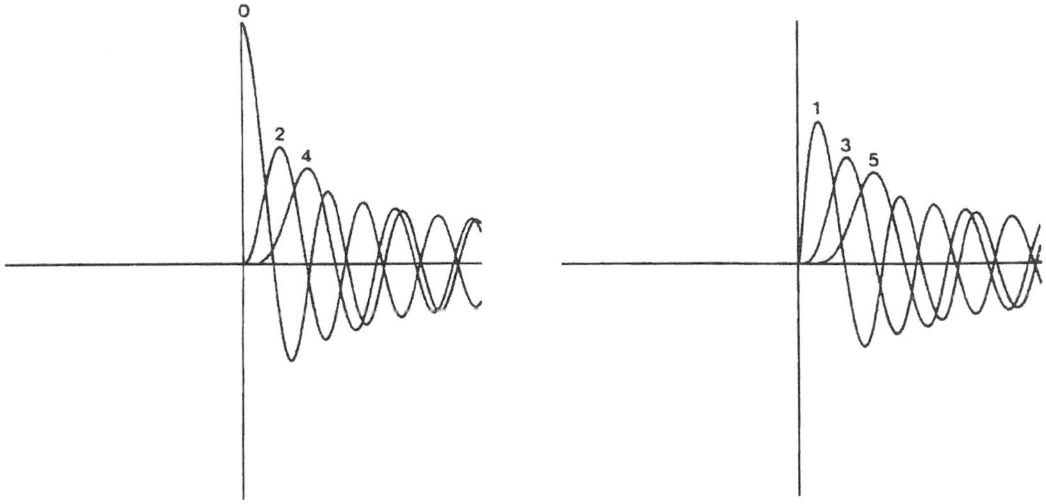

Figure 1.13 Bessel functions of the first kind.

Bessel Functions of the Second Kind, $Y_n(x)$ (Also Called *Neumann Functions* or *Weber Functions*) (Figure 1.14)

Domain: $[x > 0]$

Recurrence relation:

$$Y_{n+1}(x) = \frac{2n}{x} Y_n(x) - Y_{n-1}(x), \quad n = 0, 1, 2, \ldots$$

Symmetry: $Y_{-n}(x) = (-1)^n Y_n(x)$

0. $Y_0(20x)$ 3. $Y_3(20x)$
1. $Y_1(20x)$ 4. $Y_4(20x)$
2. $Y_2(20x)$ 5. $Y_5(20x)$

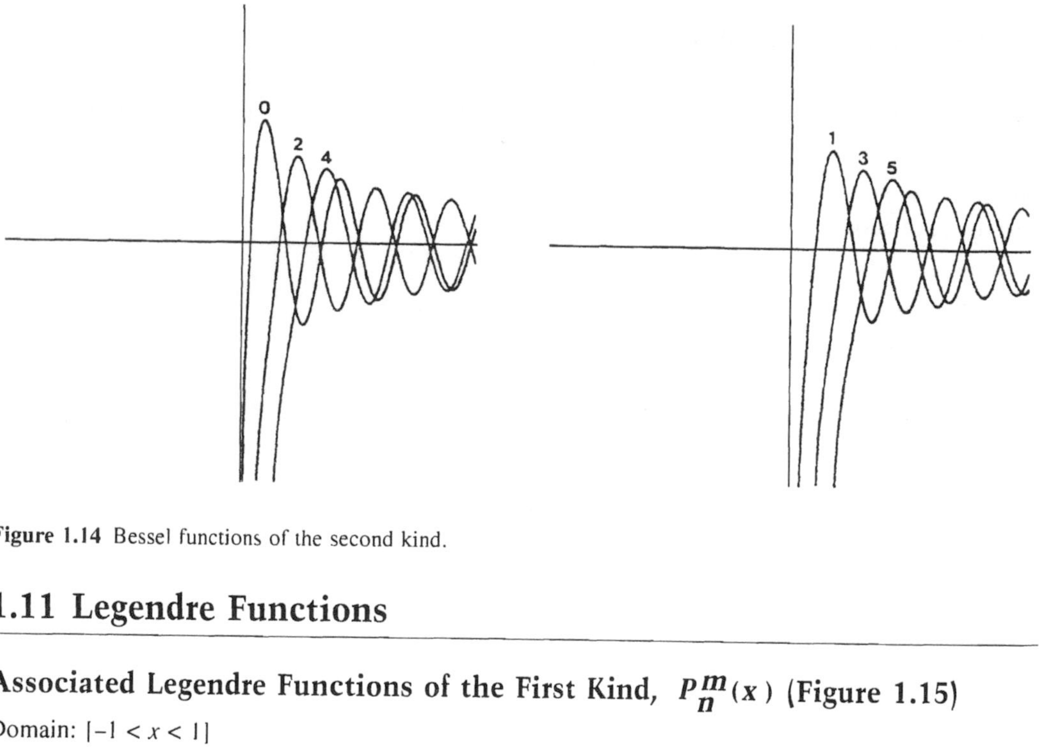

Figure 1.14 Bessel functions of the second kind.

1.11 Legendre Functions

Associated Legendre Functions of the First Kind, $P_n^m(x)$ (Figure 1.15)

Domain: $[-1 < x < 1]$

Recurrence relations:

$$P_{n+1}^m(x) = \frac{(2n+1)xP_n^m - (n+m)P_{n-1}^m(x)}{n-m+1}, \quad n = 1, 2, 3, \ldots$$

$$P_n^{m+1}(x) = \left(x^2 - 1\right)^{-1/2}\left[(n-m)xP_n^m(x) - (n+m)P_{n-1}^m(x)\right], \quad m = 0, 1, 2, \ldots$$

with

$$P_0^0 = 1 \qquad P_1^0 = x$$

Special case: P_n^0 = Legendre polynomials

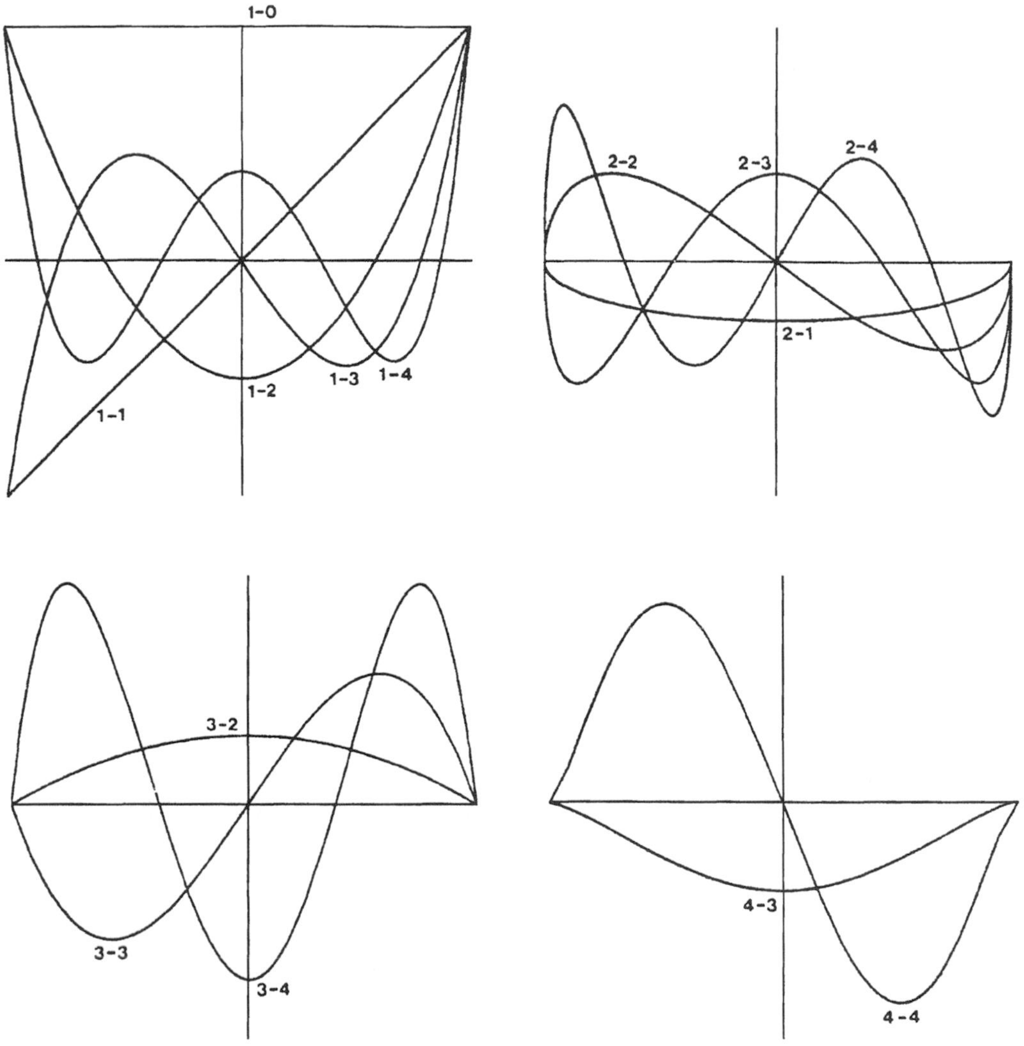

Figure 1.15 Legendre functions of the first kind.

1-0. $P_0^0(x)$
1-1. $P_1^0(x)$ 2-1. 0.25 $P_1^1(x)$
1-2. $P_2^0(x)$ 2-2. 0.25 $P_2^1(x)$ 3-2. 0.10 $P_2^2(x)$
1-3. $P_3^0(x)$ 2-3. 0.25 $P_3^1(x)$ 3-3. 0.10 $P_3^2(x)$ 4-3. 0.025 $P_3^3(x)$
1-4. $P_4^0(x)$ 2-4. 0.25 $P_4^1(x)$ 3-4. 0.10 $P_4^2(x)$ 4-4. 0.025 $P_4^3(x)$

1.12 Table of Differential Equations

Equation	Solution
1 $y' = \dfrac{dy}{dx} = f(x)$	$y = \int f(x)\,dx + c$
2. $y' + p(x)y = q(x)$	$y = \exp[-\int p(x)\,dx][c + \int \exp[\int p(x)\,dx]q(x)dx]$
3. $y' + p(x)y = q(x)y^{\alpha}$	Set $z = y^{1-\alpha} \rightarrow z' + (1-\alpha)p(x)z = (1-\alpha)q(x)$ and use 2

Equation	Solution
$\alpha \neq 0, \alpha \neq 1$	
4. $y' = f(x)g(y)$	Integrate $\dfrac{dy}{g(y)} = f(x)\,dx$ (separable)
5. $\dfrac{dy}{dx} = f(x/y)$	Set $y = xu \rightarrow u + x\dfrac{du}{dx} = f(u)$ $\displaystyle\int \frac{1}{f(u) - u}\,du = \ln\lvert x\rvert + c$
6. $y' = f\left(\dfrac{a_1x + b_1y + c_1}{a_2x + b_2y + c_2}\right)$	Set $x = X + \alpha, y = Y + \beta$ Choose $\begin{cases} a_1\alpha + b_1\beta = -c_1 \\ a_2\alpha + b_2\beta = -c_2 \end{cases} \rightarrow Y' = f\left(\dfrac{a_1X + b_1Y}{a_2X + b_2Y}\right)$ If $a_1b_2 - a_2b_1 \neq 0$, set $Y = Xu \rightarrow$ separable form $u + Xu' = f\left(\dfrac{a_1 + b_1u}{a_2 + b_2u}\right)$ If $a_1b_2 - a_2b_1 = 0$, set $u = a_1x + b_1y \rightarrow$ $\dfrac{du}{dx} = a_1 + b_1 f\left(\dfrac{u + c_1}{ku + c_2}\right)$ since $a_2x + b_2y = k(a.x + a_2y)$
7 $y'' + a^2y = 0$	$y = c_1 \cos ax + c_2 \sin ax$
8. $y'' - a^2y = 0$	$y = c_1e^{ax} + c_2e^{-ax}$
9. $y'' + ay' + by = 0$	Set $y = e^{-(a/2)x}u \rightarrow u'' + \left(b - \dfrac{a^2}{4}\right)u = 0$
10. $y'' + a(x)y' + b(x)y = 0$	Set $y = e^{-(1/2)\int a(x)dx} \rightarrow u'' + \left[b(x) - \dfrac{a^2}{4} - \dfrac{a'}{2}\right]u = 0$
11 $x^2y'' + xy' + (x^2 - a^2)\,y = 0$ $a \geq 0$ (Bessel)	i. If a is not an integer $y = c_1J_a(x) + c_2J_{-a}(x)$ (Bessel functions of first kind) ii. If a is an integer (say, n) $y = c_1J_n(x) + c_2Y_n(x)$ (Y_n is Bessel function of second kind)
12. $(1 - x^2)y'' - 2xy' + a(a + 1)y = 0$ a is real (Legendre)	$y(x) = c_1p_a(x) + c_1q_a(x)$ (Legendre functions)
13. $y' + ay^2 = bx^n$ (integrable Riccati) a, b, n real	Set $u' = ayu \rightarrow u'' - abx^nu = 0$ and use 14
14. $y'' - ax^{-1}y' + b^2x^{\mu}y = 0$	$y = x^p[c_1J_{\nu}(kx^q) + c_2J_{-\nu}(kx^q)$ where $p = (a + 1)/2,\ \nu = (a + 1)/(\mu + 2)$, $k = 2b/(\mu + 2),\ q = (\mu + 2)/2$

15. Item 13 shows that the Riccati equation is linearized by raising the order of the equation. The *Riccan chain*, which is linearizable by raising the order, is

$$u' = uy,\quad u'' = u[y' + y^2],\quad u''' = u[y'' + 3yy' + y^3],$$
$$u'''' = u[y''' + 4yy'' + 6y^2y' + 3(y')^2 + y^4]\ldots$$

To use this consider the second-order equation $y'' + 3yy' + y^3 = f(x)$. The Riccati transformation $u' = yu$ transforms this equation to the linear for $u''' = uf(x)$!

References

Kanke, E. 1956. *Differentialgleichungen Lösungsmethoden und Lösungen*. Vol. I. Akad. Verlagsges., Leipzig.

Murphy, G. M. 1960. *Ordinary Differential Equations and Their Solutions*, Van Nostrand, New York.

Zwillger, D. 1992. *Handbook of Differential Equations*, 2nd ed. Academic Press, San Diego.

2

Linear Algebra and Matrices

George Cain
Georgia Institute of Technology

2.1 Basic Definitions

A *Matrix* **A** is a rectangular array of numbers (real or complex)

$$\mathbf{A} = \begin{bmatrix} a_{11} & a_{12} & \cdots & a_{1m} \\ a_{21} & a_{22} & \cdots & a_{2m} \\ \vdots & & & \\ a_{n1} & a_{n2} & \cdots & a_{nm} \end{bmatrix}$$

The *size* of the matrix is said to be $n \times m$. The $1 \times m$ matrices $[a_{i1} \cdots a_{im}]$ are called rows of A, and the $n \times 1$ matrices

$$\begin{bmatrix} a_{1j} \\ a_{2j} \\ \vdots \\ a_{nj} \end{bmatrix}$$

are called *columns* of **A**. An $n \times m$ matrix thus consists of n rows and m columns; a_{ij} denotes the *element*, or *entry*, of **A** in the ith row and jth column. A matrix consisting of just one row is called a *row vector*, whereas a matrix of just one column is called a *column vector*. The elements of a vector are frequently called *components* of the vector. When the size of the matrix is clear from the context, we sometimes write $\mathbf{A} = (a_{ij})$.

A matrix with the same number of rows as columns is a *square* matrix, and the number of rows and columns is the *order* of the matrix. The diagonal of an $n \times n$ square matrix **A** from a_{11} to a_{nn} is called the *main*, or *principal*, *diagonal*. The word *diagonal* with no modifier usually means the main diagonal.

0-8493-0056-8/00/$0.00+$.50

The *transpose* of a matrix **A** is the matrix that results from interchanging the rows and columns of **A**. It is usually denoted by $\mathbf{A}^T$. A matrix **A** such that $\mathbf{A} = \mathbf{A}^T$ is said to be *symmetric*. The *conjugate transpose* of **A** is the matrix that results from replacing each element of $\mathbf{A}^T$ by its complex conjugate, and is usually denoted by $\mathbf{A}^H$. A matrix such that $\mathbf{A} = \mathbf{A}^H$ is said to be *Hermitian*.

A square matrix $\mathbf{A} = (a_{ij})$ is *lower triangular* if $a_{ij} = 0$ for $j > i$ and is *upper triangular* if $a_{ij} = 0$ for $j < i$. A matrix that is both upper and lower triangular is a *diagonal* matrix. The $n \times n$ *identity matrix* is the $n \times n$ diagonal matrix in which each element of the main diagonal is 1. It is traditionally denoted $\mathbf{I}_n$, or simply **I** when the order is clear from the context.

2.2 Algebra of Matrices

The sum and difference of two matrices **A** and **B** are defined whenever **A** and **B** have the same size. In that case $\mathbf{C} = \mathbf{A} \pm \mathbf{B}$ is defined by $\mathbf{C} = (c_{ij}) = (a_{ij} \pm b_{ij})$. The product $t\mathbf{A}$ of a scalar t (real or complex number) and a matrix **A** is defined by $t\mathbf{A} = (ta_{ij})$. If **A** is an $n \times m$ matrix and **B** is an $m \times p$ matrix, the product $\mathbf{C} = \mathbf{AB}$ is defined to be the $n \times p$ matrix $\mathbf{C} = (c_{ij})$ given by $c_{ij} = \sum_{k=1}^{m} a_{ik}b_{kj}$. Note that the product of an $n \times m$ matrix and an $m \times p$ matrix is an $n \times p$ matrix, and the product is defined only when the number of columns of the first factor is the same as the number of rows of the second factor. Matrix multiplication is, in general, associative: $\mathbf{A}(\mathbf{BC}) = (\mathbf{AB})\mathbf{C}$. It also distributes over addition (and subtraction):

$$\mathbf{A}(\mathbf{B} + \mathbf{C}) = \mathbf{AB} + \mathbf{AC} \quad \text{and} \quad (\mathbf{A} + \mathbf{B})\mathbf{C} = \mathbf{AC} + \mathbf{BC}$$

It is, however, not in general true that $\mathbf{AB} = \mathbf{BA}$, even in case both products are defined. It is clear that $(\mathbf{A} + \mathbf{B})^T = \mathbf{A}^T + \mathbf{B}^T$ and $(\mathbf{A} + \mathbf{B})^H = \mathbf{A}^H + \mathbf{B}^H$. It is also true, but not so obvious perhaps, that $(\mathbf{AB})^T = \mathbf{B}^T\mathbf{A}^T$ and $(\mathbf{AB})^H = \mathbf{B}^H\mathbf{A}^H$.

The $n \times n$ identity matrix **I** has the property that $\mathbf{IA} = \mathbf{AI} = \mathbf{A}$ for every $n \times n$ matrix **A**. If **A** is square, and if there is a matrix **B** such at $\mathbf{AB} = \mathbf{BA} = \mathbf{I}$, then **B** is called the *inverse* of **A** and is denoted $\mathbf{A}^{-1}$. This terminology and notation are justified by the fact that a matrix can have at most one inverse. A matrix having an inverse is said to be *invertible*, or *nonsingular*, while a matrix not having an inverse is said to be *noninvertible*, or *singular*. The product of two invertible matrices is invertible and, in fact, $(\mathbf{AB})^{-1} = \mathbf{B}^{-1}\mathbf{A}^{-1}$. The sum of two invertible matrices is, obviously, not necessarily invertible.

2.3 Systems of Equations

The system of n linear equations in m unknowns

$$a_{11}x_1 + a_{12}x_2 + a_{13}x_3 + \ldots + a_{1m}x_m = b_1$$

$$a_{21}x_1 + a_{22}x_2 + a_{23}x_3 + \ldots + a_{2m}x_m = b_2$$

$$\vdots$$

$$a_{n1}x_1 + a_{n2}x_2 + a_{n3}x_3 + \ldots + a_{nm}x_m = b_n$$

may be written $\mathbf{Ax} = \mathbf{b}$, where $\mathbf{A} = (a_{ij})$, $\mathrm{x} = [\mathrm{x}_1\ \mathrm{x}_2 \cdots \mathrm{x}_m]^T$, and $\mathrm{b} = [\mathrm{b}_1\ \mathrm{b}_2 \cdots \mathrm{b}_n]^T$. Thus A is an $n \times m$ matrix, and x and b are column vectors of the appropriate sizes.

The matrix **A** is called the *coefficient matrix* of the system. Let us first suppose the coefficient matrix is square; that is, there are an equal number of equations and unknowns. If **A** is upper triangular, it is quite easy to find all solutions of the system. The ith equation will contain only the unknowns $x_i, x_{i+1}, \ldots, x_n$, and one simply solves the equations in reverse order: the last equation is solved for x_n; the result is substituted into the $(n-1)$st equation, which is then solved for x_{n-1}; these values of x_n and x_{n-1} are substituted in the $(n-2)$th equation, which is solved for x_{n-2}, and so on. This procedure is known as *back substitution*.

The strategy for solving an arbitrary system is to find an upper-triangular system equivalent with it and solve this upper-triangular system using back substitution. First suppose the element $a_{11} \neq 0$. We may rearrange the equations to ensure this, unless, of course the first column of **A** is all 0s. In this case proceed to the next step, to be described later. For each $i \geq 2$ let $m_{i1} = a_{i1}/a_{11}$. Now replace the ith equation by the result of multiplying the first equation by m_{i1} and subtracting the new equation from the ith equation. Thus,

$$a_{i1}x_1 + a_{i2}x_2 + a_{i3}x_3 + \ldots + a_{in}x_n = b_i$$

is replaced by

$$0 \cdot x_1 + (a_{i2} + m_{i1}a_{12})x_2 + (a_{i3} + m_{i1}a_{13})x_3 + \ldots + (a_{in} + m_{i1}a_{1n})x_n = b_i + m_{i1}b_1$$

After this is done for all $i = 2, 3, \ldots, n$, there results the equivalent system

$$\begin{aligned}
a_{11}x_1 + a_{12}x_2 + a_{13}x_3 + \ldots + a_{1n}x_n &= b_1 \\
0 \cdot x_1 + a'_{22}x_2 + a'_{23}x_3 + \ldots + a'_{2n}x_n &= b'_2 \\
0 \cdot x_1 + a'_{32}x_2 + a'_{33}x_3 + \ldots + a'_{3n}x_n &= b'_3 \\
&\vdots \\
0 \cdot x_1 + a'_{n2}x_2 + a'_{n3}x_3 + \ldots + a'_{nn}x_n &= b'_n
\end{aligned}$$

in which all entries in the first column below a_{11} are 0. (Note that if all entries in the first column were 0 to begin with, then $a_{11} = 0$ also.) This procedure is now repeated for the $(n-1) \times (n-1)$ system

$$\begin{aligned}
a'_{22}x_2 + a'_{23}x_3 + \ldots + a'_{2n}x_n &= b'_2 \\
a'_{32}x_2 + a'_{33}x_3 + \ldots + a'_{3n}x_n &= b'_3 \\
&\vdots \\
a'_{n2}x_2 + a'_{n3}x_3 + \ldots + a'_{nn}x_n &= b'_n
\end{aligned}$$

to obtain an equivalent system in which all entries of the coefficient matrix below a'_{22} are 0. Continuing, we obtain an upper-triangular system Ux = c equivalent with the original system. This procedure is known as *Gaussian elimination*. The number m_{ij} are known as the *multipliers*.

Essentially the same procedure may be used in case the coefficient matrix is not square. If the coefficient matrix is not square, we may make it square by appending either rows or columns of 0s as needed. Appending rows of 0s and appending 0s to make b have the appropriate size equivalent to appending equations 0 = 0 to the system. Clearly the new system has precisely the same solutions as the original system. Appending columns of 0s and adjusting the size of x appropriately yields a new system with additional unknowns, each appearing only with coefficient 0, thus not affecting the solutions of the original system. In either case we may assume the coefficient matrix is square, and apply the Gauss elimination procedure.

Suppose the matrix **A** is invertible. Then if there were no row interchanges in carrying out the above Gauss elimination procedure, we have the *LU factorization* of the matrix **A**:

$$\mathbf{A} = \mathbf{LU}$$

where U is the upper-triangular matrix produced by elimination and L is the lower-triangular matrix given by

$$\mathbf{L} = \begin{bmatrix} 1 & 0 & \cdots & \cdots & 0 \\ m_{21} & 1 & 0 & \cdots & 0 \\ \vdots & & \ddots & & \\ m_{n1} & m_{n2} & \cdots & & 1 \end{bmatrix}$$

A *permutation* P_{ij} matrix is an $n \times n$ matrix such that $P_{ij}A$ is the matrix that results from exchanging row i and j of the matrix A. The matrix P_{ij} is the matrix that results from exchanging rows i and j of the identity matrix. A product P of such matrices P_{ij} is called a *permutation* matrix. If row interchanges are required in the Gauss elimination procedure, then we have the factorization

$$\mathbf{PA} = \mathbf{LU}$$

where P is the permutation matrix giving the required row exchanges.

2.4 Vector Spaces

The collection of all column vectors with n real components is *Euclidean n-space*, and is denoted R^n. The collection of column vectors with n complex components is denoted C^n. We shall use *vector space* to mean either R^n or C^n. In discussing the space R^n, the word *scalar* will mean a real number, and in discussing the space C^n, it will mean a complex number. A subset S of a vector space is a *subspace* such that if **u** and **v** are vectors in S, and if c is any scalar, then $\mathbf{u} + \mathbf{v}$ and $c\mathbf{u}$ are in S. We shall sometimes use the word *space* to mean a subspace. If $B = \{v_1, v_2, \ldots, v_k\}$ is a collection of vectors in a vector space, then the set S consisting of all vectors $c_1v_1 + c_2v_2 + \cdots + c_mv_m$ for all scalars $c_1, c_2, \ldots, c_m$ is a subspace, called the *span* of B. A collection $\{v_1, v_2, \ldots, v_m\}$ of vectors $c_1v_1 + c_2v_2 + \cdots + c_mv_m$ is a *linear combination* of B. If S is a subspace and $B = \{v_1, v_2, \ldots, v_m\}$ is a subset of S such that S is the span of B, then B is said to *span* S.

A collection $\{v_1, v_2, \ldots, v_m\}$ of n-vectors is *linearly dependent* if there exist scalars $c_1, c_2, \ldots, c_m$, not all zero, such that $c_1v_1 + c_2v_2 + \cdots + c_mv_m = 0$. A collection of vectors that is not linearly dependent is said to be *linearly independent*. The modifier *linearly* is frequently omitted, and we speak simply of dependent and independent collections. A linearly independent collection of vectors in a space S that spans S is a *basis* of S. Every basis of a space S contains the same number of vectors; this number is the *dimension* of S. The dimension of the space consisting of only the zero vector is 0. The collection $B = \{e_1, e_2, \ldots, e_n\}$, where $e_1 = [1,0,0,\ldots,0]^T$, $e_2 = [0,1,0,\ldots,0]^T$, and so forth (e_i has 1 as its ith component and zero for all other components) is a basis for the spaces R^n and C^n. This is the *standard basis* for these spaces. The dimension of these spaces is thus n. In a space S of dimension n, no collection of fewer than n vectors can span S, and no collection of more than n vectors in S can be independent.

2.5 Rank and Nullity

The *column space* of an $n \times m$ matrix **A** is the subspace of R^n or C^n spanned by the columns of **A**. The *row space* is the subspace of R^m or C^m spanned by the rows or **A**. Note that for any vector $x = [x_1\, x_2, \cdots x_m]^T$,

$$\mathbf{Ax} = x_1\begin{bmatrix} a_{11} \\ a_{21} \\ \vdots \\ a_{n1} \end{bmatrix} + x_2\begin{bmatrix} a_{12} \\ a_{22} \\ \vdots \\ a_{n2} \end{bmatrix} + \ldots + x_m\begin{bmatrix} a_{1m} \\ a_{2m} \\ \vdots \\ a_{nm} \end{bmatrix}$$

so that the column space is the collection of all vectors, **Ax**, and thus the system **Ax** = **b** has a solution if and only *if* **b** is a member of the column space of **A**.

The dimension of the column space is the *rank* of **A**. The row space has the same dimension as the column space. The set of all solutions of the system **Ax** = 0 is a subspace called the *null space* of **A**, and the dimension of this null space is the *nullity* of **A**. A fundamental result in matrix theory is the fact that, for an $n \times m$ matrix **A**.

$$\text{rank } \mathbf{A} + \text{nullity } \mathbf{A} = m$$

The difference of any two solutions of the linear system **Ax** = **b** is a member of the null space of **A**. Thus this system has at most one solution if and only if the nullity of **A** is zero. If the system is square (that is, if **A** is $n \times n$), then there will be a solution for every right-hand side **b** if and only if the collection of columns of **A** is linearly independent, which is the same as saying the rank of **A** is n. In this case the nullity must be zero. Thus, for any **b**, the square system **Ax** = **b** has exactly one solution if and only if rank **A** = n. In other words the $n \times n$ matrix **A** is invertible if and only if rank **A** = n.

2.6 Orthogonality and Length

The *inner product* of two vectors x and y is the scalar $x^H y$. The *length*, or *norm*, $\|x\|$, of the vector **x** is given by $\|x\| = \sqrt{\mathbf{x}^H\mathbf{x}}$. A *unit vector* is a vector of norm 1. Two vectors x and y are *orthogonal* if $x^H y = 0$. A collection of vectors $\{v_1, v_2, \ldots, v_m\}$ in a space S is said to be an *orthonormal* collection if $\mathbf{v}_i^H \mathbf{v}_j = 0$ for $i \neq j$ and $\mathbf{v}_i^H \mathbf{v}_i = 1$. An orthonormal collection is necessarily linearly independent. If S is a subspace (of R^n or C^n) spanned by the orthonormal collection $\{v_1, v_2, \ldots, v_m\}$, then the *projection* of a vector x onto S is the vector

$$\text{proj}(\mathbf{x}; \mathbf{S}) = \left(\mathbf{x}^H\mathbf{v}_1\right)\mathbf{v}_1 + \left(\mathbf{x}^H\mathbf{v}_2\right)\mathbf{v}_2 + \ldots + \left(\mathbf{x}^H\mathbf{v}_m\right)\mathbf{v}_m$$

The projection of x onto S minimizes the function $f(y) = \|x - y\|^2$ for $y \in S$. In other words the projection of x onto S is the vector in S that is "closest" to x.

If **b** is a vector and **A** is an $n \times m$ matrix, then a vector x minimizes $\|b - Ax\|^2$ if only if it is a solution of $A^H Ax = A^H b$. This system of equations is called the *system of normal equations* for the least-squares problem of minimizing $\|b - Ax\|^2$.

If **A** is an $n \times m$ matrix, and rank **A** = k, then there is a $n \times k$ matrix Q whose columns form an orthonormal basis for the column space of **A** and a $k \times m$ upper-triangular matrix **R** of rank k such that

$$\mathbf{A} = \mathbf{QR}$$

This is called the *QR factorization* of **A**. It now follows that x minimizes $\|b - Ax\|^2$ if and only if it is a solution of the upper-triangular system $Rx = Q^H b$.

If $\{w_1, w_2, \ldots, w_m\}$ is a basis for a space S, the following procedure produces an orthonormal basis $\{v_1, v_2, \ldots, v_m\}$ for S.

Set $\mathbf{v}_1 = \mathbf{w}_1/\|\mathbf{w}_1\|$.

Let $\tilde{\mathbf{v}}_2 = \mathbf{w}_2 - \text{proj}(\mathbf{w}_2; \mathbf{S}_1)$, where $\mathbf{S}_1$ is the span of $\{\mathbf{v}_1\}$; set $\mathbf{v}_2 = \tilde{\mathbf{v}}_2/\|\tilde{\mathbf{v}}_2\|$.

Next, let $\tilde{\mathbf{v}}_3 = \mathbf{w}_3 - \text{proj}(\mathbf{w}_3; \mathbf{S}_2)$, where $\mathbf{S}_2$ is the span of $\{\mathbf{v}_1, \mathbf{v}_2\}$; set $\mathbf{v}_3 = \tilde{\mathbf{v}}_3/\|\tilde{\mathbf{v}}_3\|$.

And, so on: $\tilde{\mathbf{v}}_i = w_i - \text{proj}(w_i; S_{i-1})$, where S_{i-1} is the span of $\{v_1, v_2, \ldots, v_{i-1}\}$; set $v_i = \tilde{\mathbf{v}}_i/\|\tilde{\mathbf{v}}_i\|$. This the *Gram-Schmidt procedure.*

If the collection of columns of a square matrix is an orthonormal collection, the matrix is called a *unitary matrix*. In case the matrix is a real matrix, it is usually called an *orthogonal matrix*. A unitary matrix U is invertible, and $\mathbf{U}^{-1} = \mathbf{U}^H$. (In the real case an orthogonal matrix **Q** is invertible, and $\mathbf{Q}^{-1} = \mathbf{Q}^T$.)

2.7 Determinants

The *determinant* of a square matrix is defined inductively. First, suppose the determinant det **A** has been defined for all square matrices of order < n. Then

$$\det \mathbf{A} = a_{11}\mathbf{C}_{11} + a_{12}\mathbf{C}_{12} + \ldots + a_{1n}\mathbf{C}_{1n}$$

where the numbers C_{ij} are *cofactors* of the matrix **A**:

$$\mathbf{C}_{ij} = (-1)^{i+j} \det \mathbf{M}_{ij}$$

where M_{ij} is the $(n - 1) \times (n - 1)$ matrix obtained by deleting the ith row and jth column of **A**. Now det**A** is defined to be the only entry of a matrix of order 1. Thus, for a matrix of order 2, we have

$$\det \begin{bmatrix} a & b \\ c & d \end{bmatrix} = ad - bc$$

There are many interesting but not obvious properties of determinants. It is true that

$$\det \mathbf{A} = a_{i1}\mathbf{C}_{i1} + a_{i2}\mathbf{C}_{i2} + \ldots + a_{in}\mathbf{C}_{in}$$

for any $1 \le i \le n$. It is also true that det**A** = det**A**^{T}, so that we have

$$\det \mathbf{A} = a_{1j}\mathbf{C}_{1j} + a_{2j}\mathbf{C}_{2j} + \ldots + a_{nj}\mathbf{C}_{nj}$$

for any $1 \le j \le n$.

If **A** and **B** are matrices of the same order, then det**AB** = (det**A**)(det**B**), and the determinant of any identity matrix is 1. Perhaps the most important property of the determinant is the fact that a matrix in invertible if and only if its determinant is not zero.

2.8 Eigenvalues and Eigenvectors

If **A** is a square matrix, and **Av** = λv for a scalar λ and a nonzero v, then λ is an *eigenvalue* of **A** and v is an *eigenvector* of **A** that *corresponds* to λ. Any nonzero linear combination of eigenvectors corresponding to the same eigenvalue λ is also an eigenvector corresponding to λ. The collection of all eigenvectors corresponding to a given eigenvalue λ is thus a subspace, called an *eigenspace* of **A**. A collection of eigenvectors corresponding to different eigenvalues is necessarily linear-independent. It follows that a matrix of order n can have at most n distinct eigenvectors. In fact, the eigenvalues of **A** are the roots of the nth degree polynomial equation

$$\det(\mathbf{A} - \lambda\mathbf{I}) = 0$$

called the *characteristic equation* of **A**. (Eigenvalues and eigenvectors are frequently called *characteristic values* and *characteristic vectors.*)

If the nth order matrix **A** has an independent collection of n eigenvectors, then **A** is said to have a *full set* of eigenvectors. In this case there is a set of eigenvectors of **A** that is a basis for R^n or, in the complex case, C^n. In case there are n distinct eigenvalues of **A**, then, of course, **A** has a full set of eigenvectors. If there are fewer than n distinct eigenvalues, then **A** may or may not have a full set of eigenvectors. If there is a full set of eigenvectors, then

$$\mathbf{D} = \mathbf{S}^{-1}\mathbf{A}\mathbf{S} \quad \text{or} \quad \mathbf{A} = \mathbf{S}\mathbf{D}\mathbf{S}^{-1}$$

where **D** is a diagonal matrix with the eigenvalues of **A** on the diagonal, and **S** is a matrix whose columns are the full set of eigenvectors. If **A** is symmetric, there are n real distinct eigenvalues of **A** and the corresponding eigenvectors are orthogonal. There is thus an orthonormal collection of eigenvectors that span R^n, and we have

$$\mathbf{A} = \mathbf{Q}\mathbf{D}\mathbf{Q}^T \quad \text{and} \quad \mathbf{D} = \mathbf{Q}^T\mathbf{A}\mathbf{Q}$$

where **Q** is a real orthogonal matrix and **D** is diagonal. For the complex case, if **A** is Hermitian, we have

$$\mathbf{A} = \mathbf{U}\mathbf{D}\mathbf{U}^H \quad \text{and} \quad \mathbf{D} = \mathbf{U}^H\mathbf{A}\mathbf{U}$$

where **U** is a unitary matrix and **D** is a *real* diagonal matrix. (A Hermitian matrix also has n distinct real eigenvalues.)

References

Daniel, J. W. and Nobel, B. 1988. *Applied Linear Algebra.* Prentice Hall, Englewood Cliffs, NJ.

Strang, G. 1993. *Introduction to Linear Algebra.* Wellesley-Cambridge Press, Wellesley, MA.

3

Vector Algebra and Calculus

George Cain
Georgia Institute of Technology

3.1 Basic Definitions

A vector is a directed line segment, with two vectors being equal if they have the same length and the same direction. More precisely, a *vector* is an equivalence class of directed line segments, where two directed segments are equivalent if they have the same length and the same direction. The *length* of a vector is the common length of its directed segments, and the *angle between* vectors is the angle between any of their segments. The length of a vector u is denoted $|\mathbf{u}|$. There is defined a distinguished vector having zero length, which is usually denoted $\mathbf{0}$. It is frequently useful to visualize a directed segment as an arrow; we then speak of the nose and the tail of the segment. The *sum* $\mathbf{u} + \mathbf{v}$ of two vectors $\mathbf{u}$ and $\mathbf{v}$ is defined by taking directed segments from $\mathbf{u}$ and $\mathbf{v}$ and placing the tail of the segment representing $\mathbf{v}$ at the nose of the segment representing $\mathbf{u}$ and defining $\mathbf{u} + \mathbf{v}$ to be the vector determined by the segment from the tail of the $\mathbf{u}$ representative to the nose of the $\mathbf{v}$ representative. It is easy to see that $\mathbf{u} + \mathbf{v}$ is well defined and that $\mathbf{u} + \mathbf{v} = \mathbf{v} + \mathbf{u}$. Subtraction is the inverse operation of addition. Thus the *difference* $\mathbf{u} - \mathbf{v}$ of two vectors is defined to be the vector that when added to $\mathbf{v}$ gives $\mathbf{u}$. In other words, if we take a segment from $\mathbf{u}$ and a segment from $\mathbf{v}$ and place their tails together, the difference is the segment from the nose of v to the nose of $\mathbf{u}$. The zero vector behaves as one might expect: $\mathbf{u} + \mathbf{0} = \mathbf{u}$, and $\mathbf{u} - \mathbf{u} = \mathbf{0}$. Addition is associative: $\mathbf{u} + (\mathbf{v} + \mathbf{w}) = (\mathbf{u} + \mathbf{v}) + \mathbf{w}$.

To distinguish them from vectors, the real numbers are called *scalars*. The product $t\mathbf{u}$ of a scalar t and a vector $\mathbf{u}$ is defined to be the vector having length $|t|\,|\mathbf{u}|$ and direction the same as $\mathbf{u}$ if $t > 0$, the opposite direction if $t < 0$. If $t = 0$, then $t\mathbf{u}$ is defined to be the zero vector. Note that $t(\mathbf{u} + \mathbf{v}) = t\mathbf{u} + t\mathbf{v}$, and $(t + s)\mathbf{u} = t\mathbf{u} + s\mathbf{u}$. From this it follows that $\mathbf{u} - \mathbf{v} = \mathbf{u} + (-1)\mathbf{v}$.

The *scalar product* $\mathbf{u} \cdot \mathbf{v}$ of two vectors is $|\mathbf{u}||\mathbf{v}| \cos$ **q**, where θ is the angle between $\mathbf{u}$ and $\mathbf{v}$. The scalar product is frequently called the *dot product*. The scalar product distributes over addition:

$$\mathbf{u} \cdot (\mathbf{v} + \mathbf{w}) = \mathbf{u} \cdot \mathbf{v} + \mathbf{u} \cdot \mathbf{w}$$

and it is clear that $(t\mathbf{u}) \cdot \mathbf{v} = t(\mathbf{u} \cdot \mathbf{v})$. The *vector product* $\mathbf{u} \times \mathbf{v}$ of two vectors is defined to be the vector perpendicular to both $\mathbf{u}$ and $\mathbf{v}$ and having length $\mathbf{u}||\mathbf{v}| \sin \theta$, where θ is the angle between $\mathbf{u}$ and $\mathbf{v}$. The

0-8493-0056-8/00/$0.00+$.50

direction of $\mathbf{u} \times \mathbf{v}$ is the direction a right-hand threaded bolt advances if the vector **u** is rotated to **v**. The vector is frequently called the *cross product*. The vector product is both associative and distributive, but not commutative: $\mathbf{u} \times \mathbf{v} = -\mathbf{v} \times \mathbf{u}$.

3.2 Coordinate Systems

Suppose we have a right-handed Cartesian coordinate system in space. For each vector, **u**, we associate a point in space by placing the tail of a representative of **u** at the origin and associating with **u** the point at the nose of the segment. Conversely, associated with each point in space is the vector determined by the directed segment from the origin to that point. There is thus a one-to-one correspondence between the points in space and all vectors. The origin corresponds to the zero vector. The coordinates of the point associated with a vector **u** are called *coordinates* of **u**. One frequently refers to the vector **u** and writes $\mathbf{u} = (x, y, z)$, which is, strictly speaking, incorrect, because the left side of this equation is a vector and the right side gives the coordinates of a point in space. What is meant is that (x, y, z) are the coordinates of the point associated with **u** under the correspondence described. In terms of coordinates, for $\mathbf{u} = (u_1, u_2, u_3)$ and $\mathbf{v} = (v_1, v_2, v_3)$, we have

$$\mathbf{u} + \mathbf{v} = (u_1 + v_1, u_2 + v_2, u_3 + v_3)$$

$$t\mathbf{u} = (tu_1, tu_2, tu_3)$$

$$\mathbf{u} \cdot \mathbf{v} = u_1 v_1 + u_2 v_2 + u_3 v_3$$

$$\mathbf{u} \times \mathbf{v} = (u_2 v_3 - v_2 u_3, u_3 v_1 - v_3 u_1, u_1 v_2 - v_1 u_2)$$

The *coordinate vectors* **i**, **j**, and **k** are the unit vectors i = (1, 0, 0), j = (0, 1, 0), and k = (0, 0, 1). Any vector $\mathbf{u} = (u_1, u_2, u_3)$ is thus a linear combination of these coordinate vectors: $u = u_1 i + u_2 j + u_3 k$. A convenient form for the vector product is the formal determinant

$$\mathbf{u} \times \mathbf{v} = \det \begin{bmatrix} \mathbf{i} & \mathbf{j} & \mathbf{k} \\ u_1 & u_2 & u_3 \\ v_1 & v_2 & v_2 \end{bmatrix}$$

3.3 Vector Functions

A *vector function* **F** *of one variable* is a rule that associates a vector $\mathbf{F}(t)$ with each real number t in some set, called the *domain* of **F**. The expression $\lim_{t \to t_0} \mathbf{F}(t) = \mathbf{a}$ means that for any $\varepsilon > 0$, there is a $\delta > 0$ such that $|\mathbf{F}(t) - \mathbf{a}| < \varepsilon$ whenever $0 < |t - t_0| < \delta$. If $\mathbf{F}(t) = [x(t), y(t), z(t)]$ and $\mathbf{a} = (a_1, a_2, a_3)$, then $\lim_{t \to t_0} \mathbf{F}(t) = \mathbf{a}$ if and only if

$$\lim_{t \to t_0} x(t) = a_1$$

$$\lim_{t \to t_0} y(t) = a_2$$

$$\lim_{t \to t_0} z(t) = a_3$$

A vector function **F** is *continuous* at t_0 if $\lim_{t \to t_0} \mathbf{F}(t) = \mathbf{F}(t_0)$. The vector function **F** is continuous at t_0 if and only if each of the coordinates $x(t)$, $y(t)$, and $z(t)$ is continuous at t_0.

The function **F** is *differentiable* at t_0 if the limit

$$\lim_{h\to 0}\frac{1}{h}\left[\mathbf{F}(t+h)-\mathbf{F}(t)\right]$$

exists. This limit is called the *derivative* of **F** at t_0 and is usually written $\mathbf{F}'(t_0)$, or $(d\mathbf{F}/dt)(t_0)$. The vector function **F** is differentiable at t_0 if and only if each of its coordinate functions is differentiable at t_0. Moreover, $(d\mathbf{F}/dt)(t_0) = [(dx/dt)(t_0), (dy/dt)(t_0), (dz/dt)(t_0)]$. The usual rules for derivatives of real valued functions all hold for vector functions. Thus if **F** and **G** are vector functions and s is a scalar function, then

$$\frac{d}{dt}(\mathbf{F}+\mathbf{G}) = \frac{d\mathbf{F}}{dt}+\frac{d\mathbf{G}}{dt}$$

$$\frac{d}{dt}(s\mathbf{F}) = s\frac{d\mathbf{F}}{dt}+\frac{ds}{dt}\mathbf{F}$$

$$\frac{d}{dt}(\mathbf{F}\cdot\mathbf{G}) = \mathbf{F}\cdot\frac{d\mathbf{G}}{dt}+\frac{d\mathbf{F}}{dt}\cdot\mathbf{G}$$

$$\frac{d}{dt}(\mathbf{F}\times\mathbf{G}) = \mathbf{F}\times\frac{d\mathbf{G}}{dt}+\frac{d\mathbf{F}}{dt}\times\mathbf{G}$$

If **R** is a vector function defined for t in some interval, then, as t varies, with the tail of **R** at the origin, the nose traces out some object C in space. For nice functions **R**, the object C is a *curve*. If $\mathbf{R}(t) = [x(t), y(t), z(t)]$, then the equations

$$x = x(t)$$

$$y = y(t)$$

$$z = z(t)$$

are called *parametric equations* of C. At points where **R** is differentiable, the derivative $d\mathbf{R}/dt$ is a vector *tangent* to the curve. The unit vector $\mathbf{T} = (d\mathbf{R}/dt)/|d\mathbf{R}/dt|$ is called the *unit tangent vector*. If **R** is differentiable and if the length of the arc of curve described by **R** between $\mathbf{R}(a)$ and $\mathbf{R}(t)$ is given by $s(t)$, then

$$\frac{ds}{dt} = \left|\frac{d\mathbf{R}}{dt}\right|$$

Thus the length L of the arc from $\mathbf{R}(t_0)$ to $\mathbf{R}(t_1)$ is

$$L = \int_{t_0}^{t_1}\frac{ds}{dt}\,dt = \int_{t_0}^{t_1}\left|\frac{d\mathbf{R}}{dt}\right|dt$$

The vector $d\mathbf{T}/ds = (d\mathbf{T}/dt)/(ds/dt)$ is perpendicular to the unit tangent **T**, and the number $\kappa = |d\mathbf{T}/ds|$ is the *curvature* of C. The unit vector $\mathbf{N} = (1/\kappa)(d\mathbf{T}/ds)$ is the *principal normal*. The vector $\mathbf{B} = \mathbf{T}\times\mathbf{N}$ is the *binormal*, and $d\mathbf{B}/ds = -\tau N$. The number τ is the *torsion*. Note that C is a plane curve if and only if τ is zero for all t.

A *vector function* **F** *of two variables* is a rule that assigns a vector $\mathbf{F}(s, t)$ in some subset of the plane, called the *domain* of **F**. If $\mathbf{R}(s, t)$ is defined for all (s, t) in some region D of the plane, then as the point (s, t) varies over D, with its rail at the origin, the nose of $\mathbf{R}(s, t)$ traces out an object in space. For a nice function **R**, this object is a *surface*, *S*. The partial derivatives $(\partial \mathbf{R}/\partial s)(s, t)$ and $(\partial \mathbf{R}/\partial t)(s, t)$ are tangent to the surface at $\mathbf{R}(s, t)$, and the vector $(\partial \mathbf{R}/\partial s) \times (\partial \mathbf{R}/\partial t)$ is thus *normal* to the surface. Of course, $(\partial \mathbf{R}/\partial t) \times (\partial \mathbf{R}/\partial s) = -(\partial \mathbf{R}/\partial s) \times (\partial \mathbf{R}/\partial t)$ is also normal to the surface and points in the direction opposite that of $(\partial \mathbf{R}/\partial s) \times (\partial \mathbf{R}/\partial t)$. By electing one of these normal, we are choosing an *orientation* of the surface. A surface can be oriented only if it has two sides, and the process of orientation consists of choosing which side is "positive" and which is "negative."

3.4 Gradient, Curl, and Divergence

If $f(x, y, z)$ is a scalar field defined in some region *D*, the *gradient* of **f** is the vector function

$$\text{grad } f = \frac{\partial f}{\partial x}\mathbf{i} + \frac{\partial f}{\partial y}\mathbf{j} + \frac{\partial f}{\partial z}\mathbf{k}$$

If $\mathbf{F}(x, y, z) = F_1(x, y, z)\mathbf{i} + F_2(x, y, z)\mathbf{j} + F_3(x, y, z)\mathbf{k}$ is a vector field defined in some region *D*, then the *divergence* of **F** is the scalar function

$$\text{div } \mathbf{F} = \frac{\partial F_1}{\partial x} + \frac{\partial F_2}{\partial y} + \frac{\partial F_3}{\partial z}$$

The curl is the vector function

$$\text{curl } \mathbf{F} = \left(\frac{\partial F_3}{\partial y} - \frac{\partial F_2}{\partial z}\right)\mathbf{i} + \left(\frac{\partial F_1}{\partial z} - \frac{\partial F_3}{\partial x}\right)\mathbf{j} + \left(\frac{\partial F_2}{\partial x} - \frac{\partial F_1}{\partial y}\right)\mathbf{k}$$

In terms of the vector operator *del*, $\nabla = \mathbf{i}(\partial/\partial x) + \mathbf{j}(\partial/\partial y) + \mathbf{k}(\partial/\partial z)$, we can write

$$\text{grad } f = \nabla f$$

$$\text{div } \mathbf{F} = \nabla \cdot \mathbf{F}$$

$$\text{curl } \mathbf{F} = \nabla \times \mathbf{F}$$

The *Laplacian operator* is div (grad) $= \nabla \cdot \nabla = \nabla^2 = (\partial^2/\partial x^2) + (\partial^2/\partial y^2) + (\partial^2/\partial z^2)$.

3.5 Integration

Suppose *C* is a curve from the point (x_0, y_0, z_0) to the point (x_1, y_1, z_1) and is described by the vector function $\mathbf{R}(t)$ for $t_0 \leq t \leq t_1$. If *f* f is a scalar function (sometimes called a *scalar field*) defined on *C*, then the integral of *f* over *C* is

$$\int_C f(x, y, z)\, ds = \int_{t_0}^{t_1} f[\mathbf{R}(t)] \left|\frac{d\mathbf{R}}{dt}\right| dt$$

If **F** is a vector function (sometimes called a *vector field*) defined on *C*, then the integral of **F** over *C* is

$$\int_C \mathbf{F}(x, y, z) \cdot d\mathbf{R} = \int_{t_0}^{t_1} \mathbf{F}[\mathbf{R}(t)] \frac{d\mathbf{R}}{dt}\, dt$$

These integrals are called *line integrals*.

In case there is a scalar function f such that $\mathbf{F} = \text{grad}\, f$, then the line integral

$$\int_C \mathbf{F}(x, y, z) \cdot d\mathbf{R} = f[\mathbf{R}(t_1)] - f[\mathbf{R}(t_0)]$$

The value of the integral thus depends only on the end points of the curve C and not on the curve C itself. The integral is said to be *path-independent*. The function f is called a *potential function* for the vector field $\mathbf{F}$, and $\mathbf{F}$ is said to be a *conservative field*. A vector field $\mathbf{F}$ with domain D is conservative if and only if the integral of $\mathbf{F}$ around every closed curve in D is zero. If the domain D is simply connected (that is, every closed curve in D can be continuously deformed in D to a point), then $\mathbf{F}$ is conservative if and only if curl $\mathbf{F} = 0$ in D.

Suppose S is a surface described by $\mathbf{R}(s, t)$ for (s, t) in a region D of the plane. If f is a scalar function defined on D, then the integral of f over S is given by

$$\iint_S f(x, y, z)\, dS = \iint_D f[\mathbf{R}(s, t)] \left| \frac{\partial \mathbf{R}}{\partial s} \times \frac{\partial \mathbf{R}}{\partial t} \right| ds\, dt$$

If $\mathbf{F}$ is a vector function defined on S, and if an orientation for S is chosen, then the integral $\mathbf{F}$ over S, sometimes called the flux of $\mathbf{F}$ through S, is

$$\iint_S \mathbf{F}(x, y, z) \cdot d\mathbf{S} = \iint_D \mathbf{F}[\mathbf{R}(s, t)] \left| \frac{\partial \mathbf{R}}{\partial s} \times \frac{\partial \mathbf{R}}{\partial t} \right| ds\, dt$$

3.6 Integral Thorems

Suppose $\mathbf{F}$ is a vector field with a closed domain D bounded by the surface S oriented so that the normal points out from D. Then the *divergence theorem* states that

$$\iiint_D \text{div}\, \mathbf{F}\, dV = \iint_S \mathbf{F} \cdot \mathbf{dS}$$

If S is an orientable surface bounded by a closed curve C, the orientation of the closed curve C is chosen to be consistent with the orientation of the surface S. Then we have *Stoke's theorem:*

$$\iint_S (\text{curl}\, \mathbf{F}) \cdot d\mathbf{S} = \oint_C \mathbf{F} \cdot \mathbf{ds}$$

References

Davis, H. F. and Snider, A. D. 1991. *Introduction to Vector Analysis*. 6th ed., Wm. C. Brown, Dubuque, IA.

Wylie, C. R. 1975. *Advanced Engineering Mathematics*, 4th ed., McGraw-Hill, New York.

Further Information

More advanced topics leading into the theory and applications of tensors may be found in J. G. Simmonds, *A Brief on Tensor Analysis* (1982, Springer-Verlag, New York).

4

Difference Equations

William F. Ames
Georgia Institute of Technology

Difference equations are equations involving *discrete variables*. They appear as natural descriptions of natural phenomena and in the study of discretization methods for differential equations, which have continuous variables.

Let $y_n = y(nh)$, where n is an integer and h is a real number. (One can think of measurements taken at equal intervals, h, $2h$, $3h$, ..., and y_n describes these). A typical equation is that describing the famous Fibonacci sequence — $y_{n+2} - y_{n+1} - y_n = 0$. Another example is the equation $y_{n+2} - 2zy_{n+1} + y_n = 0$, $z \in C$, which describes the Chebyshev polynomials.

4.1 First-Order Equations

The general first-order equation $y_{n+1} = f(y_n)$, $y_0 = y(0)$ is easily solved, for as many terms as are needed, by *iteration*. Then $y_1 = f(y_0)$; $y_2 = f(y_1)$,.... An example is the logistic equation $y_{n+1} = ay_n(1 - y_n) = f(y_n)$. The logistic equation has two fixed (critical or equilibrium) points where $y_{n+1} = y_n$. They are 0 and $\bar{y} = (a - 1)/a$. This has physical meaning only for $a > 1$. For $1 < a < 3$ the equilibrium $\bar{y}$ is asymptotically stable, and for $a > 3$ there are two points y_1 and y_2, called a *cycle of period two*, in which $y_2 = f(y_1)$ and $y_1 = f(y_2)$. This study leads into chaos, which is outside our interest. By iteration, with $y_0 = {}^1/_2$, we have $y_1 = (a/2)(1/2) = a/2^2$, $y_2 = a(a/2^2)(1 - a/2^2) = (a^2/2^2)(1 - a/2^2)$, ...

With a constant, the equation $y_{n+1} = ay_n$ is solved by making the assumption $y_n = A\lambda^n$ and finding λ so that the equation holds. Thus $A\lambda^{n+1} = aA\lambda^n$, and hence $\lambda = 0$ or $\lambda = a$ and A is arbitrary. Discarding the trivial solution 0 we find $y_n = Aa^{n+1}$ is the desired solution. By using a method called the *variation of constants*, the equation $y_{n+1} - ay_n = g_n$ has the solution $y_n = y_0a^n + \Sigma_{j=0}^{n-1} g_j a^{n-j-1}$, with y_0 arbitrary.

In various applications we find the first-order equation of *Riccati type* $y_ny_{n+1} + ay_n + by_{n+1} + c = 0$ where a, b, and c are real constants. This equation can be transformed to a linear second-order equation by setting $y_n = z_n/z_{n+1} - a$ to obtain $z_{n+1} + (b + a)z_n + (c - ab)z_{n-1} = 0$, which is solvable as described in the next section.

4.2 Second-Order Equations

The second-order linear equation with constant coefficients $y_{n+2} + ay_{n+1} + by_n = f_n$ is solved by first solving the homogeneous equation (with right-hand side zero) and adding to that solution any solution of the inhomogeneous equation. The *homogeneous equation* $y_{n+2} + ay_{n+1}$ $by_n = 0$ is solved by assuming $y_n = \lambda^n$, whereupon $\lambda^{n+2} + a\lambda^{n+1} + b\lambda^n = 0$ or $\lambda = 0$ (rejected) or $\lambda^2 + a\lambda + b = 0$. The roots of this

0-8493-0056-8/00/$0.00+$.50

quadratic are $\lambda_1 = {}^1/_2(-a + \sqrt{a^2 - 4b})$, $\lambda_2 = -{}^1/_2(a + \sqrt{a^2 - 4b})$ and the solution of the homogeneous equation is $y_n = c_1\lambda_1^n + c_2\lambda_2^n$. As an example consider the Fibonacci equation $y_{n+2} - y_{n+1} - y_n = 0$. The roots of $\lambda^2 - \lambda - 1 = 0$ are $\lambda_1 = {}^1/_2(1 + \sqrt{5})$, $\lambda_2 = {}^1/_2(1 - \sqrt{5})$, and the solution $y_n = c_1[(1 + \sqrt{5})/2]^n + c_2[(1 - \sqrt{5})/2]^n$ is known as the *Fibonacci sequence.*

Many of the orthogonal polynomials of differential equations and numerical analysis satisfy a second-order difference equation (recurrence relation) involving a discrete variable, say n, and a continuous variable, say z. One such is the *Chebshev equation* $y_{n+2} - 2zy_{n+1} + y_n = 0$ with the initial conditions $y_0 = 1$, $y_1 = z$ (*first-kind* Chebyshev polynomials) and $y_{n-1} = 0$, $y_0 = 1$ (second-kind Chebyshev polynomials). They are denoted $T_n(z)$ and $V_n(z)$, respectively. By iteration we find

$$T_0(z) = 1, \quad T_1(z) = z, \quad T_2(z) = 2z^2 - 1,$$

$$T_3(z) = 4z^3 - 3z, \quad T_4(z) = 8z^4 - 8z^2 + 1$$

$$V_0(z) = 0, \quad V_1(z) = 1, \quad V_2(z) = 2z,$$

$$V_3(z) = 4z^2 - 1, \quad V_4(z) = 8z^3 - 4z$$

and the general solution is $y_n(z) = c_1T_n(z) + c_2V_{n-1}(z)$.

4.3 Linear Equations with Constant Coefficients

The genral kth-order linear equation with constant coefficients is $\Sigma_{i=0}^{k} p_i y_{n+k-i} = g_n$, $p_0 = 1$. The solution to the corresponding homogeneous equation (obtained by setting $g_n = 0$) is as follows. (a) $y_n = \Sigma_{i=1}^{k} c_i\lambda_i^n$ if the λ_i are the distinct roots of the characteristic polynomial $p(\lambda) = \Sigma_{i=0}^{k} p_i\lambda^{k-i} = 0$. (b) if m_s is the multiplicity of the root λ_s, then the functions $y_{n,s} = u_s(n)\lambda_s^n$, where $u_s(n)$ are polynomials in n whose degree does not exceed $m_s - 1$, are solutions of the equation. Then the general solution of the homogeneous equation is $y_n = \Sigma_{i=1}^{d} a_i u_i(n)\lambda_i^n = \Sigma_{i=1}^{d} a_i \Sigma_{j=0}^{m_i-1} c_j n^j \lambda_i^n$. To this solution one adds any particular solution to obtain the general solution of the general equation.

Example 4.1. A model equation for the price p_n of a product, at the nth time, is $p_n + b/a(1 + \rho)p_{n-1} - (b/a)\rho p_{n-2} + (s_0 - d_0)/a = 0$. The equilibrium price is obtained by setting $p_n = p_{n-1} = p_{n-2} = p_*$, and one finds $p_* = (d_0 - s_0)/(a + b)$. The homogeneous equation has the characteristic polynomial $\lambda^2 + (b/a)(1 + \rho)\lambda - (b/a)\rho = 0$. With λ_1 and λ_2 as the roots the general solution of the full equation is $p_n = c_1\lambda_1^n + c_2\lambda_2^n + p_*$, since p_* is a solution of the full equation. This is one method for finding the solution of the nonhomogeneous equation.

4.4 Generating Function (z Transform)

An elegant way of solving linear difference equations with constant coefficients, among other applications, is by use of *generating functions* or, as an alternative, the z transform. The generating function of a sequence $\{y_n\}$, $n = 0, 1, 2, \ldots$, is the function $f(x)$ given by the formal series $f(x) = \Sigma_{n=0}^{\infty} y_n x^n$. The z transform of the same sequence is $z(x) = \Sigma_{n=0}^{\infty} y_n x^{-n}$. Clearly, $z(x) = f(1/x)$. A table of some important sequences is given in Table 4.1.

To solve the linear difference equation $\Sigma_{i=0}^{k} p_i y_{n+k-i} = 0$, $p_0 = 1$ we associate with it the two formal series $P = p_0 + p_1x + \cdots + p_kx^k$ and $Y = y_0 + y_1x + y_2x^2 \cdots$. If $p(x)$ is the characteristic polynomial then $P(x) = x^kp(1/x) = \bar{p}(x)$. The *product* of the two series is $Q = YP = q_0 + q_1x + \cdots + q_{k-1}x^{k-1} + q_kx^k + \cdots$ where $q_n = \Sigma_{i=0}^{n} p_i y_{n-i}$. Because $p_{k+1} = p_{k+2} = \cdots = 0$, it is obvious that $q_{k+1} = q_{k+2} = \cdots = 0$ — that is, Q is a polynomial (formal series with finite number of terms). Then $Y = P^{-1}Q = q(x)/\bar{p}(x) = q(x)/x^kp(1/x)$, where p is the characteristic polynomial and $q(x) = \Sigma_{i=0}^{k} q_ix^i$. The roots of $\bar{p}(x)$ are x_i^{-1} where the x_i are the roots of $p(x)$.

Table 4.1 Important Sequences

y_n	$f(x)$	Convergence Domain
1	$(1-x)^{-1}$	$\lvert x\rvert < 1$
n	$x(1-x)^{-2}$	$\lvert x\rvert < 1$
n^m	$xp_m(x)(1-x)^{-n-1}$ *	$\lvert x\rvert < 1$
k^n	$(1-kx)^{-1}$	$\lvert x\rvert < k^{-1}$
e^{an}	$(1-e^a x)^{-1}$	$\lvert x\rvert < e^{-a}$
$k^n \cos an$	$\dfrac{1-kx\cos a}{1-2kx\cos a+k^2x^2}$	$\lvert x\rvert < k^{-1}$
$k^n \sin an$	$\dfrac{kx\sin a}{1-2kx\cos a+k^2x^2}$	$\lvert x\rvert < k^{-1}$
$\binom{n}{m}$	$x^m(1-x)^{-m-1}$	$\lvert x\rvert < 1$
$\binom{k}{n}$	$(1+x)^k$	$\lvert x\rvert < 1$

* The term $p_m(z)$ is a polynomial of degree m satisfying $p_{m+1}(z) = (mz+1)\cdot p_m(z) + z(1-z)\,p'_m(x)$, $p_1 = 1$.

Theorem 1. If the roots of $p(x)$ are less than one in absolute value, then $Y(x)$ converges for $|x| < 1$. *Thorem 2*. If $p(x)$ has no roots greater than one in absolute value and those on the unit circle are simple roots, then the coefficients y_n of Y are bounded. Now $q_k = g_0$, $q_{n+k} = g_n$, and $Q(x) = Q_1(x) + x^k Q_2(x)$. Hence $\Sigma_{i=1}^{\infty} y_i x^i = [Q_1(x) + x^k Q_2(x)]/[\bar{p}(x)]$.

Example 4.2. Consider the equation $y_{n+1} + y_n = -(n+1)$, $y_0 = 1$. Here $Q_1 = 1$, $Q_2 = -\Sigma_{n=0}^{\infty}(n+1)x^n = -1/(1-x)^2$.

$$G(x) = \frac{1 - x/(1-x)^2}{1+x} = \frac{5}{4}\frac{1}{1+x} - \frac{1}{4}\frac{1}{1-x} - \frac{1}{2}\frac{x}{(1-x)^2}$$

Using the table term by term, we find $\Sigma_{n=0}^{\infty} y_n x^n = \Sigma_{n=0}^{\infty}[{}^5\!/_4(-1)^n - {}^1\!/_4 - {}^1\!/_2\, n]x^n$, so $y_n = {}^5\!/_4(-1)^n - {}^1\!/_4 - {}^1\!/_2\, n$.

References

Fort, T. 1948. *Finite Differences and Difference Equations in the Real Domain*. Oxford University Press, London.

Jordan, C. 1950. *Calculus of Finite Differences*, Chelsea, New York.

Jury, E. I. 1964. *Theory and Applications of the Z Transform Method*. John Wiley & Sons, New York.

Lakshmikantham, V. and Trigrante, D. 1988. *Theory of Difference Equations*. Academic Press, Boston, MA.

Levy, H. and Lessman, F. 1961. *Finite Difference Equations*. Macmillan, New York.

Miller, K. S. 1968. *Linear Difference Equations*, Benjamin, New York.

Wilf, W. S. 1994. *Generating Functionology*, 2nd ed. Academic Press, Boston, MA.

5

Differential Equations

William F. Ames
Georgia Institute of Technology

Any equation involving derivatives is called a *differential equation*. If there is only one independent variable the equation is termed a *total differential equation* or an *ordinary differential equation*. If there is more than one independent variable the equation is called a *partial differential equation*. If the highest-order derivative is the nth then the equation is said to be nth order. If there is no function of the dependent variable and its derivatives other than the linear one, the equation is said to be *linear*. Otherwise, it is *nonlinear*. Thus $(d^3y/dx^3) + a(dy/dx) + by = 0$ is a *linear* third-order ordinary (total) differential equation. If we replace by with by^3, the equation becomes nonlinear. An example of a second-order linear partial differential equation is the famous wave equation $(\partial^2u/\partial x^2) - a^2(\partial^2u/\partial t^2) = f(x)$. There are two independent variables x and t and $a^2 > 0$ (of course). If we replace $f(x)$ by $f(u)$ (say u^3 or $\sin u$) the equation is nonlinear. Another example of a nonlinear third-order partial differential equation is $u_t + uu_x = au_{xxx}$. This chapter uses the common subscript notation to indicate the partial derivatives.

Now we briefly indicate some methods of solution and the solution of some commonly occurring equations.

5.1 Ordinary Differential Equations

First-Order Equations

The *general* first-order equation is $f(x, y, y') = 0$. Equation capable of being written in either of the forms $y' = f(x)g(y)$ or $f(x)g(y)y' + F(x)G(y) = 0$ are *separable* equations. Their solution is obtained by using $y' = dy/dx$ and writing the equations in differential form as $dy/g(y) = f(x)dx$ or $g(y)[dy/G(y)] = -F(x)[dx/f(x)]$ and integrating. An example is the famous *logistic* equation of inhibited growth $(dy/dt) = ay(1 - y)$. The integral of $dy/y(1 - y) = adt$ is $y = 1/[1 + (y_0^{-1} - 1)e^{-at}]$ for $t \geq 0$ and $y(0) = y_0$ (the initial state called the *initial condition*).

Equations may not have unique solutions. An example is $y' = 2y^{1/2}$ with the initial condition $y(0) = 0$. One solution by separation is $y = x^2$. But there are an *infinity* of others — namely, $y_a(x) = 0$ for $-\infty < x \leq a$, and $(x - a)^2$ for $a < x < \infty$.

If the equation $P(x, y)dy + Q(x, y)dy = 0$ is reducible to

$$\frac{dy}{dx} = f\left(\frac{y}{x}\right) \quad \text{or} \quad \frac{dy}{dx} = f\left(\frac{a_1x + b_1y + c_1}{a_2x + b_2y + c_2}\right)$$

0-8493-0056-8/00/$0.00+$.50

the equation is called *homogenous* (nearly homogeneous). The first form reduces to the separable equation $u + x(du/dx) = f(u)$ with the substitution $y/x = u$. The nearly homogeneous equation is handled by setting $x = X + a$, $y = Y + \beta$, and choosing α and β so that $a_1\alpha + b_1\beta + c_1 = 0$ and $a_2\alpha + b_2\beta + c_2 = 0$. If $\begin{vmatrix} a_1 & b_1 \\ a_2 & b_2 \end{vmatrix} \neq 0$ this is always possible; the equation becomes $dY/dX = [a_1 + b_1(Y/X)]/[a_2 + b_2(Y/X)]$ and the substitution $Y = Xu$ gives a separable equation. If $\begin{vmatrix} a_1 & b_1 \\ a_2 & b_2 \end{vmatrix} = 0$ then $a_2x + b_2y = k(a_1x + b_1y)$ and the equation becomes $du/dx = a_1 + b_1(u + c_1)/(ku + c_2)$, with $u = a_1x + b_1y$. Lastly, any equation of the form $dy/dx = f(ax + by + c)$ transforms into the separable equation $du/dx = a + bf(u)$ using the change of variable $u = ax + by + c$.

The general first-order linear equation is expressible in the form $y' + f(x)y = g(x)$. It has the *general solution* (a solution with an arbitrary constant c)

$$y(x) = \exp\left[-\int f(x)\,dx\right]\left\{c + \int \exp[f(x)]g(x)\,dx\right\}$$

Two noteworthy examples of first-order equations are as follows:

1. An often-occurring nonlinear equation is the *Bernoulli equation*, $y' + p(x)y = g(x)y^\alpha$, with α real, $\alpha \neq 0$, $\alpha \neq 1$. The transformation $z = y^{1-\alpha}$ converts the equation to the linear first-order equation $z' + (1 - \alpha)p(x)z = (1 - \alpha)q(x)$.
2. The famous *Riccati equation*, $y' = p(x)y^2 + q(x)y + r(x)$, cannot in general be solved by integration. But some useful transformations are helpful. The substitution $y = y_1 + u$ leads to the equation $u' - (2py_1 + q)u = pu^2$, which is a Bernoulli equation for u. The substitution $y = y_1 + v^{-1}$ leads to the equation $v' + (2py_1 + q)v + p = 0$, which is a linear first-order equation for v. Once either of these equations has been solved, the general solution of the Riccati equation is $y = y_1 + u$ or $y = y_1 + v^{-1}$.

Second-Order Equations

The simplest of the second-order equations is $y'' + ay' + by = 0$ (a, b real), with the initial conditions $y(x_0) = y_0$, $y'(x_0) = y_0'$ or the boundary conditions $y(x_0) = y_0$, $y(x_1) = y_1$. The general solution of the equation is given as follows.

1. $a^2 - 4b > 0$, $\lambda_1 = {}^1/_2\,(-a + \sqrt{a^2 - 4b})$, $\lambda_2 = {}^1/_2\,(-a - \sqrt{a^2 - 4b})$
   $y = c_1 \exp(\lambda_1 x) + c_2 \exp(\lambda_2 x)$
2. $a^2 - 4b = 0$, $\lambda_1 = \lambda_2 = -\frac{a}{2}$, $y = (c_1 + c_2x)\exp(\lambda_1 x)$
3. $a^2 - 4b < 0$, $\lambda_1 = {}^1/_2\,(-a + i\sqrt{4b - a^2})$, $\lambda_2 = {}^1/_2\,(-a - i\sqrt{4b - a^2})$,
   $i^2 = -1$
   With $p = -a/2$ and $q = {}^1/_2\,\sqrt{4b - a^2}$,

$$y = c_1 \exp[(p + iq)x] + c_2 \exp[(p - iq)x] = \exp(px)[A \sin qx + B \cos qx]$$

The initial conditions or boundary conditions are used to evaluate the arbitrary constants c_1 and c_2 (or A and B).

Note that a linear problem with specified data may not have a solution. This is especially serious if numerical methods are employed without serious thought.

For example, consider $y'' + y = 0$ with the boundary condition $y(0) = 1$ and $y(\pi) = 1$. The general solution is $y = c_1 \sin x + c_2 \cos x$. The first condition $y(0) = 1$ gives $c_2 = 1$, and the second condition requires $y(\pi) = c_1 \sin \pi + \cos \pi$ or "$1 = -1$," which is a *contradiction*.

Example 5.1 — The Euler Strut. When a strut of uniform construction is subject to a compressive load P it exhibits no transverse displacement until P exceeds some critical value P_1. When this load is exceeded, buckling occurs and large deflections are produced as a result of small load changes. Let the rod of length ℓ be placed as shown in Figure 5.1.

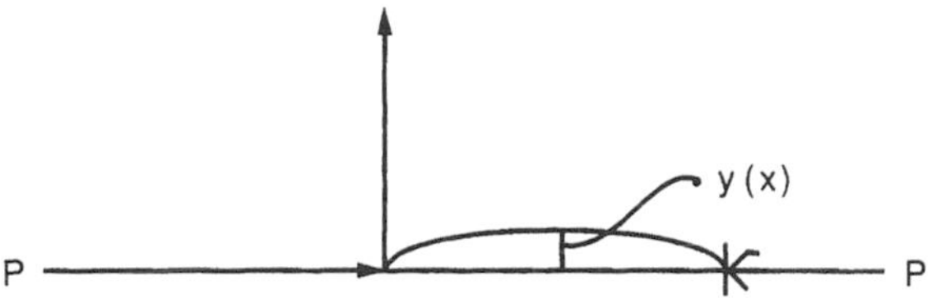

Figure 5.1

From the linear theory of elasticity (Timoshenko), the transverse displacement $y(x)$ satisfies the linear second-order equation $y'' + (Py/EI) = 0$, where E is the modulus of elasticity and I is the moment of inertia of the strut. The boundary conditions are $y(0) = 0$ and $y(a) = 0$. With $k^2 = P/EI$ the general solution is $y = c_1 \sin kx + c_2 \cos kx$. The condition $y(0) = 0$ gives $c_2 = 0$. The second condition gives $c_1 \sin ka = 0$. Since $c_1 = 0$ gives the trival solution $y = 0$ we must have $\sin ka = 0$. This occurs for $ka = n\pi$, $n = 0, 1, 2, \ldots$ (these are called *eigenvalues*). The first nontrivial solution occurs for $n = 1$ — that is, $k = \pi/a$ — whereupon $y_1 = c_1 \sin(\pi/a)$, with arbitrary c_1. Since $P = EIk^2$ the critical compressive load is $P_1 = EI\,\pi^2/a^2$. This is the buckling load. The weakness of the linear theory is its failure to model the situation when buckling occurs.

Example 5.2 — Some Solvable Nonlinear Equations. Many physical phenomena are modeled using nonlinear second-order equations. Some general cases are given here.

1. $y'' = f(y)$, *first integral* $(y')^2 = 2\int f(y)\,dy + c$.
2. $f(x, y', y'') = 0$. Set $p = y'$ and obtain a first-order equation $f(x, p, dp/dx) = 0$. Use first-order methods.
3. $f(y, y', y'') = 0$. Set $p = y'$ and then $y'' = p(dp/dy)$ so that a first-order equation $f[y, p, p(dp/dy)] = 0$ for p as a function of y is obtained.
4. The *Riccati transformation* $du/dx = yu$ leads to the Riccati chain of equations, which linearize by raising the order. Thus,

	Equation in y	**Equation in u**
1	$y' + y^2 = f(x)$	$u'' = f(x)u$
2	$y'' + 3yy' + y^3 = f(x)$	$u''' = f(x)u$
3	$y''' + 6y^2y' + 3(y')^2 + 4yy'' = f(x)$	$u'''' = f(x)u$

This method can be generalized to $u' = a(x)yu$ or $u' = a(x)f(u)y$.

Second-Order Inhomogeneous Equations

The general solution of $a_0(x)y'' + a_1(x)y' + a_2(x)y = f(x)$ is $y = y_H(x) + y_p(x)$ where $y_H(x)$ is the general solution of the homogeneous equation (with the right-hand side zero) and y_p is the particular integral of the equation. Construction of particular integrals can sometimes be done by the *method of undetermined coefficients*. See Table 5.1. This applies only to the linear constant coefficient case in which the function $f(x)$ is a linear combination of a polynomial, exponentials, sines and cosines, and some products of these functions. This method has as its base the observation that repeated differentiation of such functions gives rise to similar functions.

Example 5.3. Consider the equation $y'' + 3y' + 2y = \sin 2x$. The characteristic equation of the homogeneous equation $\lambda^2 + 3\lambda + 2 = 0$ has the two roots $\lambda_1 = -1$ and $\lambda_2 = -2$. Consequently, $y_H = c_1e^{-x}$

Table 5.1 Method of Undetermined Coefficients — Equation $L(y) = f(x)$ (Constant Coefficients)

Terms in $f(x)$	Terms To Be Included in $y_p(x)$
1 Polynomial of degree n	(i) If $L(y)$ contains y, try $y_p = a_0x^n + a_1x^{n-1} + \cdots + a_n$.
	(ii) If $L(y)$ does not contain y and lowest-order derivative is $y^{(r)}$, try $y_p = a_0x^{n+r} + \cdots + a_nx^r$
2. $\sin qx$, $\cos qx$	(i) $\sin qx$ and/or $\cos qx$ are not in y_H. $y_p = B \sin qx + C \cos qx$.
	(ii) y_H contains terms of form $x^r \sin qx$ and/or $x^r \cos qx$ for $r = 0, 1, \ldots, m$; include in y_p terms of the form $a_0x^{m+1} \sin qx + a_1x^{m+1} \cos qx$.
3. e^{ax}	(i) y_H does not contain e^{ax}; include Ae^{ax} in y_p.
	(ii) y_H contains $e^{ax}, xe^{ax}, \ldots x^ne^{ax}$; include in y_p terms of the form $Ax^{n+1}e^{ax}$,
4. $e^{px} \sin qx$, $e^{px} \cos qx$	(i) y_H does not contain these terms; in y_p include $Ae^{px} \sin qx + Be^{px} \cos qx$.
	(ii) y_H contains $x^r e^{px} \sin qx$ and/or $x^r e^{px} \cos qx$; $r = 0,1, \ldots m$ include in y_p. $Ax^{m+1}e^{px} \sin qx + Bx^{m+1} e^{px} \cos qx$.

$+ c_2e^{-2x}$. Since $\sin 2x$ is not linearly dependent on the exponentials and since $\sin 2x$ repeats after two differentiations, we assume a particular solution with undetermined coefficients of the form $y_p(x) = B \sin 2x + C \cos 2x$. Substituting into the original equation gives $-(2B + 6C) \sin 2x + (6B - 2C) \cos 2x = \sin 2x$. Consequently, $-(2B + 6C) = 1$ and $6B - 2C = 0$ to satisfy the equation. These two equations in two unknowns have the solution $B = -1/20$ and $C = -3/20$. Hence $y_p = -1/20 (\sin 2x + 3 \cos 2x)$ and $y = c_1e^{-x} + c_2e^{-2x} - 1/20 (\sin 2x + 3 \cos 2x)$.

A general method for finding $y_p(x)$ called *variation of parameters* uses as its starting point $y_H(x)$. This method applies to *all* linear differential equations irrespective of whether they have constant coefficients. But it assumes $y_H(x)$ is known. We illustrate the idea for $a(x)y'' + b(x)y' + c(x)y = f(x)$. If the solution of the homogeneous equation is $y_H(x) = c_1\phi_1(x) + c_2\phi_2(x)$, then vary the parameters c_1 and c_2 to seek $y_p(x)$ as $y_p(x) = u_1(x)\phi_1(x) + u_2(x)\phi_2(x)$. Then $y_p' = u_1 \phi_1' + u_2 \phi_2' + u_1' \phi_1 + u_2' \phi_2$ and choose $u_1' \phi_1 + u_2' \phi_2 = 0$. Calculating y_p'' and setting in the original equation gives $a(x) u_1'\phi_1' + a(x) u_2'\phi_2' = f$. Solving the last two equations for u_1' and u_2' gives $u_1' = -\phi_2 f/wa$, $u_2' = \phi_1 f/wa$, where $w = \phi_1 \phi_2' - \phi_1' \phi_2 \neq 0$. Integrating the general solution gives $y = c_1\phi_1(x) + c_2\phi_2(x) - \{\int[\phi_2 f(x)]/wa\}\phi_1(x) + [\int(\phi_1 f/wa)dx]\phi_2(x)$.

Example 5.4. Consider the equations $y'' - 4y = \sin x/(1 + x^2)$ and $y_H = c_1e^{2x} + c_2e^{-2x}$. With $\phi_1 = e^{2x}$, and $\phi_2 = e^{-2x}$, $w = 4$, so the general solution is

$$y = c_1e^{2x} + c_2e^{-2x} - \frac{e^{-2x}}{4}\int \frac{e^{2x}\sin x}{1+x^2}\,dx + \frac{e^{2x}}{4}\int \frac{e^{-2x}\sin x}{1+x^2}\,dx$$

The method of variation of parameters can be generalized as described in the references.

Higher-order systems of linear equations with constant coefficients are treated in a similar manner. Details can be found in the references.

Series Solution

The solution of differential equations can only be obtained in closed form in special cases. For all others, series or approximate or numerical solutions are necessary. In the simplest case, for an initial value problem, the solution can be developed as a Taylor series expansion about the point where the initial data are specified. The method fails in the *singular case* — that is, a point where the coefficient of the highest-order derivative is zero. The general method of approach is called the *Frobenius method*.

To understand the nonsingular case consider the equation $y'' + xy = x^2$ with $y(2) = 1$ and $y'(2) = 2$ (an initial value problem). We seek a series solution of the form $y(x) = a_0 + a_1(x - 2) + a_2(x - 2)^2 + \cdots$. To proceed, set $1 = y(2) = a_0$, which evaluates a_0. Next $y'(x) = a_1 + 2a_2(x - 2) + \cdots$, so $2 = y'(2) = a_1$ or $a_1 = 2$. Next $y''(x) = 2a_2 + 6a_3(x - 2) + \cdots$, and from the equation, $y'' = x^2 - xy$, so $y''(2) = 4 - 2y(2) = 4 - 2 = 2$. Hence $2 = 2a_2$ or $a_2 = 1$. Thus, to third-order $y(x) = 1 + 2(x - 2) + (x - 2)^2 + R_2(x)$, where the remainder $R_2(x)$ $[(x-2)^3/3]y'''(\xi)$, where $2 < \xi < x$ can be bounded for each x by finding the maximum of $y'''(x) = 2x - y - xy'$. The third term of the series follows by evaluating $y'''(2) = 4 - 1 - 2 \cdot 2 = -1$, so $6a_3 = -1$ or $a_3 = -1/6$.

By now the nonsingular process should be familiar. The algorithm for constructing a series solution about a nonsingular (ordinary) point x_0 of the equation $P(x)y'' + Q(x)y' + R(x)y = f(x)$ (note that $P(x_0) \neq 0$) is as follows:

1. Substitute into the differential equation the expressions

$$y(x) = \sum_{n=0}^{\infty} a_n (x - x_0)^n, \qquad y'(x) = \sum_{n=1}^{\infty} n a_n (x - x_0)^{n-1}, \qquad y''(x) = \sum_{n=2}^{\infty} n(n-1) a_n (x - x_0)^{n-2}$$

2. Expand $P(x)$, $Q(x)$, $R(x)$, and $f(x)$ about the point x_0 in a power series in $(x - x_0)$ and substitute these series into the equation.
3. Gather all terms involving the same power of $(x - x_0)$ to arrive at an identity of the form $\Sigma_{n=0}^{\infty} A_n(x - x_0)^n \equiv 0$.
4. Equate to zero each coefficient A_n of step 3.
5. Use the expressions of step 4 to determine a_2, a_3, ... in terms of a_0, a_1 (we need two arbitrary constants) to arrive at the general solution.
6. With the given initial conditions, determine a_0 and a_1.

If the equation has a regular singular point — that is, a point x_0 at which $P(x)$ vanishes and a series expansion is sought about that point — a solution is sought of the form $y(x) = (x - x_0)^r \Sigma_{n=0}^{\infty} a_n(x - x_0)^n$, $a_0 \neq 0$ and the index r and coefficients a_n must be determined from the equation by an algorithm analogous to that already described. The description of this Frobenius method is left for the references.

5.2 Partial Differential Equations

The study of partial differential equations is of continuing interest in applications. It is a vast subject, so the focus in this chapter will be on the most commonly occurring equations in the engineering literature — the second-order equations in two variables. Most of these are of the three basic types: elliptic, hyperbolic, and parabolic.

Elliptic equations are often called *potential equations* since they occur in potential problems where the potential may be temperature, voltage, and so forth. They also give rise to the steady solutions of parabolic equations. They require boundary conditions for the complete determination of their solution.

Hyperbolic equations are often called *wave equations* since they arise in the propagation of waves. For the development of their solutions, initial and boundary conditions are required. In principle they are solvable by the method of characteristics.

Parabolic equations are usually called *diffusion equations* because they occur in the transfer (diffusion) of heat and chemicals. These equations require initial conditions (for example, the initial temperature) and boundary conditions for the determination of their solutions.

Partial differential equations (PDEs) of the second order in two independent variables (x, y) are of the form $a(x, y)u_{xx} + b(x, y)u_{xy} + c(x, y)u_{yy} = E(x, y, u, u_x, u_y)$. If $E = E(x, y)$ the equation is linear; if E depends also on u, u_x, and u_y, it is said to be *quasilinear*, and if E depends only on x, y, and u, it is *semilinear*. Such equations are classified as follows: If $b^2 - 4ac$ is less than, equal to, or greater than zero at some point (x, y), then the equation is elliptic, parabolic, or hyperbolic, respectively, at that point. A PDE of this form can be transformed into canonical (standard) forms by use of new variables. These standard forms are most useful in analysis and numerical computations.

For hyperbolic equations the standard form is $u_{\xi\eta} = \phi(u, u_\eta, u_\xi, \eta, \xi)$, where $\xi_x/\xi_y = (-b + \sqrt{b^2 - 4ac})/2a$, and $\eta_x/\eta_y = (-b - \sqrt{b^2 - 4ac})/2a$. The right-hand sides of these equations determine the so-called characteristics $(dy/dx)|_+ = (-b + \sqrt{b^2 - 4ac})/2a$, $(dy/dx)|_- = (-b - \sqrt{b^2 - 4ac})/2a$.

Example 5.5. Consider the equation $y^2u_{xx} - x^2u_{yy} = 0$. $\xi_y/\xi_x = -x/y$, $\eta_y/\eta_x = x/y$, so $\xi = y^2 - x^2$ and $\eta = y^2 + x^2$. In these new variables the equation becomes $u_{\xi\eta} = (\xi u_\eta - \eta u_\xi)/2(\xi^2 - \eta^2)$.

For parabolic equations the standard form is $u_{\xi\xi} = \phi(u, u_\eta, u_\xi, \eta, \xi)$ or $u_{\eta\eta} = \phi(u, u_\eta, u_\xi, \xi, \eta)$, depending upon how the variables are defined. In this case $\xi_y/\xi_x = -b/2a$ if $a \neq 0$, and $\xi_y/\xi_x = -b/2c$ if $c \neq 0$. Only ξ must be determined (there is only one characteristic) and η can be chosen as any function that is linearly independent of ξ.

Example 5.6. Consider the equation $y^2u_{xx} - 2xyu_{xy} + x^2u_{yy} + u_y = 0$. Clearly, $b^2 - 4ac = 0$. Neither a nor c is zero so either path can be chosen. With $\xi_y/\xi_x = -b/2a = x/y$, there results $\xi = x^2 + y^2$. With $\eta = x$, the equation becomes $u_{\eta\eta} = [2(\xi + \eta)u_\xi + u_\eta]/(\xi - \eta^2)$.

For *elliptic equations* the standard form is $u_{\alpha\alpha} + u_{\beta\beta} = \phi(u, u_\alpha, u_\beta, \alpha, \beta)$, where ξ and η are determined by solving the ξ and η equations of the hyperbolic system (they are complex) and taking $\alpha = (\eta + \xi)/2$, $\beta = (\eta - \xi)/2i(i^2 = -1)$. Since ξ and η are complex conjugates, both α and β are real.

Example 5.7. Consider the equation $y^2u_{xx} + x^2u_{yy} = 0$. Clearly, $b^2 - 4ac < 0$, so the equation is elliptic. Then $\xi_y/\xi_x = -ix/y$, $\eta_y/\eta_x = ix/y$, so $\alpha = (\eta + \xi)/2 = y^2$ and $\beta = (\eta - \xi)/2i = x^2$. The standard form is $u_{\alpha\alpha} + u_{\beta\beta} = -(u_\alpha/2\alpha + u_\beta/2\beta)$.

Methods of Solution

Separation of Variables. Perhaps the most elementary method for solving linear PDEs with homogeneous boundary conditions is the method of *separation of variables.* To illustrate, consider $u_t - u_{xx} = 0$, $u(x, 0) = f(x)$ (the initial condition) and $u(0, t) = u(1, t) = 0$ for $t > 0$ (the boundary conditions). A solution is assumed in "separated form" $u(x, t) = X(x)T(t)$. Upon substituting into the equation we find $\dot{T}/T = X''/X$ (where $\dot{T} = dT/dt$ and $X'' = d^2X/dx^2$). Since $T = T(t)$ and $X = X(x)$, the ratio must be constant, and for finiteness in t the constant must be negative, say $-\lambda^2$. The solutions of the separated equations $X'' + \lambda^2X = 0$ with the boundary conditions $X(0) = 0$, $X(1) = 0$, and $\dot{T} = -\lambda^2T$ are $X = A \sin \lambda x + B \cos \lambda x$ and $T = Ce^{-\lambda^2 t}$, where A, B, and C are arbitrary constants. To satisfy the boundary condition $X(0) = 0$, $B = 0$. An infinite number of values of λ (eigenvalues), say $\lambda_n = n\pi (n = 1, 2, 3, \ldots)$, permit all the eigenfunctions $X_n = b_n \sin \lambda_n x$ to satisfy the other boundary condition $X(1) = 0$. The solution of the equation and boundary conditions (not the initial condition) is, by superposition, $u(x, t) = \sum_{n=1}^{\infty} b_n e^{-n^2\pi^2 t} \cdot \sin n\pi x$ (a Fourier sine series), where the b_n are arbitrary. These values are obtained from the initial condition using the orthogonality properties of the trigonometric function (e.g., $\int_{-\pi}^{\pi} \sin mx \sin n x \, d x$ is 0 for $m \neq n$ and is π for $m = n \neq 0$) to be $b_n = 2\int_0^1 f(r) \sin n \pi r \, d r$. Then the solution of the problem is $u(x, t) = \sum_{n=1}^{\infty} [2\int_0^1 f(r) \sin n \pi r \, d r] e^{-n^2\pi^2 t} \sin n \pi x$, which is a Fourier sine series.

If $f(x)$ is a piecewise smooth or a piecewise continuous function defined for $a \leq x \leq b$, then its Fourier series within $a \leq x \leq b$ as its fundamental interval (it is extended periodically ouside that interval) is

$$f(x) \sim \tfrac{1}{2}a_0 + \sum_{n=1}^{\infty} a_n \cos[2n\pi x/(b-a)] + b_n \sin[2n\pi x/(b-a)]$$

where

$$a_n = \left[\frac{2}{(b-a)}\right]\int_a^b f(x)\cos[2n\pi x/(b-a)]\,dx, \quad n = 0, 1, \ldots$$

$$b_n = \left[\frac{2}{(b-a)}\right]\int_a^b f(x)\sin[2n\pi x/(b-a)]\,dx, \quad n = 1, 2, \ldots$$

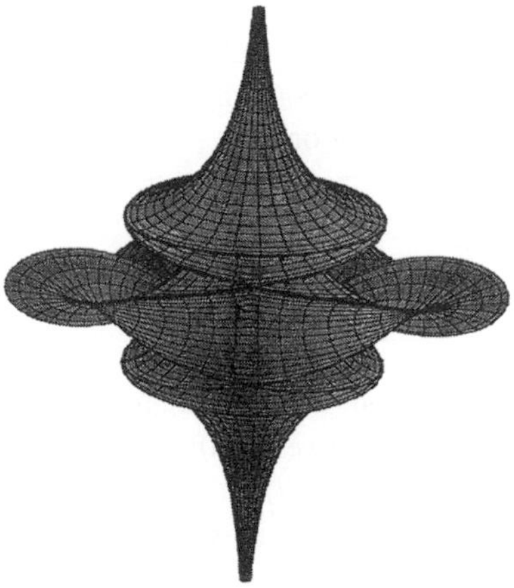

Figure 5.2

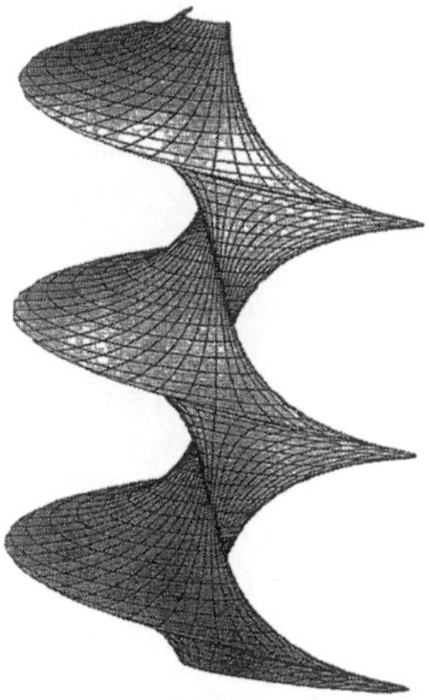

Figure 5.3

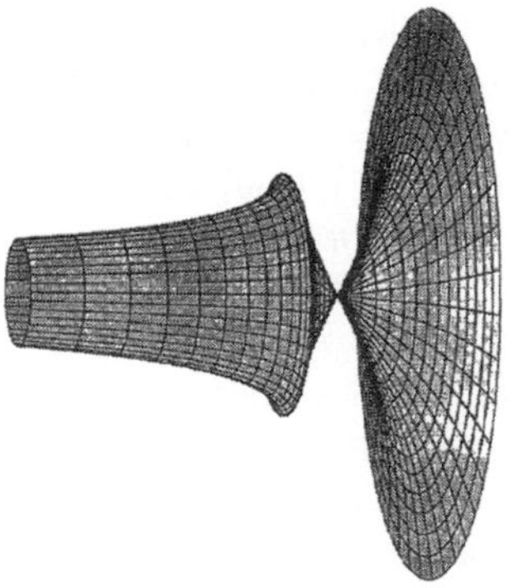

Figure 5.4

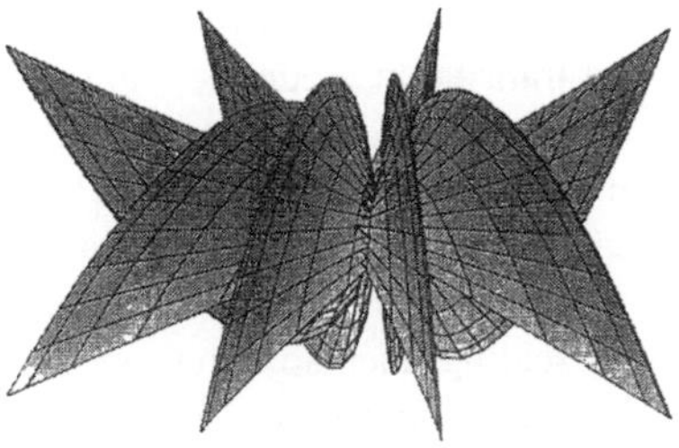

Figure 5.5

Figure 5.2 to 5.5 The mathematical equations used to generate these three-dimensional figures are worth a thousand words. The figures shown illustrate some of the nonlinear ideas of engineering, applied physics, and chemistry. Figure 5.2 represents a breather soliton surface for the sine-Gordon equation $w_{uv} = \sin w$ generated by a Backlund transformation. A single-soliton surface for the sine-Gordon equation $w_{uv} = \sin w$ is illustrated in Figure 5.3. Figure 5.4 represents a single-soliton surface for the Tzitzecia-Dodd-Bullough equation associated with an integrable anisentropic gas dynamics system. Figure 5.5 represents a single-soliton Bianchi surface.
The solutions to the equations were developed by W. K. Schief and C. Rogers at the Center for Dynamical Systems and Nonlinear Studies at the Georgia Institute of Technology and the University of New South Wales in Sydney, Australia. All of these three-dimensional projections were generated using the MAPLE software package. (Figures courtesy of Schief and Rogers).

The Fourier sine series has $a_n \equiv 0$, and the Fourier cosine series has $b_n \equiv 0$. The symbol ~ means that the series converges to $f(x)$ at points of continuity, and at the (allowable) points of finite discontinuity the series converges to the *average value* of the discontinuous values.

Caution: This method *only* applies to linear equations with homogeneous boundary conditions. Linear equations with variable coefficients use other orthogonal functions, such as the Bessel functions, Laguerre functions, Chebyshev functions, and so forth.

Some inhomogeneous boundary value problems can be transformed into homogeneous ones. Consider the problem $u_t - u_{xx} = 0$, $0 \le x \le 1$, $0 \le t < \infty$ with initial condition $u(x, 0) = f(x)$, and boundary conditions $u(0, t) = g(t)$, $u(1, t) = h(t)$. To homogenize the boundary conditions set $u(x, t) = w(x, t) + x[h(t) - g(t)] + g(t)$ and then solve $w_t - w_{xx} = [\dot{g}(t) - \dot{h}(t)]x - \dot{g}(t)$ with the initial condition $w(x, 0) = f(x) - x[h(0) - g(0)] + g(0)$ and $w(0, t) = w(1, t) = 0$.

Operational Methods. A number of integral transforms are useful for solving a variety of linear problems. To apply the Laplace transform to the problem $u_t - u_{xx} = \delta(x)\,\delta(t)$, $-\infty < x < \infty$, $0 \le t$ with the initial condition $u(x, 0^-) = 0$, where δ is the Dirac delta function, we multiply by e^{-st} and integrate with respect to t from 0 to ∞. With the Laplace transform of $u(x, t)$ denoted by $U(x, s)$ — that is, $U(x, s) = \int_0^\infty e^{-st}\, u(x, t)\, dt$ — we have $sU - U_{xx} = \delta(x)$, which has the solution

$$U(x,s) = A(s)e^{-\sqrt{s}\,x} + B(s)e^{\sqrt{s}\,x} \qquad \text{for } x > 0$$

$$U(x,s) = C(s)e^{-\sqrt{s}\,x} + D(s)e^{\sqrt{s}\,x} \qquad \text{for } x < 0$$

Clearly, $B(s) = C(s) = 0$ for bounded solutions as $|x| \to \infty$. Then, from the boundary condition, $U(0^+, s) - U(0^-, s) = 0$ and integration of $sU - U_{xx} = \delta(x)$ from 0^- to 0^+ gives $U_x(0^+, s) - U(0^-, s) = -1$, so $A = D = 1/2\ \sqrt{s}$. Hence, $U(x, s) = (1/2\ \sqrt{s})e^{-\sqrt{s}|x|}$ and the inverse is $u(x, t) = (1/2\ \pi i)\int_\Gamma e^{st}U(x, s)\, ds$, where Γ is a Bromwich path, a vertical line taken to the right of all singularities of U on the sphere.

Similarity (Invariance). This very useful approach is related to dimensional analysis; both have their foundations in group theory. The three important transformations that play a basic role in Newtonian mechanics are translation, scaling, and rotations. Using two independent variables x and t and one dependent variable $u = u(x, t)$, the *translation group* is $\bar{x} = x + \alpha a$, $\bar{t} = t + \beta a$, $\bar{u} = u + \gamma a$; the *scaling group* is $\bar{x} = a^\alpha x$, $\bar{t} = a^\beta t$, and $\bar{u} = a^\gamma u$; the *rotation group* is $\bar{x} = x \cos a + t \sin a$, $\bar{t} = t \cos a - x \sin a$, $\bar{u} = u$, with a nonnegative real number a. Important in which follows are the *invariants* of these groups. For the translation group there are two $\eta = x - \lambda t$, $\lambda = \alpha/\beta$, $f(\eta) = u - \varepsilon t$, $\varepsilon = \gamma/\beta$ or $f(\eta) = u - \theta x$, $\theta = \gamma/\alpha$; for the scaling group the invariants are $\eta = x/t^{\alpha/\beta}$ (or $t/x^{\beta/\alpha}$) and $f(\eta) = u/t^{\gamma/\beta}$ (or $u/x^{\gamma/\alpha}$); for the rotation group the invariants are $\eta = x^2 + t^2$ and $u = f(\eta) = f(x^2 + t^2)$.

If a PDE and its data (initial and boundary conditions) are left invariant by a transformation group, then similar (invariant) solutions are sought using the invariants. For example, if an equation is left invariant under scaling, then solutions are sought of the form $u(x, t) = t^{\gamma/\beta} f(\eta)$, $\eta = xt^{-\alpha/\beta}$ or $u(x, t) = x^{\gamma/\alpha} f(tx^{-\beta/\alpha})$; invariance under translation gives solutions of the form $u(x, t) = f(x - \lambda t)$; and invariance under rotation gives rise to solutions of the form $u(x, t) = f(x^2 + t^2)$.

Examples of invariance include the following:

1. The equation $u_{xx} + u_{yy} = 0$ is invariant under rotation, so we search for solutions of the form $u = f(x^2 + y^2)$. Substitution gives the ODE $f' + \eta f'' = 0$ or $(\eta f')' = 0$. The solution is $u(x, t) = c \ln \eta = c \ln(x^2 + t^2)$, which is the (so-called) fundamental solution of Laplace's equation.
2. The nonlinear diffusion equation $u_t = (u^n u_x)_x$ $(n > 0)$, $0 \le x$, $0 \le t$, $u(0, t) = ct^n$ is invariant under scaling with the similar form $u(x, t) = t^n f(\eta)$, $\eta = xt^{-(n+1)/2}$. Substituting into the PDE gives the equation $(f^n f')' + ((n + 1)/2)\eta f' - nf = 0$, with $f(0) = c$ and $f(\infty) = 0$. Note that the equation is an ODE.
3. The wave equation $u_{xx} - u_{tt} = 0$ is invariant under translation. Hence, solutions exist of the form $u = f(x - \lambda t)$. Substitution gives $f''(1 - \lambda^2) = 0$. Hence, $\lambda = \pm 1$ or f is linear. Rejecting the trivial linear solution we see that $u = f(x - t) + g(x + t)$, which is the general (d'Alembert) solution of the wave equation; the quantities $x - t = \alpha$, $x + t = \beta$ are the characteristics of the next section.

The construction of all transformations that leave a PDE invariant is a solved problem left for the references.

The study of "solitons" (solitary traveling waves with special properties) has benefited from symmetry considerations. For example, the nonlinear third-order (Korteweg-de Vries) equation $u_t + uu_x - au_{xxx} = 0$ is invariant under translation. Solutions are sought of the form $u = f(x - \lambda t)$, and f satisfies the ODE, in $\eta = x - \lambda t$, $-\lambda f' + ff' - af''' = 0$.

Characteristics. Using the characteristics the solution of the hyperbolic problem $u_{tt} - u_{xx} = p(x, t)$, $-\infty < x < \infty$, $0 \le t$, $u(x, 0) = f(x)$, $u_t(x, 0) = h(x)$ is

$$u(x,t) = \tfrac{1}{2}\int_0^t d\tau \int_{x-(t-\tau)}^{x+(t-\tau)} p(\xi,\tau)\, d\xi + \tfrac{1}{2}\int_{x-t}^{x+t} h(\xi)\, d\xi + \tfrac{1}{2}\left[f(x+t) + f(x-t)\right]$$

The solution of $u_{tt} - u_{xx} = 0$, $0 \le x < \infty$, $0 \le t < \infty$, $u(x, 0) = 0$, $u_t(x, 0) = h(x)$, $u(0, t) = 0$, $t > 0$ is $u(x, t) = {}^1/_2 \int_{-x+t}^{x+t} h(\xi)\, d\xi$.

The solution of $u_{tt} - u_{xx} = 0$, $0 \le x < \infty$, $0 \le t < \infty$, $u(x, 0) = 0$, $u_t(x, 0) = 0$, $u(0, t) = g(t)$, $t > 0$ is

$$u(x,t) = \begin{cases} 0 & \text{if } t < x \\ g(t-x) & \text{if } t > x \end{cases}$$

From time to time, lower-order derivatives appear in the PDE in use. To remove these from the equation $u_{tt} - u_{xx} + au_x + bu_t + cu = 0$, where a, b, and c are constants, set $\xi = x + t$, $\mu = t - x$, whereupon $u(x, t) = u[(\xi - \mu)/2, (\xi + \mu)/2] = U(\xi, \mu)$, where $U_{\xi\mu} + [(b + a)/4]\, U_\xi + [(b - a)/4]\, U_\mu + (c/4)U = 0$. The transformation $U(\xi, \mu) = W(\xi, \mu) \exp[-(b - a)\xi/4 - (b + a)\mu/4]$ reduces to satisfying $W_{\xi\mu} + \lambda W = 0$, where $\lambda = (a^2 - b^2 + 4c)/16$. If $\lambda \ne 0$, we lose the simple d'Alembert solution. But the equation for W is still easier to handle.

In linear problems discontinuities propagate along characteristics. In nonlinear problems the situation is usually different. The characteristics are often used as new coordinates in the numerical method of characteristics.

Green's Function. Consider the diffusion problem $u_t - u_{xx} = \delta(t)\delta(x - \xi)$, $0 \le x < \infty$, $\xi > 0$, $u(0, t) = 0$, $u(x, 0) = 0$ $[u(\infty, t) = u(\infty, 0) = 0]$, a problem that results from a unit source somewhere in the domain subject to a homogeneous (zero) boundary condition. The solution is called a *Green's function of the first kind*. For this problem there is $G_1(x, \xi, t) = F(x - \xi, t) - F(x + \xi, t)$, where $F(x, t) = e^{-x^2/4t}/\sqrt{4\pi t}$ is the *fundamental* (invariant) *solution*. More generally, the solution of $u_t - u_{xx} = \delta(x - \xi)\, \delta(t - \tau)$, $\xi > 0$, $\tau > 0$, with the same conditions as before, is the Green's function of the first kind.

$$G_1(x,\xi,t-\tau) = \frac{1}{\sqrt{4\pi(t-\tau)}}\left[e^{-(x-\xi)^2/4(t-\tau)} - e^{-(x-\xi)^2/4(t-\tau)}\right]$$

for the semi-infinite interval.

The solution of $u_t - u_{xx} = p(x, t)$, $0 \le x < \infty$, $0 \le t < \infty$, with $u(x, 0) = 0$, $u(0, t) = 0$, $t > 0$ is $u(x, t) = \int_0^t d\tau \int_0^\infty p(\xi, \tau) G_1(x, \xi, t - \tau)]\, d\,\xi$, which is a superposition. Note that the Green's function and the desired solution must both satisfy a zero boundary condition at the origin for this solution to make sense.

The solution of $u_t - u_{xx} = 0$, $0 \le x < \infty$, $0 \le t < \infty$, $u(x, 0) = f(x)$, $u(0, t) = 0$, $t > 0$ is $u(x, t) = \int_0^\infty f(\xi) G_1(x, \xi, t)\, d\,\xi$.

The solution of $u_t - u_{xx} = 0$, $0 \le x < \infty$, $0 \le t < \infty$, $u(x, 0) = 0$, $u(0, t) = g(t)$, $t > 0$ (nonhomogeneous) is obtained by transforming to a new problem that has a homogeneous boundary condition. Thus, with $w(x, t) = u(x, t) - g(t)$ the equation for w becomes $w_t - w_{xx} = -\dot{g}(t) - g(0)\, \delta(t)$ and $w(x, 0) = 0$, $w(0, t) = 0$. Using G_1 above, we finally obtain $u(x, t) = (x/\sqrt{4\pi}) \int_0^t g(t - \tau) e^{-x^2/4\tau}/\tau^{3/2}\, d\tau$.

The Green's function approach can also be employed for elliptic and hyperbolic problems.

Equations in Other Spatial Variables. The sperically symmetric wave equation $u_{rr} + 2u_r/r - u_{tt} = 0$ has the general solution $u(\tau, t) = [f(t - r) + g(t + r)]/r$.

The Poisson-Euler-Darboux equation, arising in gas dynamics,

$$u_{rs} + N(u_r + u_s)/(r + s) = 0$$

where N is a positive integer ≥ 1, has the general solution

$$u(r,s) = k + \frac{\partial^{N-1}}{\partial r^{N-1}}\left[\frac{f(r)}{(r+s)^N}\right] + \frac{\partial^{N-1}}{\partial s^{N-1}}\left[\frac{g(s)}{(r+s)^N}\right]$$

Here, k is an arbitrary constant and f and g are arbitrary functions whose form is determined from the problem initial and boundary conditions.

Conversion to Other Orthogonal Coordinate Systems. Let (x^1, x^2, x^3) be rectangular (Cartesian) coordinates and (u^1, u^2, u^3) be any orthogonal coordinate system related to the rectangular coordinates by $x^i = x^i(u^1, u^2, u^3)$, $i = 1, 2, 3$. With $(ds)^2 = (dx^1)^2 + (dx^2)^2 + (dx^3)^2 = g_{11}(du^1)^2 + g_{22}(du^2)^2 + g_{33}(du^3)^2$, where $g_{ii} = (\partial x^1/\partial u^i)^2 + (\partial x^2/\partial u^i)^2 + (\partial x^3/\partial u^i)^2$. In terms of these "metric" coefficients the basic operations of applied mathematics are expressible. Thus (with $g = g_{11}g_{22}g_{33}$)

$$dA = (g_{11}g_{22})^{1/2}\, du^1\, du^2; \qquad dV = (g_{11}g_{22}g_{33})^{1/2}\, du^1\, du^2\, du^3$$

$$\text{grad}\,\phi = \frac{\vec{a}_1}{(g_{11})^{1/2}}\frac{\partial\phi}{\partial u^1} + \frac{\vec{a}_2}{(g_{22})^{1/2}}\frac{\partial\phi}{\partial u^2} + \frac{\vec{a}_3}{(g_{33})^{1/2}}\frac{\partial\phi}{\partial u^3}$$

($\vec{a}_i$ are unit vectors in direction i);

$$\text{div}\,\vec{E} = g^{-1/2}\left\{\frac{\partial}{\partial u^1}\left[(g_{22}g_{33})^{1/2}E_1\right] + \frac{\partial}{\partial u^2}\left[(g_{11}g_{33})^{1/2}E_2\right] + \frac{\partial}{\partial u^3}\left[(g_{11}g_{22})^{1/2}E_3\right]\right\}$$

[here $\vec{E} = (E_1, E_2, E_3)$];

$$\begin{aligned}\text{curl}\,\vec{E} = g^{-1/2}\Big\{&\vec{a}_1(g_{11})^{1/2}\left(\frac{\partial}{\partial u^2}\left[(g_{33})^{1/2}E_3\right] - \frac{\partial}{\partial u^3}\left[(g_{22})^{1/2}E_2\right]\right)\\ &+ \vec{a}_2(g_{22})^{1/2}\left(\frac{\partial}{\partial u^3}\left[(g_{11})^{1/2}E_1\right] - \frac{\partial}{\partial u^1}\left[(g_{33})^{1/2}E_3\right]\right)\\ &+ \vec{a}_3(g_{33})^{1/2}\left(\frac{\partial}{\partial u^1}\left[(g_{22})^{1/2}E_2\right] - \frac{\partial}{\partial u^2}\left[(g_{11})^{1/2}E_1\right]\right)\Big\}\end{aligned}$$

$$\text{div}\,\text{grad}\,\psi = \nabla^2\psi = \text{Laplacian of } \psi = g^{-1/2}\sum_{i=1}^{3}\frac{\partial}{\partial u^i}\left[\frac{g^{1/2}}{g_{ii}}\frac{\partial\psi}{\partial u^i}\right]$$

Table 5.2 shows some coordinate systems.

Table 5.2 Some Coordinate Systems

Coordinate System	Metric Coefficients	
	Circular Cylindrical	
$x = r\cos\theta$	$u^1 = r$	$g_{11} = 1$
$y = r\sin\theta$	$u^2 = \theta$	$g_{22} = r^2$
$z = z$	$u^3 = z$	$g_{33} = 1$
	Spherical	
$x = r\sin\psi\cos\theta$	$u^1 = r$	$g_{11} = 1$
$y = r\sin\psi\sin\theta$	$u^2 = \psi$	$g_{22} = r^2$

Table 5.2 (continued) Some Coordinate Systems

Coordinate System	Metric Coefficients	
	Circular Cylindrical	
$z = r \cos \psi$	$u^3 = \theta$	$g_{33} = r^2 \sin^2 \psi$
	Parabolic Coordinates	
$x = \mu \nu \cos \theta$	$u^1 = \mu$	$g_{11} = \mu^2 + \nu^2$
$y = \mu \nu \sin \theta$	$u^2 = \nu$	$g_{22} = \mu^2 + \nu^2$
$z = 1/2 (\mu^2 - \nu^2)$	$u^3 = \theta$	$g_{33} = \mu^2\nu^2$

Other metric coefficients and so forth can be found in Moon and Spencer [1961].

References

Ames, W. F. 1965. *Nonlinear Partial Differential Equations in Science and Engineering, Volume 1.* Academic Press, Boston, MA.

Ames, W. F. 1972. *Nonlinear Partial Differential Equations in Science and Engineering, Volume 2.* Academic Press, Boston, MA.

Brauer, F. and Nohel, J. A. 1986. *Introduction to Differential Equations with Applications*, Harper & Row, New York.

Jeffrey, A. 1990. *Linear Algebra and Ordinary Differential Equations*, Blackwell Scientific, Boston, MA.

Kevorkian, J. 1990. *Partial Differential Equations.* Wadsworth and Brooks/Cole, Belmont, CA.

Moon, P. and Spencer, D. E. 1961. *Field Theory Handbook*, Springer, Berlin.

Rogers, C. and Ames, W. F. 1989. *Nonlinear Boundary Value Problems in Science and Engineering.* Academic Press, Boston, MA.

Whitham, G. B. 1974. *Linear and Nonlinear Waves.* John Wiley & Sons, New York.

Zauderer, E. 1983. *Partial Differential Equations of Applied Mathematics.* John Wiley & Sons, New York.

Zwillinger, D. 1992. *Handbook of Differential Equations.* Academic Press, Boston, MA.

Further Information

A collection of solutions for linear and nonlinear problems is found in E. Kamke, *Differential-gleichungen-Lösungsmethoden und Lösungen*. Akad. Verlagsges. Leipzig, 1956. Also see G. M. Murphy, *Ordinary Differential Equations and Their Solutions.* Van Nostrand, Princeton, NJ, 1960 and D. Zwillinger, *Handbook of Differential Equations*, Academic Press, Boston, MA, 1992. For nonlinear problems see

Ames, W. F. 1968. *Ordinary Differential Equations in Transport Phenomena.* Academic Press, Boston, MA.

Cunningham, W. J. 1958. *Introduction to Nonlinear Analysis.* McGraw-Hill, New York.

Jordan, D. N. and Smith, P. 1977. *Nonlinear Ordinary Differential Equations.* Clarendon Press, Oxford, UK.

McLachlan, N. W. 1955. *Ordinary Non Linear Differential Equations in Engineering and Physical Sciences*, 2nd ed. Oxford University Press, London.

Zwillinger, D. 1992.

6

Integral Equations

William F. Ames
Georgia Institute of Technology

6.1 Classification and Notation

Any equation in which the unknown function $u(x)$ appears under the integral sign is called an *integral equation*. If $f(x)$, $K(x, t)$, a, and b are known then the integral equation for u, $\int_a^b K(x, t), u(t)\, dt = f(x)$ is called a *linear integral equation of the first kind of Fredholm type*. $K(x, t)$ is called the *kernel function* of the equation. If b is replaced by x (the independent variable) the equation is an equation of *Volterra type of the first kind.*

An equation of the form $u(x) = f(x) + \lambda \int_a^b K(x, t)u(t)\, dt$ is said to be a linear integral equation of *Fredholm type of the second kind.* If b is replaced by x it is of *Volterra type*. If $f(x)$ is not present the equation is homogeneous.

The equation $\phi(x)\, u(x) = f(x) + \lambda \int_a^{b(x)} K(x, t)u(t)\, dt$ is the *third kind equation* of Fredholm or Volterra type. If the unknown function u appears in the equation in any way other than to the first power then the integral equation is said to be *nonlinear*. Thus, $u(x) = f(x) + \int_a^b K(x, t) \sin u(t)\, dt$ is nonlinear. An integral equation is said to be *singular* when either or both of the limits of integration are infinite or if $K(x, t)$ becomes infinite at one or more points of the integration interval.

Example 6.1. Consider the singular equations $u(x) = x + \int_0^\infty \sin (x\, t)\, u(t)\, dt$ and $f(x) = \int_0^x [u(t)/(x - t)^2]\, dt$.

6.2 Relation to Differential Equations

The *Leibniz rule* $(d/dx) \int_{a(x)}^{b(x)} F(x, t)\, dt \int_{a(x)}^{b(x)} (\partial F/\partial x)\, dt + F[x, b(x)](db/dx) - F[x, a(x)] \times (da/dx)$ is useful for differentiation of an integral involving a parameter (x in this case). With this, one can establish the relation

$$I_n(x) = \int_a^x (x - t)^{n-1} f(t)\, dt = (n-1)! \underbrace{\int_a^x \cdots \int_a^x}_{n \text{ times}} f(x) \underbrace{dx \ldots dx}_{n \text{ times}}$$

This result will be used to establish the relation of the second-order initial value problem to a Volterra integral equation.

The second-order differential equation $y''(x) + A(x)y'(x) + B(x)y = f(x)$, $y(a) = y_0$, $y'(a) = y_0'$ is equivalent to the integral equations

0-8493-0056-8/00/$0.00+$.50

$$y(x) = -\int_a^x \{A(t) + (x-t)[B(t) - A'(t)]\} y(t)\, dt + \int_a^x (x-t) f(t)\, dt + [A(a) y_0 + y_0'](x-a) + y_0$$

which is of the type $(x)y = \int_a^x K(x, t)y(t)\, dt + F(x)$ where $K(x, t) = (t - x)[B(t) - A'(t)] - A(t)$ and $F(x)$ includes the rest of the terms. Thus, this initial value problem is equivalent to a Volterra integral equation of the second kind.

Example 6.2. Consider the equation $y'' + x^2y' + xy = x$, $y(0) = 1$, $y'(0) = 0$. Here $A(x) = x^2$, $B(x) = x$, $f(x) = x$, $a = 0$, $y_0 = 1$, $y_0' = 0$. The integral equation is $y(x) = \int_0^x t(x - 2t)y(t)\, dt + (x^3/6) + 1$.

The expression for $I_n(x)$ can also be useful in converting boundary value problems to integral equations. For example, the problem $y''(x) + \lambda y = 0$, $y(0) = 0$, $y(a) = 0$ is equivalent to the Fredholm equation $y(x) = \lambda \int_0^a K(x, t)y(t)\, dt$, where $K(x, t) = (t/a)(a - x)$ when $t < x$ and $K(x, t) = (x/a)(a - t)$ when $t > x$.

In both cases the differential equation can be recovered from the integral equation by using the Leibnitz rule.

Nonlinear differential equations can also be transformed into integral equations. In fact this is one method used to establish properties of the equation and to develop approximate and numerical solutions. For example, the "forced pendulum" equation $y''(x) + a^2 \sin y(x) = f(x)$, $y(0) = y(1) = 0$ transforms into the nonlinear Fredholm equation.

$$y(x) = \int_0^1 K(x,t)\left[a^2 \sin y(t) - f(t)\right] dt$$

with $K(x, t) = x(1 - t)$ for $0 < x < t$ and $K(x, t) = t(1 - x)$ for $t < x < 1$.

6.3 Methods of Solution

Only the simplest integral equations can be solved exactly. Usually approximate or numerical methods are employed. The advantage here is that integration is a "smoothing operation," whereas differentiation is a "roughening operation." A few exact and approximate methods are given in the following sections. The numerical methods are found under Section 12.

Convolution Equations

The special convolution equation $y(x) = f(x) + \lambda \int_0^x K(x - t)y(t)\, dt$ is a special case of the Volterra equation of the second kind. $K(x - t)$ is said to be a *convolution kernel*. The integral part is the convolution integral discussed under Chapter 8. The solution can be accomplished by transforming with the Laplace transform: $L[y(x)] = L[f(x)] + \lambda L[y(x)]L[K(x)]$ or $y(x) = L^{-1}\{L[f(x)]/(1 - \lambda L[K(x)])\}$.

Abel Equation

The Volterra equation $f(x) = \int_0^x y(t)/(x - t)^\alpha\, dt$, $0 < \alpha < 1$ is the (singular) Abel equation. Its solution is $y(x) = (\sin \alpha\pi/\pi)(d/dx) \int_0^x F(t)/(x - t)^{1-\alpha} dt$.

Approximate Method (Picard's Method)

This method is one of successive approximations that is described for the equation $y(x) = f(x) + \lambda \int_a^x K(x, t)y(t)\, dt$. Beginning with an initial guess $y_0(t)$ (often the value at the initial point a) generate the next approximation with $y_1(x) = f(x) + \int_a^x K(x, t)y_0(t)\, dt$ and continue with the general iteration

$$y_n(x) = f(x) + \lambda \int_a^x K(x,t) y_{n-1}(t)\, dt$$

Then, by iterating, one studies the convergence of this process, as is described in the literature.

Example 6.3. Let $y(x) = 1 + \int_0^x xt[y(t)]^2\,dt$, $y(0) = 1$. With $y_0(t) = 1$ we find $y_1(x) = 1 + \int_0^x xt\,dt = 1 + (x^3/2)$ and $y_2(x) = 1 + \int_0^x xt[1 + (t^3/2)^2 dt$, and so forth.

References

Jerri, A. J. 1985. *Introduction to Integral Equations with Applications.* Marcel Dekker, New York.

Tricomi, F. G. 1958. *Integral Equations.* Wiley-Interscience, New York.

Yosida, K. 1960. *Lectures on Differential and Integral Equations.* Wiley-Interscience, New York.

7

Approximation Methods

William F. Ames
Georgia Institute of Technology

The term *approximation methods* usually refers to an analytical process that generates a symbolic approximation rather than a numerical one. Thus, $1 + x + x^2/2$ is an approximation of e^x for small x. This chapter introduces some techniques for approximating the solution of various operator equations.

7.1 Perturbation

Regular Perturbation

This procedure is applicable to *some* equations in which a small parameter, ε, appears. Use this procedure with care; the procedure involves expansion of the dependent variables and data in a power series in the small parameter. The following example illustrates the procedure.

Example 7.1. Consider the equation $y'' + \varepsilon y' + y = 0$, $y(0) = 1$, $y'(0) = 0$. Write $y(x; \varepsilon) = y_0(x) + \varepsilon y_1(x) + \varepsilon^2 y_2(x) + \cdots$, and the initial conditions (data) become

$$y_0(0) + \varepsilon y_1(0) + \varepsilon^2 y_2(0) + \cdots = 1$$

$$y_0'(0) + \varepsilon y_1'(0) + \varepsilon^2 y_2'(0) + \cdots - 0$$

Equating like powers of ε in all three equations yields the sequence of equations

$$O(\varepsilon^0): y_0'' + y_0 = 0, \qquad y_0(0) = 1, \qquad y_0'(0) = 0$$

$$O(\varepsilon^1): y_1'' + y_1 = -y_0', \qquad y_1(0) = 0, \qquad y_1'(0) = 0$$

$$\vdots$$

The solution for y_0 is $y_0 = \cos x$ and using this for y_1 we find $y_1(x) = 1/2\,(\sin x - x\cos x)$. So $y(x; \varepsilon) = \cos x + \varepsilon(\sin x - x\cos x)/2 + O(\varepsilon^2)$. Appearance of the term $x \cos x$ indicates a *secular term* that becomes arbitrarily large as $x \to \infty$. Hence, this approximation is valid only for $x \ll 1/\varepsilon$ and for small ε. If an approximation is desired over a larger range of x then the method of multiple scales is required.

0-8493-0056-8/00/$0.00+$.50

Singular Perturbation

The *method of multiple scales* is a singular method that is *sometimes* useful if the regular perturbation method fails. In this case the assumption is made that the solution depends on *two* (or more) different length (or time) scales. By trying various possibilities, one can determine those scales. The scales are treated as dependent variables when transforming the given ordinary differential equation into a partial differential equation, but then the scales are treated as independent variables when solving the equations.

Example 7.2. Consider the equation $\varepsilon y'' + y' = 2$, $y(0) = 0$, $y(1) = 1$. This is singular since (with $\varepsilon = 0$) the resulting first-order equation cannot satisfy both boundary conditions. For the problem the proper length scales are $u = x$ and $v = x/\varepsilon$. The second scale can be ascertained by substituting $\varepsilon^n x$ for x and requiring $\varepsilon y''$ and y' to be of the same order in the transformed equation. Then

$$\frac{d}{dx} = \frac{\partial}{\partial u}\frac{du}{dx} + \frac{\partial}{\partial v}\frac{dv}{dx} = \frac{\partial}{\partial u} + \frac{1}{\varepsilon}\frac{\partial}{\partial v}$$

and the equation becomes

$$\varepsilon\left(\frac{\partial}{\partial u} + \frac{1}{\varepsilon}\frac{\partial}{\partial v}\right)^2 y + \left(\frac{\partial}{\partial u} + \frac{1}{\varepsilon}\frac{\partial}{\partial v}\right) y = 2$$

With $y(x; \varepsilon) = y_0(u, v) + \varepsilon y_1(u, v) + \varepsilon^2 y_2(u, v) + \cdots$ we have terms

$$O(\varepsilon^{-1}): \frac{\partial^2 y_0}{\partial v^2} + \frac{\partial y_0}{\partial v} = 0 \qquad \text{(actually ODEs with parameter } u\text{)}$$

$$O(\varepsilon^{0}): \frac{\partial^2 y_1}{\partial v^2} + \frac{\partial y_1}{\partial v} = 2 - 2\frac{\partial^2 y_0}{\partial u \partial v} - \frac{\partial y_0}{\partial u}$$

$$O(\varepsilon^{1}): \frac{\partial^2 y_2}{\partial v^2} + \frac{\partial y_2}{\partial v} = -2\frac{\partial^2 y_1}{\partial u \partial v} - \frac{\partial y_1}{\partial u} - \frac{\partial^2 y_0}{\partial u^2}$$

$$\vdots$$

Then $y_0(u, v) = A(u) + B(u)e^{-v}$ and so the second equation becomes $\partial^2 y_1/\partial v^2 + \partial y_1/\partial v = 2 - A'(u) + B'(u)e^{-v}$, with the solution $y_1(u, v) = [2 - A'(u)]v + vB'(u)e^{-v} + D(u) + E(u)e^{-v}$. Here A, B, D and E are still arbitrary. Now the solvability condition — "higher order terms must vanish no slower (as $\varepsilon \to 0$) than the previous term" (Kevorkian and Cole, 1981) — is used. For y_1 to vanish no slower than y_0 we must have $2 - A'(u) = 0$ and $B'(u) = 0$. If this were not true the terms in y_1 would be larger than those in y_0 ($v \gg 1$). Thus $y_0(u, v) = (2u + A_0) + B_0 e^{-v}$, or in the original variables $y(x; \varepsilon) \approx (2x + A_0) + B_0 e^{-x/\varepsilon}$ and matching to both boundary conditions gives $y(x; \varepsilon) \approx 2x - (1 - e^{-x/\varepsilon})$.

Boundary Layer Method

The boundary layer method is applicable to regions in which the solution is *rapidly varying*. See the references at the end of the chapter for detailed discussion.

7.2 Iterative Methods

Taylor Series

If it is known that the solution of a differential equation has a power series in the independent variable (t), then we may proceed from the initial data (the easiest problem) to compute the Taylor series by differentiation.

Example 7.3. Consider the equation $(d^2x/dt) = -x - x^2$, $x(0) = 1$, $x'(0) = 1$. From the differential equation, $x''(0) = -2$, and, since $x''' = -x' - 2xx'$, $x'''(0) = -1 - 2 = -3$, so the four term approximation for $x(t) \approx 1 + t - (2t^2/2!) - (3t^3/3!) = 1 + t - t^2 - t^3/2$. An estimate for the error at $t = t_1$, (see a discussion of series methods in any calculus text) is not greater than $|d^4x/dt^4|_{max}[(t_1)^4/4!]$, $0 \le t \le t_1$.

Picard's Method

If the vector differential equation $x' = f(t, x)$, $x(0)$ given, is to be approximated by Picard iteration, we begin with an initial guess $x_0 = x(0)$ and calculate iteratively $x_i' = f(t, x_{i-1})$.

Example 7.4. Consider the equation $x' = x + y^2$, $y' = y - x^3$, $x(0) = 1$, $y(0) = 2$. With $x_0 = 1$, $y_0 = 2$, $x_1' = 5$, $y_1' = 1$, so $x_1 = 5t + 1$, $y_1 = t + 2$, since $x_i(0) = 1$, $y_i(0) = 2$ for $i \ge 0$. To continue, use $x_{i+1}' = x_i + y_i^2$, $y_{i+1}' = y_i - x_i^3$. A modification is the utilization of the first calculated term immediately in the second equation. Thus, the calculated value of $x_1 = 5t + 1$, when used in the second equation, gives $y_1' = y_0 - (5t + 1)^3 = 2 - (125t^3 + 75t^2 + 15t + 1)$, so $y_1 = 2t - (125t^4/4) - 25t^3 - (15t^2/2) - t + 2$. Continue with the iteration $x_{i+1}' = x_i + y_i^3$, $y_{i+1}' = y_i - (x_{i+1})^3$.

Another variation would be $x_{i+1}' = x_{i-1} + (y_i)^3$, $y_{i+1}' = y_{i-1} - (x_{i-1})^3$.

References

Ames, W. F. 1965. *Nonlinear Partial Differential Equations in Science and Engineering, Volume I.* Academic Press, Boston, MA.

Ames, W. F. 1968. *Nonlinear Ordinary Differential Equations in Transport Processes.* Academic Press, Boston, MA.

Ames, W. F. 1972. *Nonlinear Partial Differential Equations in Science and Engineering, Volume II.* Academic Press, Boston, MA.

Kevorkian, J. and Cole, J. D. 1981. *Perturbation Methods in Applied Mathematics.* Springer, New York.

Miklin, S. G. and Smolitskiy, K. L. 1967. *Approximate Methods for Solutions of Differential and Integral Equations.* Elsevier, New York.

Nayfeh, A. H. 1973. *Perturbation Methods.* John Wiley & Sons, New York.

Zwillinger, D. 1992. *Handbook of Differential Equations.* 2nd ed. Academic Press, Boston, MA.

8

Integral Transforms

William F. Ames
Georgia Institute of Technology

All of the integral transforms are special cases of the equation $g(s) = \int_a^b K(s, t)f(t)dt$, in which $g(s)$ is said to be the *transform* of $f(t)$, and $K(s, t)$ is called the *kernel* of the transform. Table 8.1 shows the more important kernels and the corresponding intervals (a, b).

Details for the first three transforms listed in Table 8.1 are given here. The details for the other are found in the literature.

8.1 Laplace Transform

The Laplace transform of $f(t)$ is $g(s) = \int_0^\infty e^{-st} f(t)\,dt$. It may be thought of as transforming one class of functions into another. The advantage in the operation is that under certain circumstances it replaces complicated functions by simpler ones. The notation $L[f(t)] = g(s)$ is called the *direct transform* and $L^{-1}[g(s)] = f(t)$ is called the *inverse transform*. Both the direct and inverse transforms are tabulated for many often-occurring functions. In general $L^{-1}[g(s)] = (1/2\pi i)\int_{\alpha-i\infty}^{\alpha+i\infty} e^{st}g(s)\,ds$, and to evaluate this integral requires a knowledge of complex variables, the theory of residues, and contour integration.

Properties of the Laplace Transform

Let $L[f(t)] = g(s)$, $L^{-1}[g(s)] = f(t)$.

1. The Laplace transform may be applied to a function $f(t)$ if $f(t)$ is continuous or piecewise continuous; if $t^n|f(t)|$ is finite for all t, $t \to 0$, $n < 1$; and if $e^{-at}|f(t)|$ is finite as $t \to \infty$ for some value of a, $a > 0$.
2. L and L^{-1} are unique.
3. $L[af(t) + bh(t)] = aL[f(t)] + bL[h(t)]$ (linearity).
4. $L[e^{at}f(t)] = g(s - a)$ (shift theorem).
5. $L[(-t)^k f(t)] = d^k g/ds^k$; k a positive integer.

Example 8.1. $L[\sin at] = \int_0^\infty e^{-st} \sin at\, dt = a/(s^2 + a^2)$, $s > 0$. By property 5,

$$\int_0^\infty e^{-st} t \sin at\, dt = L[t \sin at] = \frac{2as}{s^2 + a^2}$$

0-8493-0056-8/00/$0.00+$.50

Table 8.1 Kernels and Intervals of Various Integral Transforms

Name of Transform	(a, b)	$K(s, t)$
Laplace	$(0, \infty)$	e^{-st}
Fourier	$(-\infty, \infty)$	$\frac{1}{\sqrt{2\pi}} e^{-ist}$
Fourier cosine	$(0, \infty)$	$\sqrt{\frac{2}{\pi}} \cos st$
Fourier sine	$(0, \infty)$	$\sqrt{\frac{2}{\pi}} \sin st$
Mellin	$(0, \infty)$	t^{s-1}
Hankel	$(0, \infty)$	$tJ_v(st),\ v \geq -\frac{1}{2}$

6.
$$L[f'(t)] = sL[f(t)] - f(0)$$
$$L[f''(t)] = s^2 L[f(t)] - sf(0) - f'(0)$$
$$\vdots$$
$$L[f^{(n)}(t)] = s^n L[f(t)] - s^{n-1} f(0) - \cdots - sf^{(n-2)}(0) - f^{(n-1)}(0)$$

In this property it is apparent that the initial data are automatically brought into the computation.

Example 8.2. Solve $y'' + y = e^t$, $y(0) = 1$, $y'(0) = 1$. Now $L[y''] = s^2L[y] - sy(0) - y'(0) = s^2L[y] - s - 1$. Thus, using the linear property of the transform (property 3), $s^2L[y] + L[y] - s - 1 = L[e^t] = 1/(s - 1)$. Therefore, $L[y] = s^2/[(s - 1)(s^2 + 1)]$.

With the notations $\Gamma(n + 1) = \int_0^\infty x^n e^{-x}\, dx$ (gamma function) and $J_n(t)$ the Bessel function of the first kind of order n, a short table of Laplace transforms is given in Table 8.2.

7. $$L\left[\int_a^t f(t)\, dt\right] = \frac{1}{s} L[f(t)] + \frac{1}{s}\int_a^0 f(t)\, dt.$$

Example 8.3. Find $f(t)$ if $L[f(t)] = (1/s^2)[1/(s^2 - a^2)]$. $L[1/a \sinh a\, t] = 1/(s^2 - a^2)$. Therefore, $f(t) = \int_0^t [\int_0^t \frac{1}{a} \sinh a\, t\, d\, t] d\, t = 1/a^2[(\sinh a\, t)/a - t]$.

$$L\left[\frac{f(t)}{t}\right] = \int_s^\infty g(s)\, ds; \qquad L\left[\frac{f(t)}{t^k}\right] = \underbrace{\int_s^\infty \cdots \int_s^\infty}_{k \text{ integrals}} g(s)(ds)^k$$

Example 8.4. $L[(\sin a\, t)/t] = \int_s^\infty L[\sin a\, t] d\, s = \int_s^\infty [a\, d\, s/(s^2 + a^2)] = \cot^{-1}(s/a)$.

9. The *unit step function* $u(t - a) = 0$ for $t < a$ and 1 for $t > a$. $L[u(t - a)] = e^{-as}/s$.
10. The *unit impulse function* is $\delta(a) = u'(t - a) = 1$ at $t = a$ and 0 elsewhere. $L[u'(t - a)] = e^{-as}$.
11. $L^{-1}[e^{-as} g(s)] = f(t - a)u(t - a)$ (second shift theorem).
12. If $f(t)$ is *periodic* of period b — that is, $f(t + b) = f(t)$ — then $L[f(t)] = [1/(1 - e^{-bs})] \times \int_0^b e^{-st} f(t)\, dt$.

Example 8.5. The equation $\partial^2 y/(\partial t \partial x) + \partial y/\partial t + \partial y/\partial x = 0$ with $(\partial y/\partial x)(0, x) = y(0, x) = 0$ and $y(t, 0) + (\partial y/\partial t)(t, 0) = \delta(0)$ (see property 10) is solved by using the Laplace transform of y with respect to t. With $g(s, x) = \int_0^\infty e^{-st} y(t, x)\, dt$, the transformed equation becomes

Table 8.2 Some Laplace Transforms

$f(t)$	$g(s)$	$f(t)$	$g(s)$
1	$\frac{1}{s}$	$e^{-at}(1-at)$	$\frac{s}{(s+a)^2}$
t^n, n is a + integer	$\frac{n!}{s^{n+1}}$	$\frac{t \sin at}{2a}$	$\frac{s}{(s^2+a^2)^2}$
t^n, $n \neq$ a + integer	$\frac{\Gamma(n+1)}{s^{n+1}}$	$\frac{1}{2a^2}\sin at \sinh at$	$\frac{s}{s^4+4a^4}$
$\cos at$	$\frac{s}{s^2+a^2}$	$\cos at \cosh at$	$\frac{s^3}{s^4+4a^4}$
$\sin at$	$\frac{a}{s^2+a^2}$	$\frac{1}{2a}(\sinh at + \sin at)$	$\frac{s^2}{s^4-a^4}$
$\cosh at$	$\frac{s}{s^2-a^2}$	$\frac{1}{2}(\cosh at + \cos at)$	$\frac{s^3}{s^4-a^4}$
$\sinh at$	$\frac{a}{s^2-a^2}$	$\frac{\sin at}{t}$	$\tan^{-1}\frac{a}{s}$
e^{-at}	$\frac{1}{s+a}$	$J_0(at)$	$\frac{1}{\sqrt{s^2+a^2}}$
$e^{-bt}\cos at$	$\frac{s+b}{(s+b)^2+a^2}$	$\frac{n}{a^n}\frac{J_n(at)}{t}$	$\frac{1}{\left(\sqrt{s^2+a^2}+s\right)^n}$
$e^{-bt}\sin at$	$\frac{a}{(s+b)^2+a^2}$	$J_0\left(2\sqrt{at}\right)$	$\frac{1}{s}e^{-a/s}$

$$s\frac{\partial g}{\partial x} - \frac{\partial y}{\partial x}(0,x) + sg - y(0,x) + \frac{\partial g}{\partial x} = 0$$

or

$$(s+1)\frac{\partial g}{\partial x} + sg = \frac{\partial y}{\partial x}(0,x) + y(0,x) = 0$$

The second (boundary) condition gives $g(s, 0) + sg(s, 0) - y(0, 0) = 1$ or $g(s, 0) = 1/(1 + s)$. A solution of the preceding ordinary differential equation consistent with this condition is $g(s, x) = [1/(s+1)]e^{-sx/(s+1)}$. Inversion of this transform gives $y(t, x) = e^{-(t+x)}I_0(2\sqrt{tx})$, where I_0 is the zero-order Bessel function of an imaginary argument.

8.2 Convolution Integral

The *convolution integral* (*faltung*) of two functions $f(t)$, $r(t)$ is $x(t) = f(t)*r(t) = \int_0^t f(\tau)r(t-\tau)\,d\tau$.

Example 8.6. $t * \sin t = \int_0^t \tau \sin(t-\tau)\,d\tau = t - \sin t$.

13. $L[f(t)]L[h(t)] = L[f(t) * h(t)]$.

8.3 Fourier Transform

The *Fourier transform* is given by $F[f(t)] = (1/\sqrt{2\pi})\int_{-\infty}^{\infty} f(t)e^{-ist}\,dt = g(s)$ and its *inverse* by $F^{-1}[g(s)] = (1/\sqrt{2\pi})\int_{-\infty}^{\infty} g(s)e^{ist}\,dt = f(t)$. In brief, the condition for the Fourier transform to exist is that $\int_{-\infty}^{\infty}|f(t)|\,dt < \infty$, although certain functions may have a Fourier transform even if this is violated.

Example 8.7. The function $f(t) = 1$ for $-a \le t \le a$ and $= 0$ elsewhere has

$$F[f(t)] = \int_{-a}^{a} e^{-ist}\,dt = \int_0^{a} e^{ist}\,dt + \int_0^{a} e^{-ist}\,dt = 2\int_0^{a} \cos st\,dt = \frac{2\sin sa}{s}$$

Properties of the Fourier Transform

Let $F[f(t)] = g(s)$; $F^{-1}[g(s)] = f(t)$.

1. $F[f^{(n)}(t)] = (i\,s)^n F[f(t)]$
2. $F[af(t) + bh(t)] = aF[f(t)] + bF[h(t)]$
3. $F[f(-t)] = g(-s)$
4. $F[f(at)] = 1/a\; g(s/a)$, $a > 0$
5. $F[e^{-iwt}f(t)] = g(s + w)$
6. $F[f(t + t_1)] = e^{ist_1}\, g(s)$
7. $F[f(t)] = G(i\,s) + G(-i\,s)$ if $f(t) = f(-t)$($f(t)$ even)
   $F[f(t)] = G(i\,s) - G(-i\,s)$ if $f(t) = -f(-t)$(f odd)

where $G(s) = L[f(t)]$. This result allows the use of the Laplace transform tables to obtain the Fourier transforms.

Example 8.8. Find $F[e^{-a|t|}]$ by property 7. The term $e^{-a|t|}$ is even. So $L[e^{-at}] = 1/(s + a)$. Therefore, $F[e^{-a|t|}] = 1/(i\,s + a) + 1/(-i\,s + a) = 2a/(s^2 + a^2)$.

8.4 Fourier Cosine Transform

The *Fourier cosine transform* is given by $F_c[f(t)] = g(s) = \sqrt{(2/\pi)}\int_0^{\infty} f(t)\cos s\,t\,d\,t$ and its *inverse* by $F_c^{-1}[g(s)] = f(t) = \sqrt{(2/\pi)}\int_0^{\infty} g(s)\cos s\,t\,d\,s$. The *Fourier sine transform* F_s is obtainable by replacing the cosine by the sine in the above integrals.

Example 8.9. $F_c[f(t)]$, $f(t) = 1$ for $0 < t < a$ and 0 for $a < t < \infty$. $F_c[f(t)] = \sqrt{(2/\pi)}\int_0^{a}\cos s\,t\,d\,t = \sqrt{(2/\pi)}\,(\sin a\,s)/s$.

Properties of the Fourier Cosine Transform

$F_c[f(t)] = g(s)$.

1. $F_c[af(t) + bh(t)] = aF_c[f(t)] + bF_c[h(t)]$
2. $F_c[f(at)] = (1/a)\; g\,(s/a)$
3. $F_c[f(at)\cos bt] = 1/2a\,[g\,((s + b)/a) + g((s - b)/a)]$, $a, b > 0$
4. $F_c[t^{2n}f(t)] = (-1)^n(d^{2n}g)/(d\,s^{2n})$
5. $F_c[t^{2n+1}f(t)] = (-1)^n(d^{2n+1})/(d\,s^{2n+1})\;F_s[f(t)]$

Table 8.3 presents some Fourier cosine transforms.

Example 8.10. The temperature θ in the semiinfinite rod $0 \le x < \infty$ is determined by the differential equation $\partial\theta/\partial t = k(\partial^2\theta/\partial x^2)$ and the condition $\theta = 0$ when $t = 0$, $x \ge 0$; $\partial\theta/\partial x = -\mu$ = constant when $x = 0$, $t > 0$. By using the Fourier cosine transform, a solution may be found as $\theta(x, t) = (2\mu/\pi)\int_0^{\infty}(\cos px/p)(1 - e^{-kp^2t})\,d\,p$.

Table 8.3 Fourier Cosine Transforms

$f(t)$	$\frac{g(s)}{\sqrt{2/\pi}}$
$\begin{cases} t & 0 < t < 1 \\ 2 - t & 1 < t < 2 \\ 0 & 2 < t < \infty \end{cases}$	$\frac{1}{s^2}[2\cos s - 1 - \cos 2s]$
$t^{-1/2}$	$\pi^{1/2}(s)^{-1/2}$
$\begin{cases} 0 & 0 < t < a \\ (t-a)^{-1/2} & a < t < \infty \end{cases}$	$\pi^{1/2}(s)^{-1/2}[\cos a s - \sin a s]$
$(t^2 + a^2)^{-1}$	$\frac{1}{2}\pi a^{-1}e^{-as}$
e^{-at}, $a > 0$	$\frac{a}{s^2 + a^2}$
e^{-at^2}, $a > 0$	$\frac{1}{2}\pi^{1/2}a^{-1/2}e^{-s^2/4a}$
$\frac{\sin at}{t}$ $a > 0$	$\begin{cases} \pi/2 & s < a \\ \pi/4 & s = a \\ 0 & s > a \end{cases}$

References

Churchill, R. V. 1958. *Operational Mathematics*. McGraw-Hill, New York.

Ditkin, B. A. and Proodnikav, A. P. 1965. *Handbook of Operational Mathematics* (in Russian). Nauka, Moscow.

Doetsch, G. 1950–1956. *Handbuch der Laplace Transformation*, vols. I-IV (in German). Birkhauser, Basel.

Nixon, F. E. 1960. *Handbook of Laplace Transforms*. Prentice-Hall, Englewood Cliffs, NJ.

Sneddon, I. 1951. *Fourier Transforms*. McGraw-Hill, New York.

Widder, D. 1946. *The Laplace Transform*. Princeton University Press, Princeton, NJ.

Further Information

The references citing G. Doetsch, *Handbuch der Laplace Transformation*, vols. I-IV, Birkhauser, Basel, 1950–1956 (in German) *and* B. A. Ditkin and A. P. Prodnikav, *Handbook of Operational Mathematics*, Moscow, 1965 (in Russian) are the most extensive tables known. The latter reference is 485 pages.

9

Calculus of Variations

William F. Ames
Georgia Institute of Technology

The basic problem in the *calculus of variations* is to determine a function such that a certain *functional*, often an integral involving that function and certain of its derivatives, takes on *maximum or minimum values*. As an example, find the function $y(x)$ such that $y(x_1) = y_1$, $y(x_2) = y_2$ and the integral (functional) $I = 2\pi \int_{x_1}^{x_2} y[1 + y')^2]^{1/2}\, dx$ is a minimum. A second example concerns the transverse deformation $u(x, t)$ of a beam. The energy functional $I = \int_{t_1}^{t_2} \int_0^L [1/2\, \rho\, (\partial u/\partial t)^2 - 1/2\, EI\, (\partial^2 u/\partial x^2)^2 + fu]\, dx\, dt$ is to be minimized.

9.1 The Euler Equation

The elementary part of the theory is concerned with a *necessary* condition (generally in the form of a differential equation with boundary conditions) that the required function must satisfy. To show mathematically that the function obtained actually maximizes (or minimizes) the integral is much more difficult than the corresponding problems of the differential calculus.

The *simplest case* is to determine a function $y(x)$ that makes the integral $I = \int_{x_1}^{x_2} F(x, y, y')\, dx$ stationary and that satisfies the prescribed end conditions $y(x_1) = y_1$ and $y(x_2) = y_2$. Here we suppose F has continuous second partial derivatives with respect to x, y, and $y' = dy/dx$. If $y(x)$ is such a function, then it must satisfy the *Euler equation* $(d/dx)(\partial F/\partial y') - (\partial F/\partial y) = 0$, which is the required necessary condition. The indicated partial derivatives have been formed by treating x, y, and y' as independent variables. Expanding the equation, the equivalent form $F_{y'y'}y'' + F_{y'y}y' + (F_{y'x} - F_y) = 0$ is found. This is second order in y unless $F_{y'y'} = (\partial^2 F)/[(\partial y')^2] = 0$. An alternative form $1/y'[d/dx(F - (\partial F/\partial y')(dy/dx)) - (\partial F/\partial x)] = 0$ is useful. Clearly, if F does not involve x explicitly $[(\partial F/\partial x) = 0]$ a first integral of Euler's equation is $F - y'(\partial F/\partial y') = c$. If F does not involve y explicitly $[(\partial F/\partial y) = 0]$ a first integral is $(\partial F/\partial y') = c$.

The Euler equation for $I = 2\pi \int_{x_1}^{x_2} y[1 + (y')^2]^{1/2}\, dx$, $y(x_1) = y_1$, $y(x_2) = y_2$ is $(d/dx)[yy'/[1 + (y')^2]^{1/2}] - [1 + (y')^2]^{1/2} = 0$ or after reduction $yy'' - (y')^2 - 1 = 0$. The solution is $y = c_1 \cosh(x/c_1 + c_2)$, where c_1 and c_2 are integration constants. Thus the required minimal surface, if it exists, must be obtained by revolving a catenary. Can c_1 and c_2 be chosen so that the solution passes through the assigned points? The answer is found in the solution of a transcendental equation that has two, one, or no solutions, depending on the prescribed values of y_1 and y_2.

9.2 The Variation

If $F = F(x, y, y')$, with x independent and $y = y(x)$, then the *first variation* δF of F is defined to be $\delta F = (\partial F/\partial x)\, \delta y + (\partial F/\partial y)\, \delta y'$ and $\delta y' = \delta\, (dy/dx) = (d/dx)\, (\delta y)$ — that is, they commute. Note that the first variation, δF, of a functional is a first-order change from curve to curve, whereas the differential of a

0-8493-0056-8/00/$0.00+$.50

function is a first-order approximation to the change in that function along a *particular curve*. The laws of δ are as follows: $\delta(c_1F + c_2G) = c_1\delta F + c_2\delta G$; $\delta(FG) = F\delta G + G\delta F$; $\delta(F/G) = (G\delta F - F\delta G)/G^2$; if x is an independent variable, $\delta x = 0$; if $u = u(x, y)$; $(\partial/\partial x)(\delta u) = \delta(\partial u/\partial x)$, $(\partial/\partial y)(\delta u) = \delta(\partial u/\partial y)$.

A necessary condition that the integral $I = \int_{x_1}^{x_2} F(x, y, y')\,dx$ be stationary is that its (first) variation vanish — that is, $\delta I = \delta \int_{x_1}^{x_2} F(x, y, y')\,dx = 0$. Carrying out the variation and integrating by parts yields of $\delta I = \int_{x_1}^{x_2} [(\partial F/\partial y) - (d/dx)(\partial F/\partial y')]\,\delta y\,dx + [(\partial F/\partial y')\,\partial y]_{x_1}^{x_2} = 0$. The arbitrary nature of δy means the square bracket must vanish and the last term constitutes the *natural boundary conditions*.

Example. The *Euler equation* of $\int_{x_1}^{x_2} F(x, y, y', y'')\,dx$ is $(d^2/dx^2)(\partial F/\partial y'') - (d/dx)(\partial F/\partial y') + (\partial F/\partial y) = 0$, with natural boundary conditions $\{[(d/dx)(\partial F/\partial y'') - (\partial F/\partial y')]\,\delta y\}_{x_1}^{x_2} = 0$ and $(\partial F/\partial y'')\,\delta y'|_{x_1}^{x_2} = 0$. The Euler equation of $\int_{x_1}^{x_2}\int_{y_1}^{y_2} F(x, y, u, u_x, u_y, u_{xx}, u_{xy}, u_{yy})\,dx\,dy$ is $(\partial^2/\partial x^2)(\partial F/\partial u_{xx}) + (\partial^2/\partial x\partial y)(\partial F/u_{xy}) + (\partial^2/\partial y^2)(\partial F/\partial u_{yy}) - (\partial/\partial x)(\partial F/\partial u_x) - (\partial/\partial y)(\partial F/\partial u_y) + (\partial F/\partial u)$, and the natural boundary conditions are

$$\left[\left(\frac{\partial}{\partial x}\left(\frac{\partial F}{\partial u_{xx}}\right) + \frac{\partial}{\partial y}\left(\frac{\partial F}{\partial u_{xy}}\right) - \frac{\partial F}{\partial u_x}\right)\delta u\right]_{x_1}^{x_2} = 0, \qquad \left[\left(\frac{\partial F}{\partial u_{xx}}\right)\delta u_x\right]_{x_1}^{x_2} = 0$$

$$\left[\left(\frac{\partial}{\partial y}\left(\frac{\partial F}{\partial u_{yy}}\right) + \frac{\partial}{\partial x}\left(\frac{\partial F}{\partial u_{xy}}\right) - \frac{\partial F}{\partial u_y}\right)\delta u\right]_{y_1}^{y_2} = 0, \qquad \left[\left(\frac{\partial F}{\partial u_{yy}}\right)\delta u_y\right]_{y_1}^{y_2} = 0$$

In the more general case of $I = \iint_R F(x, y, u, v, u_x, u_y, v_x, v_y)\,dx\,dy$, the condition $\delta I = 0$ gives rise to the two Euler equations $(\partial/\partial x)(\partial F/\partial u_x) + (\partial/\partial y)(\partial F/\partial u_y) - (\partial F/\partial u) = 0$ and $(\partial/\partial x)(\partial F/\partial v_x) + (\partial/\partial y)(\partial F/\partial v_y) - (\partial F/\partial v) = 0$. These are two PDEs in u and v that are linear or quasi-linear in u and v. The Euler equation for $I = \iiint_R (u_x^2 + u_y^2 + u_z^2)\,dx\,dy\,dz$, from $\delta I = 0$, is Laplace's equation $u_{xx} + u_{yy} + u_{zz} = 0$.

Variational problems are easily derived from the differential equation and associated boundary conditions by multiplying by the variation and integrating the appropriate number of times. To illustrate, let $F(x)$, $\rho(x)$, $p(x)$, and w be the tension, the linear mass density, the natural load, and (constant) angular velocity of a rotating string of length L. The equation of motion is $(d/dx)[F\,(dy/dx)] + \rho w^2 y + p = 0$. To formulate a corresponding variational problem, multiply all terms by a variation δy and integrate over $(0, L)$ to obtain

$$\int_0^L \frac{d}{dx}\left(F\frac{dy}{dx}\right)\delta y\,dx + \int_0^L \rho w^2 y\delta y\,dx + \int_0^L p\delta y\,dx = 0$$

The second and third integrals are the variations of $1/2\ \rho w^2 y^2$ and py, respectively. To treat the first integral, integrate by parts to obtain

$$\left[F\frac{dy}{dx}\delta y\right]_0^L - \int_0^L F\frac{dy}{dx}\delta\frac{dy}{dx}\,dx = \left[F\frac{dy}{dx}\delta y\right]_0^L - \int_0^L \frac{1}{2}F\delta\left(\frac{dy}{dx}\right)^2 dx = 0$$

So the variation formulation is

$$\delta\int_0^L\left[\frac{1}{2}\rho w^2 y^2 + py - \frac{1}{2}F\left(\frac{dy}{dx}\right)^2\right]dx + \left[F\frac{dy}{dx}\delta y\right]_0^L = 0$$

The last term represents the *natural boundary conditions*. The term $1/2\ \rho w^2 y^2$ is the kinetic energy per unit length, the term $-py$ is the potential energy per unit length due to the radial force $p(x)$, and the term $1/2\ F(dy/dx)^2$ is a first approximation to the potential energy per unit length due to the tension $F(x)$ in the string. Thus the integral is often called the *energy integral*.

9.3 Constraints

The variations in some cases cannot be arbitrarily assigned because of one or more auxiliary conditions that are usually called *constraints*. A typical case is the functional $\int_{x_1}^{x_2} F(x, u, v, u_x, v_x)\, dx$ with a constraint $\phi(u, v) = 0$ relating u and v. If the variations of u and v (δu and δv) vanish at the end points, then the variation of the integral becomes

$$\int_{x_1}^{x_2} \left\{ \left[\frac{\partial F}{\partial u} - \frac{d}{dx}\left(\frac{\partial F}{\partial u_x} \right) \right] \delta u + \left[\frac{\partial F}{\partial v} - \frac{d}{dx}\left(\frac{\partial F}{\partial v_x} \right) \right] \delta v \right\} dx = 0$$

The variation of the constraint $\phi(u, v) = 0$, $\phi_u \delta u + \phi_v \delta v = 0$ means that the variations cannot both be assigned arbitrarily inside (x_1, x_2), so their coefficients need not vanish separately. Multiply $\phi_u \delta u + \phi_v \delta v = 0$ by a Lagrange multiplier λ (may be a function of x) and integrate to find $\int_{x_1}^{x_2} (\lambda\phi_u \delta u + \lambda\phi_v \delta v)\, dx = 0$. Adding this to the previous result yields

$$\int_{x_1}^{x_2} \left\{ \left[\frac{\partial F}{\partial u} - \frac{d}{dx}\left(\frac{\partial F}{\partial u_x} \right) + \lambda\phi_u \right] \delta u + \left[\frac{\partial F}{\partial v} - \frac{d}{dx}\left(\frac{\partial F}{\partial v_x} \right) + \lambda\phi_v \right] \delta v \right\} dx = 0$$

which must hold for any λ. Assign λ so the first square bracket vanishes. Then δv can be assigned to vanish inside (x_1, x_2) so the two systems

$$\frac{d}{dx}\left[\frac{\partial F}{\partial u_x} \right] - \frac{\partial F}{\partial u} - \lambda\phi_u = 0, \qquad \frac{d}{dx}\left[\frac{\partial F}{\partial v_x} \right] - \frac{\partial F}{\partial v} - \lambda\phi_v = 0$$

plus the constraint $\phi(u, v) = 0$ are three equations for u, v and λ.

References

Gelfand, I. M. and Fomin, S. V. 1963. *Calculus of Variations*. Prentice Hall, Englewood Cliffs, NJ.

Lanczos, C. 1949. *The Variational Principles of Mechanics*. Univ. of Toronto Press, Toronto.

Schechter, R. S. 1967. *The Variational Method in Engineering*. McGraw-Hill, New York.

Vujanovic, B. D. and Jones, S. E. 1989. *Variational Methods in Nonconservative Phenomena*. Academic Press, New York.

Weinstock, R. 1952. *Calculus of Variations, with Applications to Physics and Engineering*. McGraw-Hill, New York.

10

Optimization Methods

George Cain
Georgia Institute of Technology

10.1 Linear Programming

Let **A** be an $m \times n$ matrix, **b** a column vector with m components, and **c** a column vector with n components. Suppose $m < n$, and assume the rank of **A** is m. The standard linear programming problem is to find, among all nonnegative solutions of $\mathbf{Ax} = \mathbf{b}$, one that minimizes

$$\mathbf{c}^T\mathbf{x} = c_1x_1 + c_2x_2 + \cdots + c_nx_n$$

This problem is called a *linear* program. Each solution of the system $\mathbf{Ax} = \mathbf{b}$ is called a *feasible* solution, and the *feasible set* is the collection of all *feasible solutions.* The function $\mathbf{c}^T\mathbf{x} = c_1x_1 + c_2x_2 + \cdots + c_nx_n$ is the cost function, or the objective function. A solution to the linear program is called an *optimal feasible solution.*

Let **B** be an $m \times n$ submatrix of **A** made up of m linearly independent columns of **A**, and let **C** be the $m \times (n - m)$ matrix made up of the remaining columns of **A**. Let $\mathbf{x}_B$ be the vector consisting of the components of **x** corresponding to the columns of **A** that make up **B**, and let $\mathbf{x}_C$ be the vector of the remaining components of **x**, that is, the components of **x** that correspond to the columns of **C**. Then the equation $\mathbf{Ax} = \mathbf{b}$ may be written $\mathbf{Bx}_B + \mathbf{Cx}_C = \mathbf{b}$. A solution of $\mathbf{Bx}_B = \mathbf{b}$ together with $\mathbf{x}_C = \mathbf{0}$ gives a solution **x** of the system $\mathbf{Ax} = \mathbf{b}$. Such a solution is called a *basic solution*, and if it is, in addition, nonnegative, it is a *basic feasible solution.* If it is also optimal, it is an *optimal basic feasible solution.* The components of a basic solution are called *basic variables.*

The Fundamental Theorem of Linear Programming says that if there is a feasible solution, there is a basic feasible solution, and if there is an optimal feasible solution, there is an optimal basic feasible solution. The linear programming problem is thus reduced to searching among the set of basic solutions for an optimal solution. This set is, of course, finite, containing as many as $n!/[m!(n - m)!]$ points. In practice, this will be a very large number, making it imperative that one use some efficient search procedure in seeking an optimal solution. The most important of such procedures is the *simplex method*, details of which may be found in the references.

The problem of finding a solution of $\mathbf{Ax} \le \mathbf{b}$ that minimizes $\mathbf{c}^T\mathbf{x}$ can be reduced to the standard problem by appending to the vector **x** an additional m nonnegative components, called *slack variables*. The vector **x** is replaced by **z**, where $\mathbf{z}^T = [x_1,x_2 \ldots x_n\ s_1,s_2 \ldots x_m]$, and the matrix **A** is replaced by $\mathbf{B} = [\mathbf{A}\ \mathbf{I}]$, where **I** is the $m \times m$ identity matrix. The equation $\mathbf{Ax} = \mathbf{b}$ is thus replaced by $\mathbf{Bz} = \mathbf{Ax} + s = \mathbf{b}$, where $s^T = [s_1,s_2,\ldots,s_m]$. Similarly, if inequalities are reversed so that we have $\mathbf{Ax} \le \mathbf{b}$, we simply append $-s$ to the vector x. In this case, the additional variables are called *surplus variables.*

0-8493-0056-8/00/$0.00+$.50

Associated with every linear programming problem is a corresponding dual problem. If the *primal* problem is to minimize $\mathbf{c}^T\mathbf{x}$ subject to $\mathbf{Ax} \geq \mathbf{b}$, and $\mathbf{x} \geq 0$, the corresponding *dual* problem is to maximize $\mathbf{y}^T\mathbf{b}$ subject to $\mathbf{t}^T\mathbf{A} \leq \mathbf{c}^T$. If either the primal problem or the dual problem has an optimal solution, so also does the other. Moreover, if $\mathbf{x}_p$ is an optimal solution for the primal problem and $\mathbf{y}_d$ is an optimal solution for the corresponding dual problem $\mathbf{c}^T\mathbf{x}_p = \mathbf{y}_d^T\mathbf{b}$.

10.2 Unconstrained Nonlinear Programming

The problem of minimizing or maximizing a sufficiently smooth nonlinear function $f(x)$ of n variables, $\mathbf{x}^T = [x_1, x_2 \ldots x_n]$, with no restrictions on x is essentially an ordinary problem in calculus. At a minimizer or maximizer $\mathbf{x}^*$, it must be true that the gradient of f vanishes:

$$\nabla f(\mathbf{x}^*) = 0$$

Thus $\mathbf{x}^*$ will be in the set of all solutions of this system of n generally nonlinear equations. The solution of the system can be, of course, a nontrivial undertaking. There are many recipes for solving systems of nonlinear equations. A method specifically designed for minimizing f is the *method of steepest descent*. It is an old and honorable algorithm, and the one on which most other more complicated algorithms for unconstrained optimization are based. The method is based on the fact that at any point x, the direction of maximum decrease of f is in the direction of $-\nabla f(x)$. The algorithm searches in this direction for a minimum, recomputes $\nabla f(x)$ at this point, and continues iteratively. Explicitly:

1. Choose an initial point x_0.
2. Assume x_k has been computed; then compute $y_k = \nabla f(x_k)$, and let $t_k \geq 0$ be a local minimum of $g(t) = f(x_k - ty_k)$. Then $x_{k+1} = x_k - t_k y_k$.
3. Replace k by $k + 1$, and repeat step 2 until t_k is small enough.

Under reasonably general conditions, the sequence (x_k) converges to a minimum of f.

10.3 Constrained Nonlinear Programming

The problem of finding the maximum or minimum of a function $f(x)$ of n variables, subject to the constraints

$$\mathbf{a}(\mathbf{x}) = \begin{bmatrix} a_1(x_1, x_2, \ldots, x_n) \\ a_2(x_1, x_2, \ldots, x_n) \\ \vdots \\ a_m(x_1, x_2, \ldots, x_n) \end{bmatrix} = \begin{bmatrix} b_1 \\ b \\ \vdots \\ b_m \end{bmatrix} = \mathbf{b}$$

is made into an unconstrained problem by introducing the new function $L(x)$:

$$L(\mathbf{x}) = f(\mathbf{x}) + \mathbf{z}^T\mathbf{a}(\mathbf{x})$$

where $\mathbf{z}^T = [\lambda_1, \lambda_2, \ldots, \lambda_m]$ is the vector of *Lagrange multipliers*. Now the requirement that $\nabla L(x) = 0$, together with the constraints $a(x) = \mathbf{b}$, give a system of $n + m$ equations

$$\nabla f(\mathbf{x}) + \mathbf{z}^T \nabla \mathbf{a}(\mathbf{x}) = 0$$

$$\mathbf{a}(\mathbf{x}) = \mathbf{b}$$

for the $n + m$ unknowns $x_1, x_2, \ldots, x_n, \lambda_1 \lambda_2, \ldots, \lambda_m$ that must be satisfied by the minimizer (or maximizer) x.

The problem of inequality constraints is significantly more complicated in the nonlinear case than in the linear case. Consider the problem of minimizing $f(x)$ subject to m equality constraints $\mathbf{a}(\mathbf{x}) = \mathbf{b}$, and p inequality constraints $\mathbf{c}(\mathbf{x}) \le \mathbf{d}$ [thus $\mathbf{a}(\mathbf{x})$ and $\mathbf{b}$ are vectors of m components, and $\mathbf{c}(\mathbf{x})$ and $\mathbf{d}$ are vectors of p components.] A point $\mathbf{x}^*$ that satisfies the constraints is a *regular point* if the collection

$$\left\{\nabla a_1(\mathbf{x}^*), \nabla a_2(\mathbf{x}^*), \ldots, \nabla a_m(\mathbf{x}^*)\right\} \cup \left\{\nabla c_j(\mathbf{x}^*) : j \in J\right\}$$

where

$$J = \left\{j : c_j(\mathbf{x}^*) = d_j\right\}$$

is linearly independent. If $\mathbf{x}^*$ is a local minimum for the constrained problem and if it is a regular point, there is a vector $\mathbf{z}$ with m components and a vector $\mathbf{w} \ge \mathbf{0}$ with p components such that

$$\nabla f(\mathbf{x}^*) + \mathbf{z}^T \nabla \mathbf{a}(\mathbf{x}^*) + \mathbf{w}^T \nabla \mathbf{c}(\mathbf{x}^*) = 0$$

$$\mathbf{w}^T \left(\mathbf{c}(\mathbf{x}^*) - \mathbf{d}\right) = 0$$

These are the *Kuhn-Tucker conditions.* Note that in order to solve these equations, one needs to know for which j it is true that $c_j(\mathbf{x}^*) = 0$. (Such a constraint is said to be *active.*)

References

Luenberger, D. C. 1984. *Linear and Nonlinear Programming.* 2nd ed. Addison-Wesley, Reading, MA.

Peressini, A. L. Sullivan, F. E., and Uhl, J. J., Jr. 1988. *The Mathematics of Nonlinear Programming.* Springer-Verlag, New York.

11

Engineering Statistics

Y. L. Tong
Georgia Institute of Technology

11.1 Introduction

In most engineering experiments, the outcomes (and hence the observed data) appear in a random and on deterministic fashion. For example, the operating time of a system before failure, the tensile strength of a certain type of material, and the number of defective items in a batch of items produced are all subject to random variations from one experiment to another. In engineering statistics, we apply the theory and methods of statistics to develop procedures for summarizing the data and making statistical inferences, thus obtaining useful information with the presence of randomness and uncertainty.

11.2 Elementary Probability

Random Variables and Probability Distributions

Intuitively speaking, a random variable (denoted by X, Y, Z, etc.) takes a numerical value that depends on the outcome of the experiment. Since the outcome of an experiment is subject to random variation, the resulting numerical value is also random. In order to provide a stochastic model for describing the probability distribution of a random variable X, we generally classify random variables into two groups: the discrete type and the continuous type. The discrete random variables are those which, technically speaking, take a finite number or a countably infinite number of possible numerical values. (In most engineering applications they take nonnegative integer values.) Continuous random variables involve outcome variables such as time, length or distance, area, and volume. We specify a function $f(x)$, called the probability density function (p.d.f.) of a random variable X, such that the random variable X takes a value in a set A (or real numbers) as given by

$$P[X \in A] = \begin{cases} \sum_{x \in A} f(x) & \text{for all sets } A \text{ if } X \text{ is discrete} \\ \int_A f(x)\, dx & \text{for all intervals } A \text{ if } X \text{ is continuous} \end{cases} \tag{11.1}$$

0-8493-0056-8/00/$0.00+$.50

By letting A be the set of all values that are less than or equal to a fixed number t, i.e., $A = (-\infty, t)$, the probability function $P[X \leq t]$, denoted by $F(t)$, is called the distribution function of X. We note that, by calculus, if X is a continuous random variable and if $F(x)$ is differentiable, then $f(x) = \frac{d}{dx} F(x)$.

Expectations

In many applications the "payoff" or "reward" of an experiment with a numerical outcome X is a specific function of X ($u(X)$, say). Since X is a random variable, $u(X)$ is also a random variable. We define the expected value of $u(X)$ by

$$Eu(X) = \begin{cases} \sum_x u(x) f(x) & \text{if } X \text{ is discrete} \\ \int_{-\infty}^{\infty} u(x) f(x)\, dx & \text{if } X \text{ is continuous} \end{cases} \tag{11.2}$$

provided of course, that, the sum or the integral exists. In particular, if $u(x) = x$, the $EX \equiv \mu$ is called the mean of X (of the distribution) and $E(X - \mu)^2 \equiv \sigma^2$ is called the variance of X (of the distribution). The mean is a measurement of the central tendency, and the variance is a measurement of the dispersion of the distribution.

Some Commonly Used Distributions

Many well-known distributions are useful in engineering statistics. Among the discrete distributions, the hypergeometric and binomial distributions have applications in acceptance sampling problems and quality control, and the Poisson distribution is useful for studying queuing theory and other related problems. Among the continuous distributions, the uniform distribution concerns random numbers and can be applied in simulation studies, the exponential and gamma distributions are closely related to the Poisson distribution, and they, together with the Weibull distribution, have important applications in life testing and reliability studies. All of these distributions involve some unknown parameter(s), hence their means and variances also depend on the parameter(s). The reader is referred to textbooks in this area for details. For example, Hahn and Shapiro (1967, pp. 163–169 and pp. 120–134) contains a comprehensive listing of these and other distributions on their p.d.f.'s and the graphs, parameter(s), means, variances, with discussions and examples of their applications.

The Normal Distribution

Perhaps *the* most important distribution in statistics and probability is the normal distribution (also known as the Gaussian distribution). This distribution involves two parameters: μ and σ^2, and its p.d.f. is given by

$$f(x) = f(x;\mu,\sigma^2) = \frac{1}{\sqrt{2\pi}\sigma} e^{-\frac{1}{2\sigma^2}(x-\mu)^2} \tag{11.3}$$

for $-\infty < \mu < \infty$, $\sigma^2 > 0$, and $-\infty < \chi < \infty$. It can be shown analytically that, for a p.d.f. of this form, the values of μ and σ^2 are, respectively, that of the mean and the variance of the distribution. Further, the quantity, $\sigma = \sqrt{\sigma^2}$ is called the standard deviation of the distribution. We shall use the symbol $X \sim N(\mu, \sigma^2)$ to denote that X has a normal distribution with mean μ and variance σ^2. When plottting the p.d.f. $f(x;\mu,\sigma^2)$ given in Equation (11.3) we see that the resulting graph represents a bell-shaped curve symmetric about μ, as shown in Figure 11.1.

If a random variable Z has an $N(0,1)$ distribution, then the p.d.f. of Z is given by (from Equation (11.3))

$$\phi(z) = \frac{1}{\sqrt{2\pi}} e^{-\frac{1}{2}z^2} \qquad -\infty < z < \infty \tag{11.4}$$

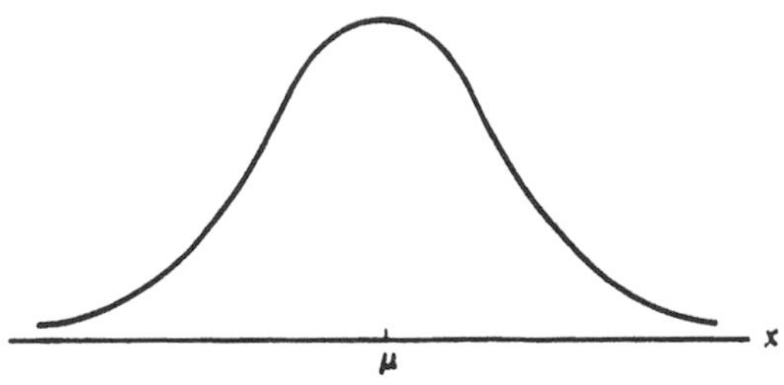

Figure 11.1 The normal curve with mean μ and variance σ^2.

The distribution function of Z,

$$\Phi(z) = \int_{-\infty}^{z} \phi(u)\, du \qquad -\infty < z < \infty \tag{11.5}$$

cannot be given in a closed form, hence it has been tabulated. The table of $\Phi(z)$ can be found in most textbooks in statistics and probability, including those listed in the references at the end of this section. (We note in passing that, by the symmetry property, $\Phi(z) + \Phi(-z) = 1$ holds for all z.)

11.3 Random Sample and Sampling Distributions

Random Sample and Related Statistics

As noted in Box et al., (1978), the design and analysis of engineering experiments usually involves the following steps:

1. The choice of a suitable stochastic model by assuming that the observations follow a certain distribution. The functional form of the distribution (or the p.d.f.) is assumed to be known, except the value(s) of the parameters(s).
2. Design of experiments and collection of data.
3. Summarization of data and computation of certain statistics.
4. Statistical inference (including the estimation of the parameters of the underlying distribution and the hypothesis-testing problems).

In order to make statistical inference concerning the parameter(s) of a distribution, it is essential to first study the sampling distributions. We say that $X_1, X_2, \ldots, X_n$ represent a random sample of size n if they are independent random variables and each of them has the same p.d.f., $f(x)$. (Due to space limitations, the notion of independence will not be carefully discussed here. Nevertheless, we say that $X_1, X_2, \ldots, X_n$ are independent if

$$P\left[X_1 \in A_1, X_2 \subset A_2, \ldots, X_n \subset A_n\right] = \prod_{i=1}^{n} P\left[X_i \subset A_i\right] \tag{11.6}$$

holds for all sets $A_1, A_2, \ldots, A_n$.) Since the parameter(s) of the population is (are) unknown, the population mean μ and the population variance σ^2 are unknown. In most commonly used distributions μ and σ^2 can be estimated by the sample mean $\bar{X}$ and the sample variance S^2, respectively, which are given by

$$\bar{X} = \frac{1}{n}\sum_{i=1}^{n} X_i, \qquad S^2 = \frac{1}{n-1}\sum_{i=1}^{n}\left(X_i - \bar{X}\right)^2 = \frac{1}{n-1}\left[\sum_{i=1}^{n} X_i^2 - n\bar{X}^2\right] \tag{11.7}$$

(The second equality in the formula for S^2 can be verified algebraically.) Now, since $X_1, X_2, \ldots, X_n$ are random variables $\bar{X}$ and S^2 are also random variables. Each of them is called a statistic and has a probability distribution which also involves the unknown parameter(s). In probability theory there are two fundamental results concerning their distributional properties.

Theorem 1. (Weak Law of Large Numbers). As the sample size n becomes large, $\bar{X}$ converges to μ in probability and S^2 converges to σ^2 in probability. More precisely, for every fixed positive number $\varepsilon > 0$ we have

$$P\left[\left|\bar{X} - \mu\right| \le \varepsilon\right] \to 1, \qquad P\left[\left|S^2 - \sigma^2\right| \le \varepsilon\right] \to 1 \tag{11.8}$$

as $n \to \infty$.

Theorem 2. (Central Limit Theorem). As n becomes large, the distribution of the random variable

$$Z = \frac{\bar{X} - \mu}{\sigma \sqrt{n}} = \frac{\sqrt{n}\left(\bar{X} - \mu\right)}{\sigma} \tag{11.9}$$

has approximately an $N(0,1)$ distribution. More precisely,

$$P[Z \le z] \to \Phi(z) \text{ for every fixed } z \text{ as } n \to \infty \tag{11.10}$$

11.4 Normal Distribution-Related Sampling Distributions

One-Sample Case

Additional results exist when the observations come from a normal population. If $X_1, X_2, \ldots, X_n$ represent a random sample of size n from an $N(\mu,\sigma^2)$ population, then the following sample distributions are useful:

Fact 3. For every fixed n the distribution of Z given in Equation (11.9) has *exactly* an N(0,1) distribution.

Fact 4. The distribution of the statistic $T = \sqrt{n}(\bar{X} - \mu)/S$, where $S = \sqrt{S^2}$ is the sample standard deviation, is called a Student's t distribution with $v = n - 1$ degrees of freedom, in symbols, $t(n - 1)$.

This distribution is useful for making inference on μ when σ^2 is unknown; a table of the percentiles can be found in most statistics textbooks.

Fact 5. The distribution of the statistic $W = (n - 1)S^2/\sigma^2$ is called a chi-squared distribution with $v = n - 1$ degrees of freedom, in symbols $\chi^2(v)$.

Such a distribution is useful in making inference on σ^2; a table of the percentiles can also be found in most statistics books.

Two-Sample Case

In certain applications we may be interested in the comparisons of two different treatments. Suppose that independent samples from treatments T_1 and T_2 are to be observed as shown in Table 11.1.

The difference of the population means $(\mu_1 - \mu_2)$ and the ratio of the population variances can be estimated, respectively, by $(\bar{X}_1 - \bar{X}_2)$ and S_1^2/S_2^2. The following facts summarize the distributions of these statistics:

Fact 6. Under the assumption of normality, $(\bar{X}_1 - \bar{X}_2)$ has an $N(\mu_1 - \mu_2, (\sigma_1^2/n_1) + (\sigma_2^2/n_2))$ distribution; or equivalently, for all n_1, n_2 the statistic

Table 11.1 Summarization of Data for a Two-Sample Problem

Treatment	Observations	Distribution	Sample Size	Sample Mean	Sample Variance
T_1	$X_{11}, X_{12}, \ldots, X_{1n_1}$	$\mathcal{N}(\mu_1, \sigma_1^2)$	n_1	$\bar{X}_1$	S_1^2
T_2	$X_{21}, X_{22}, \ldots, X_{2n_2}$	$\mathcal{N}(\mu_2, \sigma_2^2)$	n_2	$\bar{X}_2$	S_2^2

$$Z = \left[\left(\bar{X}_1 - \bar{X}_2\right) - \left(\mu_1 - \mu_2\right)\right] \Big/ \left(\sigma_1^2/n_1 + \sigma_2^2/n_2\right)^{1/2} \qquad (11.11)$$

has an $N(0,1)$ distribution.

Fact 7. When $\sigma_1^2 = \sigma_2^2 \equiv \sigma^2$, the common population variance is estimated by

$$S_p^2 = \left(n_1 + n_2 - 2\right)^{-1}\left[\left(n_1 - 1\right)S_1^2 + \left(n_2 - 1\right)S_2^2\right] \qquad (11.12)$$

and $(n_1 + n_2 - 2)S_p^2/\sigma^2$ has a $\chi^2(n_1 + n_2 - 2)$ distribution.

Fact 8. When $\sigma_1^2 = \sigma_2^2$, the statistic

$$T = \left[\left(\bar{X}_1 - \bar{X}_2\right) - \left(\mu_1 - \mu_2\right)\right] \Big/ S_p\left(1/n_1 + 1/n_2\right)^{1/2} \qquad (11.13)$$

has a $t(n_1 + n_2 - 2)$ distribution, where $S_p = \sqrt{S_p^2}$.

Fact 9. The distribution of $F = (S_1^2/\sigma_1^2)/(S_2^2/\sigma_2^2)$ is called an F distribution with degrees of freedom $(n_1 - 1, n_2 - 1)$, in symbols, $F(n_1 - 1, n_2 - 1)$.

The percentiles of this distribution have also been tabulated and can be found in statistics books.

In the following two examples we illustrate numerically how to find probabilities and percentiles using the existing tables for the normal, Student's *t*, chi-squared, and *F* distributions.

Example 10. Suppose that in an experiment four observations are taken, and that the population is assumed to have a normal distribution with mean μ and variance σ^2. Let $\bar{X}$ and S^2 be the sample mean and sample variance as given in Equation (11.7).

(a) If, based on certain similar experiments conducted in the past, we know that $\sigma^2 = 1.8^2 \times 10^{-6}$ ($\sigma = 1.8 \times 10^{-3}$), then from $\Phi(-1.645) = 0.05$ and $\Phi(1.96) = 0.975$ we have

$$P\left[-1.645 \le \frac{\bar{X} - \mu}{1.8 \times 10^{-3}/\sqrt{4}} \le 1.96\right] = 0.975 - 0.05 = 0.925$$

or equivalently,

$$P\left[-1.645 \times 0.9 \times 10^{-3} \le \bar{X} - \mu \le 1.96 \times 0.9 \times 10^{-3}\right] = 0.925$$

(b) The statistic $T = 2(\bar{X} - \mu)/S$ has a Student's *t* distribution with 3 degrees of freedom (in symbols, $t(3)$). From the *t* table we have

$$P\left[-3.182 \le 2\left(\bar{X} - \mu\right)/S \le 3.182\right] = 0.95$$

which yields

$$P\left[-3.182\times\frac{S}{2}\leq \bar{X}-\mu\leq 3.182\times\frac{S}{2}\right]=0.95$$

or equivalently,

$$P\left[\bar{X}-3.182\times\frac{S}{2}\leq \mu\leq \bar{X}+3.182\times\frac{S}{2}\right]=0.95$$

This is, in fact, the basis for obtaining the confidence interval for μ given in Equation (11.17) when σ^2 is unknown.

(c) The statistic $3S^2/\sigma^2$ has a chi-squared distribution with 3 degrees of freedom (in symbols, $\chi^2(3)$). Thus from the chi-squared table we have $P[0.216 \leq 3S^2/\sigma^2 \leq 9.348] = 0.95$, which yields

$$P\left[\frac{3S^2}{9.348}\leq \sigma^2\leq \frac{3S^2}{0.216}\right]=0.95$$

and it forms the basis for obtaining a confidence interval for σ^2 as given in Equation (11.18).

Example 11. Suppose that in Table 11.1 (with two treatments) we have $n_1 = 4$ and $n_2 = 5$, and we let $\bar{X}_1$, $\bar{X}_2$ and S_1^2, S_2^2 denote the corresponding sample means and sample variances, respectively.

(a) Assume that $\sigma_1^2 = \sigma_2^2$ where the common variance is unknown and is estimated by S_p^2 given in Equation (11.12). Then the statistic

$$T=\left[(\bar{X}_1-\bar{X}_2)-(\mu_1-\mu_2)\right]\big/S_p\left(\tfrac{1}{4}+\tfrac{1}{5}\right)^{1/2}$$

has a $t(7)$ distribution. Thus from the t table we have

$$P=\left[-2.998\leq\left[(\bar{X}_1-\bar{X}_2)-(\mu_1-\mu_2)\right]\big/S_p\left(\tfrac{1}{4}+\tfrac{1}{5}\right)^{1/2}\leq 2.998\right]=0.98$$

which is equivalent to saying that

$$P\left[-2.998S_p\left(\tfrac{1}{4}+\tfrac{1}{5}\right)^{1/2}\leq\mu_1-\mu_2\leq(\bar{X}_1-\bar{X}_2)+2.998S_p\left(\tfrac{1}{4}+\tfrac{1}{5}\right)^{1/2}\right]=0.98$$

(b) The statistic $F = (S_1^2/\sigma_1^2)/(S_2^2/\sigma_2^2)$ has an $F(3,4)$ distribution. Thus from the F-table we have

$$P\left[\left(\frac{\sigma_2^2}{\sigma_1^2}\right)\left(\frac{S_1^2}{S_2^2}\right)\leq 6.59\right]=0.95$$

or equivalently,

$$P\left[\frac{\sigma_2^2}{\sigma_1^2}\leq 6.59\frac{S_2^2}{S_1^2}\right]=0.95$$

The distributions listed above (normal, Student's t, chi-squared, and F) form an integral part of the classical statistical inference theory, and they are developed under the assumption that the observations follow a normal distribution. When the distribution of the population is not normal and inference on the populations means is to be made, we conclude that (1) if the sample sizes n_1, n_2 are large, then the statistic Z in Equation (11.11) has an approximate $N(0,1)$ distribution and (2) in the small-sample case, the exact distribution of $\bar{X}$ (of $(\bar{X}_1 - \bar{X}_2)$) depends on the population p.d.f. There are several analytical methods for obtaining it, and those methods can be found in statistics textbooks.

11.5 Confidence Intervals

A method for estimating the population parameters based on the sample mean(s) and sample variance(s) involves the confidence intervals for the parameters.

One-Sample Case

1. Confidence Interval for μ When σ^2 is Known. Consider the situation in which a random sample of size n is taken from an $N(\mu,\sigma^2)$ population and σ^2 is known. An interval, I_1, of the form $I_1 = (\bar{X} - d, \bar{X} + d)$ (with width $2d$) is to be constructed as a "confidence interval or μ." If we make the assertion that μ is in this interval (i.e., μ is bounded below by $\bar{X} - d$ and bounded above by $\bar{X} + d$), then sometimes this assertion is correct and sometimes it is wrong, depending on the value of $\bar{X}$ in a given experiment. If for a fixed α value we would like to have a confidence probability (called confidence coefficient) such that

$$P[\mu \in I_1] = P[\bar{X} - d < \mu < \bar{X} + d] = 1 - \alpha \tag{11.14}$$

then we need to choose the value of d to satisfy $d = z_{\alpha/2}\dfrac{\sigma}{\sqrt{n}}$, i.e.,

$$I_1 = \left(\bar{X} - z_{\alpha/2}\frac{\sigma}{\sqrt{n}}, \bar{X} + z_{\alpha/2}\frac{\sigma}{\sqrt{n}}\right) \tag{11.15}$$

where $z_{\alpha/2}$ is the $(1 - \alpha/2)$th percentile of the $N(0,1)$ distribution such that $\Phi(z_{\alpha/2}) = 1 - \alpha/2$. To see this, we note that from the sampling distribution of $\bar{X}$ (Fact 3) we have

$$\begin{aligned} P\left[\bar{X} - z_{\alpha/2}\frac{\sigma}{\sqrt{n}} < \mu < \bar{X} + z_{\alpha/2}\frac{\sigma}{\sqrt{n}}\right] &= P\left[\frac{|\bar{X} - \mu|}{\sigma/\sqrt{n}} \le z_{\alpha/2}\right] \\ &= \Phi(z_{\alpha/2}) - \Phi(-z_{\alpha/2}) = 1 - \alpha \end{aligned} \tag{11.16}$$

We further note that, even when the original population is not normal, by Theorem 2 the confidence probability is approximately $(1 - \alpha)$ when the sample size is reasonably large.

2. Confidence Interval for μ When σ^2 is Unknown. Assume that the observations are from an $N(\mu,\sigma^2)$ population. When σ^2 is unknown, by Fact 4 and a similar argument we see that

$$I_2 = \left(\bar{X} - t_{\alpha/2}(n-1)\frac{S}{\sqrt{n}}, \bar{X} + t_{\alpha/2}(n-1)\frac{S}{\sqrt{n}}\right) \tag{11.17}$$

is a confidence interval for μ with confidence probability $1 - \alpha$, where $t_{\alpha/2}(n - 1)$ is the $(1 - \alpha/2)$th percentile of the $t(n - 1)$ distribution.

3. Confidence Interval for σ^2. If, under the same assumption of normality, a confidence interval for σ^2 is needed when μ is unknown, then

$$I_3 = \left((n-1)S^2/\chi^2_{1-\alpha/2}(n-1),\ (n-1)S^2/\chi^2_{\alpha/2}(n-1)\right) \tag{11.18}$$

has a confidence probability $1 - \alpha$, when $\chi^2_{1-\alpha/2}(n-1)$ and $\chi^2_{\alpha/2}(n-1)$ are the $(\alpha/2)$th and $(1-\alpha/2)$th percentiles, respectively, of the $\chi^2(n-1)$ distribution.

Two-Sample Case

1. Confidence Intervals for $\mu_1 - \mu_2$ When $\sigma_1^2 = \sigma_2^2$ are Known. Consider an experiment that involves the comparison of two treatments, T_1 and T_2, as indicated in Table 11.1. If a confidence interval for $\delta = \mu_1 - \mu_2$ is needed when σ_1^2 and σ_2^2 are unknown, then by Fact 6 and a similar argument, the confidence interval

$$I_4 = \left((\bar{X}_1 - \bar{X}_2) - z_{\alpha/2}\sqrt{\sigma_1^2/n_1 + \sigma_2^2/n_2},\ (\bar{X}_1 - \bar{X}_2) + z_{\alpha/2}\sqrt{\sigma_1^2/n_1 + \sigma_2^2/n_2}\right) \tag{11.19}$$

has a confidence probability $1 - \alpha$.

2. Confidence Interval for $\mu_1 - \mu_2$ when σ_1^2, σ_2^2 are Unknown but Equal. Under the additional assumption that $\sigma_1^2 = \sigma_2^2$, but the common variance is unknown, then by Fact 8 the confidence interval

$$I_5 = \left((\bar{X}_1 - \bar{X}_2) - d,\ (\bar{X}_1 - \bar{X}_2) + d\right) \tag{11.20}$$

has a confidence probability $1 - \alpha$, where

$$d = t_{\alpha/2}(n_1 + n_2 - 2)S_p(1/n_1 + 1/n_2)^{1/2} \tag{11.21}$$

3. Confidence Interval for σ_2^2/σ_1^2. A confidence interval for the ratio of the variances σ_2^2/σ_1^2 can be obtained from the F distribution (see Fact 9), and the confidence interval

$$I_6 = \left(F_{1-\alpha/2}(n_1 - 1, n2_1 - 1)\frac{S_2^2}{S_1^2},\ F_{\alpha/2}(n_1 - 1, n2_1 - 1)\frac{S_2^2}{S_1^2}\right) \tag{11.22}$$

has a confidence probability $1 - \alpha$, where $F_{1-\alpha/2}(n_1 - 1, n_2 - 1)$ and $F_{\alpha/2}(n_1 - 1, n_2 - 1)$ are, respectively, the $(\alpha/2)$th and $(1 - \alpha/2)$th percentiles of the $F(n_1 - 1, n_2 - 1)$ distribution.

11.6 Testing Statistical Hypotheses

A statistical hypothesis concerns a statement or assertion about the true value of the parameter in a given distribution. In the two-hypothesis problems, we deal with a null hypothesis and an alternative hypothesis, denoted by H_0 and H_1, respectively. A decision is to be made, based on the data of the experiment, to either accept H_0 (hence reject H_1) or reject H_0 (hence accept H_1). In such a two-action problem, we may commit two types of errors: the type I error is to reject H_0 when it is true, and the type II error is to accept H_0 when it is false. As a standard practice, we do not reject H_0 unless there is significant evidence indicating that it may be false. (In doing so, the burden of proof that H_0 is false is on the experimenter.) Thus we usually choose a small fixed number, α (such as 0.05 or 0.01), such that the probability of committing a type I error is at most (or equal to) α. With such a given α, we can then determine the region in the data space for the rejection of H_0 (called the critical region).

One-Sample Case

Suppose that $X_1, X_2, \ldots, X_n$ represent a random sample of size n from an $N(\mu,\sigma^2)$ population, and $\bar{X}$ and S^2 are, respectively, the sample mean and sample variance.

1. Test for Mean. In testing

$$H_0 : \mu = \mu_0 \text{ vs. } H_1 : \mu = \mu_1 (\mu_1 > \mu_0) \text{ or } H_1 : \mu > \mu_0$$

when σ^2 is known, we reject H_0 when $\bar{X}$ is large. To determine the cut-off point, we note (by Fact 3) that the statistic $Z_0 = (\bar{X} - \mu_0)/(\sigma/\sqrt{n})$ has an $N(0,1)$ distribution under H_0. Thus, if we decide to reject H_0 when $Z_0 > z_\alpha$, then the probability of committing a type I error is α. As a consequence, we apply the decision rule

$$d_1 : \text{reject } H_0 \text{ if and only if } \bar{X} > \mu_0 + z_\alpha \frac{\sigma}{\sqrt{n}}$$

Similarly, from the distribution of Z_0 under H_0 we can obtain the critical region for the other types of hypotheses. When σ^2 is unknown, then by Fact 4 $T_0 = \sqrt{n}(\bar{X} - \mu_0)/S$ has a $t(n-1)$ distribution under H_0. Thus the corresponding tests can be obtained by substituting $t_\alpha(n-1)$ for z_α and S for σ. The tests for the various one-sided and two-sided hypotheses are summarized in Table 11.2 below. For each set of hypotheses, the critical region given on the first line is for the case when σ^2 is known, and that given on the second line is for the case when σ^2 is unknown. Furthermore, t_α and $t_{\alpha/2}$ stand for $t_\alpha(n-1)$ and $t_{\alpha/2}(n-1)$, respectively.

Table 11.2 One-Sample Tests for Mean

Null Hypothesis H_0	Alternative Hypothesis H_1	Critical Region
$\mu = \mu_0$ or $\mu \le \mu_0$	$\mu = \mu_1 > \mu_0$ or $\mu > \mu_0$	$\bar{X} > \mu_0 + z_\alpha \frac{\sigma}{\sqrt{n}}$
		$\bar{X} > \mu_0 + t_\alpha \frac{S}{\sqrt{n}}$
$\mu = \mu_0$ or $\mu \ge \mu_0$	$\mu = \mu_1 < \mu_0$ or $\mu < \mu_0$	$\bar{X} < \mu_0 - z_\alpha \frac{\sigma}{\sqrt{n}}$
		$\bar{X} < \mu_0 - t_\alpha \frac{S}{\sqrt{n}}$
$\mu = \mu_0$	$\mu \ne \mu_0$	$\lvert\bar{X} - \mu_0\rvert > z_{\alpha/2} \frac{\sigma}{\sqrt{n}}$
		$\lvert\bar{X} - \mu_0\rvert > t_{\alpha/2} \frac{S}{\sqrt{n}}$

2. Test for Variance. In testing hypotheses concerning the variance σ^2 of a normal distribution, use Fact 5 to assert that, under H_0: $\sigma^2 = \sigma_0^2$, the distribution of $w_0 = (n-1)S^2/\sigma_0^2$ is $\chi^2(n-1)$. The corresponding tests and critical regions are summarized in the following table (χ_α^2 and $\chi_{\alpha/2}^2$ stand for $\chi_\alpha^2(n-1)$ and $\chi_{\alpha/2}^2(n-1)$, respectively):

Two-Sample Case

In comparing the means and variances of two normal populations, we once again refer to Table 11.1 for notation and assumptions.

Table 11.3 One-Sample Tests for Variance

Null Hypothesis H_0	Alternative Hypothesis H_1	Critical Region
$\sigma^2 = \sigma_0^2$ or $\sigma^2 \leq \sigma_0^2$	$\sigma^2 = \sigma_1^2 > \sigma_0^2$ or $\sigma^2 > \sigma_0^2$	$\left(S^2/\sigma_0^2\right) > \frac{1}{n-1}\chi_\alpha^2$
$\sigma^2 = \sigma_0^2$ or $\sigma^2 \geq \sigma_0^2$	$\sigma^2 = \sigma_1^2 < \sigma_0^2$ or $\sigma^2 < \sigma_0^2$	$\left(S^2/\sigma_0^2\right) < \frac{1}{n-1}\chi_{1-\alpha}^2$
$\sigma^2 = \sigma_0^2$	$\sigma^2 \neq \sigma_0^2$	$\left(S^2/\sigma_0^2\right) > \frac{1}{n-1}\chi_{\alpha/2}^2$
		or $\left(S^2/\sigma_0^2\right) < \frac{1}{n-1}\chi_{1-\alpha/2}^2$

1. Test for Difference of Two Means. Let $\delta = \mu_1 - \mu_2$ be the difference of the two population means. In testing H_0: $\delta = \delta_0$ vs. a one-sided or two-sided alternative hypothesis, we note that, for

$$\tau = \left(\sigma_1^2/n_1 + \sigma_2^2/n_2\right)^{1/2} \tag{11.23}$$

and

$$\nu = S_p\left(1/n_1 + 1/n_2\right)^{1/2} \tag{11.24}$$

$Z_0 = [\,(\bar{X}_1 - \bar{X}_2) - \delta_0]/\tau$ has an $N(0,1)$ distribution under H_0 and $T_0 = [\,(\bar{X}_1 - \bar{X}_2) - \delta_0]/\nu$ has a $t(n_1 + n_2 - 2)$ distribution under H_0 when $\sigma_1^2 = \sigma_2^2$. Using these results, the corresponding critical regions for one-sided and two-sided tests can be obtained, and they are listed below. Note that, as in the one-sample case, the critical region given on the first line for each set of hypotheses is for the case of known variances, and that given on the second line is for the case in which the variances are equal but unknown. Further, t_α and $t_{\alpha/2}$ stand for $t_\alpha(n_1 + n_2 - 2)$ and $t_{\alpha/2}(n_1 + n_2 - 2)$, respectively.

Table 11.4 Two-Sample Tests for Difference of Two Means

Null Hypothesis H_0	Alternative Hypothesis H_1	Critical Region
$\delta = \delta$ or $\delta \leq \delta_0$	$\delta = \delta_1 > \delta_0$ or $\delta > \delta_0$	$\left(\bar{X}_1 - \bar{X}_2\right) > \delta_0 + z_\alpha\tau$
		$\left(\bar{X}_1 - \bar{X}_2\right) > \delta_0 + t_\alpha\nu$
$\delta = \delta_0$ or $\delta \geq \delta_0$	$\delta = \delta_1 < \delta_0$ or $\delta < \delta_0$	$\left(\bar{X}_1 - \bar{X}_2\right) < \delta_0 - z_\alpha\tau$
		$\left(\bar{X}_1 - \bar{X}_2\right) < \delta_0 - t_\alpha\nu$
$\delta = \delta_0$	$\delta \neq \delta_0$	$\left|\left(\bar{X}_1 - \bar{X}_2\right) - \delta_0\right| > z_{\alpha/2}\tau$
		$\left|\left(\bar{X}_1 - \bar{X}_2\right) - \delta_0\right| > t_{\alpha/2}\nu$

11.7 A Numerical Example

In the following we provide a numerical example for illustrating the construction of confidence intervals and hypothesis-testing procedures. The example is given along the line of applications in Wadsworth (1990, p. 4.21) with artificial data.

Suppose that two processes (T_1 and T_2) manufacturing steel pins are in operation, and that a random sample of 4 pins (or 5 pins) was taken from the process T_1 (the process T_2) with the following results (in units of inches):

$$T_1 : 0.7608, 0.7596, 0.7622, 0.7638$$

$$T_2 : 0.7546, 0.7561, 0.7526, 0.7572, 0.7565$$

Simple calculation shows that the observed values of sample means sample variances, and sample standard deviations are:

$$\bar{X}_1 = 0.7616, \quad S_1^2 = 3.280 \times 10^{-6}, \quad S_1 = 1.811 \times 10^{-3}$$

$$\bar{X}_2 = 0.7554, \quad S_2^2 = 3.355 \times 10^{-6}, \quad S_2 = 1.832 \times 10^{-3}$$

One-Sample Case

Let us first consider confidence intervals for the parameters of the first process, T_1, only.

1. Assume that, based on previous knowledge of processes of this type, the variance is known to be $\sigma_1^2 = 1.80^2 \times 10^{-6}$ ($\sigma_1 = 0.0018$). Then from the normal table (see, e.g., Ross (1987, p. 482) we have $z_{0.025} = 1.96$. Thus a 95% confidence interval for μ_1 is

$$\left(0.7616 - 1.96 \times 0.0018/\sqrt{4}, 0.7616 + 1.96 \times 0.0018/\sqrt{4}\right)$$

   or (0.7598, 0.7634) (after rounding off to the 4th decimal place).
2. If σ_1^2 is unknown and a 95% confidence interval for μ_1 is needed then, for $t_{0.023}(3) = 3.182$ (see, e.g., Ross, 1987, p. 484) the confidence interval is

$$\left(0.7616 - 3.182 \times 0.001811/\sqrt{4}, 0.7616 + 3.182 \times 0.001811/\sqrt{4}\right)$$

   or (0.7587, 0.7645)
3. From the chi-squared table with 4 – 1 = 3 degrees of freedom, we have (see, e.g., Ross, 1987, p. 483) $\chi^2_{0.975} = 0.216$, $\chi^2_{0.025} = 9.348$. Thus a 95% confidence interval for σ_1^2 is $(3 \times 3.280 \times 10^{-6}/9.348, 3 \times 3.280 \times 10^{-6}/0.216)$, or $(1.0526 \times 10^{-6}, 45.5556 \times 10^{-6})$.
4. In testing the hypotheses

$$H_0 : \mu_1 = 0.76 \text{ vs. } H_1 : \mu_1 > 0.76$$

   with a = 0.01 when σ_1^2 is unknown, the critical region is $\bar{x}_1 > 0.76 + 4.541 \times 0.001811/\sqrt{4} = 0.7641$. Since the observed value $\bar{x}_1$ is 0.7616, H_0 is accepted. That is, we assert that there is no significant evidence to call for the rejection of H_0.

Two-Sample Case

If we assume that the two populations have a common unknown variance, we can use the Student's t distribution (with degree of freedom $\nu = 4 + 5 - 2 = 7$) to obtain confidence intervals and to test hypotheses for $\mu_1 - \mu_2$. We first note that the data given above yield

$$S_p^2 = \frac{1}{7}(3 \times 3.280 + 4 \times 3.355) \times 10^{-6}$$

$$= 3.3229 \times 10^{-6}$$

$$S_p = 1.8229 \times 10^{-3} \qquad \nu = S_p\sqrt{1/4 + 1/5} = 1.2228 \times 10^{-3}$$

and $\bar{X}_1 - \bar{X}_2 = 0.0062$.

1. A 98% confidence interval for $\mu_1 - \mu_2$ is (0.0062 – 2.998ν, 0.0062 + 2.998ν) or (0.0025, 0.0099).
2. In testing the hypotheses H_0: $\mu_1 = \mu_2$ (i.e., $\mu_1 - \mu_2 = 0$) vs. H_1: $\mu_1 > \mu_2$ with $\alpha = 0.05$, the critical region is $(\bar{X}_1 - \bar{X}_2) > 1.895\nu = 2.3172 \times 10^{-3}$. Thus H_0 is rejected; i.e., we conclude that there is significant evidence to indicate that $\mu_1 > \mu_2$ may be true.
3. In testing the hypotheses H_0: $\mu_1 = \mu_2$ vs. $\mu_1 \neq \mu_2$ with $\alpha = 0.02$, the critical region is $|\bar{X}_1 - \bar{X}_2| > 2.998\nu = 3.6660 \times 10^{-3}$. Thus H_0 is rejected. We note that the conclusion here is consistent with the result that, with confidence probability $1 - \alpha = 0.98$, the confidence interval for $(\mu_1 - \mu_2)$ does not contain the origin.

11.8 Concluding Remarks

The history of probability and statistics goes back to the days of the celebrated mathematicians K. F. Gauss and P. S. Laplace. (The normal distribution, in fact, is also called the Gaussian distribution.) The theory and methods of classical statistical analysis began its developments in the late 1800s and early 1900s when F. Galton and R.A. Fisher applied statistics to their research in genetics, when Karl Pearson developed the chi-square goodness-of-fit method for stochastic modeling, and when E.S. Pearson and J. Neyman developed the theory of hypotheses testing. Today statistical methods have been found useful in analyzing experimental data in biological science and medicine, engineering, social sciences, and many other fields. A non-technical review on some of the applications is Hacking (1984).

Applications of statistics in engineering include many topics. In addition to those treated in this section, other important ones include sampling inspection and quality (process) control, reliability, regression analysis and prediction, design of engineering experiments, and analysis of variance. Due to space limitations, these topics are not treated here. The reader is referred to textbooks in this area for further information. There are many well-written books that cover most of these topics, the following short list consists of a small sample of them.

References

Box, G.E.P., Hunter, W.G., and Hunter, J.S. 1978. *Statistics for Experimenters*. John Wiley & Sons, New York.

Bowker, A.H. and Lieberman, G.J. 1972. *Engineering Statistics*, 2nd ed. Prentice-Hall, Englewood Cliffs, NJ.

Hacking, I. 1984. Trial by number, *Science*, 84(5), 69–70.

Hahn, G.J. and Shapiro, S.S. 1967. *Statistical Models in Engineering*. John Wiley & Sons, New York.

Hines, W.W. and Montgomery, D.G. 1980. *Probability and Statistics in Engineering and Management Science*. John Wiley & Sons, New York.

Hogg, R.V. and Ledolter, J. 1992. *Engineering Statistics*. Macmillan, New York.

Ross, S.M. 1987. *Introduction to Probability and Statistics for Engineers and Scientists*. John Wiley & Sons, New York.

Wadsworth, H.M., Ed. 1990. *Handbook of Statistical Methods for Engineers and Scientists*. John Wiley & Sons, New York.

12

Numerical Methods

William F. Ames
Georgia Institute of Technology

12.1 Introduction

Since many mathematical models of physical phenomena are not solvable by available mathematical methods one must often resort to approximate or numerical methods. These procedures do not yield exact results in the mathematical sense. This inexact nature of numerical results means we must pay attention to the errors. The two errors that concern us here are *round-off errors* and *truncation errors.*

Round-off errors arise as a consequence of using a number specified by m correct digits to approximate a number which requires more than m digits for its exact specification. For example, using 3.14159 to approximate the irrational number π. Such errors may be especially serious in matrix inversion or in any area where a very large number of numerical operations are required. Some attempts at handling these errors are called *enclosure methods.* (Adams and Kulisch, 1993).

Truncation errors arise from the substitution of a finite number of steps for an infinite sequence of steps (usually an iteration) which would yield the exact result. For example, the iteration $y_n(x) = 1 + \int_0^1 xty_{n-1}(t)dt$, $y(0) = 1$ is only carried out for a *few steps*, but it converges in *infinitely* many steps.

The study of some errors in a computation is related to the theory of probability. In what follows, a relation for the error will be given in certain instances.

12.2 Linear Algebra Equations

A problem often met is the determination of the solution vector $u = (u_1, u_2, \ldots, u_n)^T$ for the set of linear equations $Au = v$ where A is the $n \times n$ square matrix with coefficients, a_{ij} $(i, j = 1, \ldots, n)$, $v = (v_1, \ldots, v_n)^T$ and i denotes the row index and j the column index.

0-8493-0056-8/00/$0.00+$.50

There are many numerical methods for finding the solution, u, of $Au = v$. The direct inversion of A is usually too expensive and is not often carried out unless it is needed elsewhere. We shall only list a few methods. One can check the literature for the many methods and computer software available. Some of the software is listed in the References section at the end of this chapter. The methods are usually subdivided into *direct* (once through) or *iterative* (repeated) procedures.

In what follows, it will often be convenient to partition the matrix A into the form $A = U + D + L$, where U, D, and L are matrices having the same elements as A, respectively, above the main diagonal, on the main diagonal, and below the main diagonal, and zeros elsewhere. Thus,

$$U = \begin{bmatrix} 0 & a_{12} & & \cdots & a_{1n} \\ 0 & 0 & a_{23} & \cdots & a_{2n} \\ \vdots & \cdots & & \cdots & \\ 0 & 0 & \cdots & \cdots & 0 \end{bmatrix}$$

We also assume the u_is are not all zero and det $A \neq 0$ so the solution is unique.

Direct Methods

Gauss Reduction. This classical method has spawned many variations. It consists of dividing the first equation by a_{11} (if $a_{11} = 0$, reorder the equations to find an $a_{11} \neq 0$) and using the result to eliminate the terms in u_1 from each of the succeeding equations. Next, the modified second equation is divided by a'_{22} (if $a'_{22} = 0$, a reordering of the modified equations may be necessary) and the resulting equation is used to eliminate all terms in u_2 in the succeeding modified equations. This elimination is done n times resulting in a triangular system:

$$\begin{aligned} u_1 + a'_{12}u_2 + \cdots + a'_{1n}u_n &= v'_1 \\ 0 + u_2 + \cdots + a'_{2n}u_n &= v'_2 \\ &\cdots \\ 0 + \cdots + u_{n-1} + a'_{n-1,n}u_n &= v'_{n-1} \\ u_n &= v'_n \end{aligned}$$

where a'_{ij} and v'_i represent the specific numerical values obtained by this process. The solution is obtained by working backward from the last equation. Various modifications, such as the Gauss-Jordan reduction, the Gauss-Doolittle reduction, and the Crout reduction, are described in the classical reference authored by Bodewig (1956). Direct methods prove very useful for sparse matrices and banded matrices that often arise in numerical calculation for differential equations. Many of these are available in computer packages such as IMSL, Maple, Matlab, and Mathematica.

The Tridiagonal Algorithm. When the linear equations are tridiagonal, the system

$$\begin{aligned} b_1u_1 + c_1u_2 &= d_1 \\ a_iu_{i-1} + b_iu_i + c_iu_{i+1} &= d_i \\ a_nu_{n-1} + b_nu_n &= d_n, \quad i = 2, 3, \ldots, n-1 \end{aligned}$$

can be solved explicitly for the unknown, thereby eliminating any matrix operations.

The Gaussian elimination process transforms the system into a simpler one of *upper bidiagonal* form. We designate the coefficients of this new system by a_i', b_i' c_i' and d_i', and we note that

$$a_i' = 0, \quad i = 2, 3, \ldots, n$$

$$b_i' = 1, \quad i = 1, 2, \ldots, n$$

The coefficients c_i' and d_i' are calculated successively from the relations

$$c_1' = \frac{c_1}{b_1} \qquad d_1' = \frac{d_1}{b_1}$$

$$c_{i+1}' = \frac{c_{i+1}}{b_{i+1} - a_{i+1}c_i'}$$

$$d_{i+1}' = \frac{d_{i+1} - a_{i+1}d_i'}{b_{i+1} - a_{i+1}c_i'}, \quad i = 1, 2, \ldots, n-1$$

and, of course, $c_n = 0$.

Having completed the elimination we examine the new system and see that the nth equation is now

$$u_n = d_n'$$

Substituting this value into the $(n-1)$st equation,

$$u_{n-1} + c_{n-1}'u_n = d_{n-1}'$$

we have

$$u_{n-1} = d_{n-1}' - c_{n-1}'u_n$$

Thus, starting with u_n, we have successively the solution for u_i as

$$u_i = d_i' - c_i'u_{i+1}, \quad i = n-1, n-2, \ldots, 1$$

Algorithm for Pentadiagonal Matrix. The equations to be solved are

$$a_iu_{i-2} + b_iu_{i-1} + c_iu_i + d_iu_{i+1} + e_iu_{i+2} = f_i$$

for $1 \le i \le R$ with $a_1 = b_1 = a_2 = e_{R-1} = d_R = e_R = 0$.

The algorithm is as follows. First, compute

$$\delta_1 = d_1/c_1$$

$$\lambda_1 = e_1/c_1$$

$$\gamma_1 = f_1/c_1$$

and

$$\mu_2 = c_2 - b_2\delta_1$$
$$\delta_2 = (d_2 - b_2\lambda_1)/\mu_2$$
$$\lambda_2 = e_2/\mu_2$$
$$\gamma_2 = (f - b_2\gamma_1)/\mu_2$$

Then, for $3 \le i \le R-2$, compute

$$\beta_i = b_i - a_i\delta_{i-2}$$
$$\mu_i = c_i - \beta_i\delta_{i-1} - a_i\lambda_{i-2}$$
$$\delta_i = (d_i - \beta_i\lambda_{i-1})/\mu_i$$
$$\lambda_i = e_i/\mu_i$$
$$\gamma_i = (f_i - \beta_i\gamma_{i-1} - a_i\gamma_{i-2})/\mu_i$$

Next, compute

$$\beta_{R-1} = b_{R-1} - a_{R-1}\delta_{R-3}$$
$$\mu_{R-1} = c_{R-1} - \beta_{R-1}\delta_{R-2} - a_{R-1}\lambda_{R-3}$$
$$\delta_{R-1} = (d_{R-1} - \beta_{R-1}\lambda_{R-2})/\mu_{R-1}$$
$$\gamma_{R-1} = (f_{R-1} - \beta_{R-1}\gamma_{R-2} - a_{R-1}\gamma_{R-3})/\mu_{R-1}$$

and

$$\beta_R = b_R - a_R\delta_{R-2}$$
$$\mu_R = c_R - \beta_R\delta_{R-1} - a_R\lambda_{R-2}$$
$$\gamma_R = (f_R - \beta_R\gamma_{R-1} - a_R\gamma_{R-2})/\mu_R$$

The β_i and μ_i are used only to compute δ_i, λ_i, and γ_i, and need not be stored after they are computed. The δ_i, λ_i, and γ_i, must be stored, as they are used in the back solution. This is

$$u_R = \gamma_R$$
$$u_{R-1} = \gamma_{R-1} - \delta_{R-1}u_R$$

and

$$u_i = \gamma_i - \delta_i u_{i+1} - \lambda_i u_{i+2}$$

for $R-2 \ge i \ge 1$.

General Band Algorithm. The equations are of the form

$$A_j^{(M)}X_{j-M} + A_j^{(M-1)}X_{j-M+1} + \cdots + A_j^{(2)}X_{j-2} + A_j^{(1)}X_{j-1} + B_jX_j$$

$$+ C_j^{(1)}X_{j+1} + C_j^{(2)}X_{j+2} + \cdots + C_j^{(M-1)}X_{j+M-1} + C_j^{(M)}X_{j+M} = D_j$$

for $1 \le j \le N$, $N \ge M$. The algorithm used is as follows:

$$\alpha_j^{(k)} = A_j^{(k)} = 0, \quad \text{for } k \ge j$$

$$C_j^{(k)} = 0, \quad \text{for } k \ge N + 1 - j$$

The forward solution ($j = 1, \ldots, N$) is

$$\alpha_j^{(k)} = A_j^{(k)} - \sum_{p=k+1}^{p=M} \alpha_j^{(p)} W_{j-p}^{(p-k)}, \quad k = M, \ldots, 1$$

$$\beta_j = B_j - \sum_{p=1}^{M} \alpha_j^{(p)} W_{j-p}^{(p)}$$

$$W_j^{(k)} = \left(C_j^{(k)} - \sum_{p=k+1}^{p=M} \alpha_j^{(p-k)} W_{j-(p-k)}^{(p)} \right) \Big/ \beta_j, \quad k = 1, \ldots, M$$

$$\gamma_j = \left(D_j - \sum_{p=1}^{M} \alpha_j^{(p)} \gamma_{j-p} \right) \Big/ \beta_j$$

The back solution ($j = N, \ldots, 1$) is

$$X_j = \gamma_j - \sum_{p=1}^{M} W_j^{(p)} X_{j+p}$$

Cholesky Decomposition. When the matrix A is a symmetric and positive definite, as it is for many discretizations of self-adjoint positive definite boundary value problems, one can improve considerably on the band procedures by using the Cholesky decomposition. For the system $Au = v$, the Matrix A can be written in the form

$$A = (I + L)D(I + U)$$

where L is lower triangular, U is upper triangular, and D is diagonal. If $A = A'$ (A' represents the transpose of A), then

$$A = A' = (I + U)' D(I + L)'$$

Hence, because of the uniqueness of the decomposition,

$$I + L = (I + U)' = I + U'$$

and therefore,

$$A = (I + U)' D(I + U)$$

that is,

$$A = B'B, \text{ where } B = \sqrt{D}(I + U)$$

The system $Au = v$ is then solved by solving the two triangular system

$$B'w = v$$

followed by

$$Bu = w$$

To carry out the decomposition $A = B'B$, all elements of the first row of A, and of the derived system, are divided by the square root of the (positive) leading coefficient. This yields smaller rounding errors than the banded methods because the relative error of $\sqrt{a}$ is only half as large as that of a itself. Also, taking the square root brings numbers nearer to each other (i.e., the new coefficients do not differ as widely as the original ones do). The actual computation of $B = (b_{ij})$, $j > i$, is given in the following:

$$b_{11} = (a_{11})^{1/2}, \qquad b_{1j} = a_{1j}/b_{11}, \quad j \geq 2$$

$$b_{22} = (a_{22} - b_{12}^2)^{1/2}, \qquad b_{2j} = (a_{2j} - b_{12}b_{1j})/b_{22}$$

$$b_{33} = (a_{33} - b_{13}^2 - b_{23}^2)^{1/2}, \qquad b_{3j} = (a_{3j} - b_{13}b_{1j} - b_{23}b_{2j})/b_{33}$$

$$\vdots$$

$$b_{ii} = \left(a_{ii} - \sum_{k=1}^{i-1} b_{ki}^2\right)^{1/2}, \qquad b_{ij} = \left(a_{ij} - \sum_{k=1}^{i-1} b_{ki}b_{kj}\right)\Big/ b_{ii}, \qquad i \geq 2, j \geq 2$$

Iterative Methods

Iterative methods consist of repeated application of an often simple algorithm. They yield the exact answer only as the limit of a sequence. They can be programmed to take care of zeros in A and are self-correcting. Their structure permits the use of convergence accelerators, such as overrelaxation, Aitkins acceleration, or Chebyshev acceleration.

Let $a_{ii} > 0$ for all i and $\det A \neq 0$. With $A = U + D + L$ as previously described, several iteration methods are described for $(U + D + L)u = v$.

Jacobi Method (Iteration by total steps). Since $u = -D^{-1}[U + L]u + D^{-1}v$, the iteration $u^{(k)}$ is $u^{(k)} = -D^{-1}[U + L]u^{(k-1)} + D^{-1}v$. This procedure has a slow convergent rate designated by R, $0 < R \ll 1$.

Gauss-Seidel Method (Iteration by single steps). $u^{(k)} = -(L + D)^{-1}Uu^{(k-1)} + (L + D)^{-1}v$. Convergence rate is $2R$, twice as fast as that of the Jacobi method.

Gauss-Seidel with Successive Overrelaxation (SOR). Let $\bar{u}_i^{(k)}$ be the ith components of the Gauss-Seidel iteration. The SOR technique is defined by

$$u_i^{(k)} = (1 - \omega)u_i^{(k-1)} + \omega\bar{u}_i^{(k)}$$

where $1 < \omega < 2$ is the overrelaxation parameter. The full iteration is $u^{(k)} = (D + \omega L)^{-1}\{[(1 - \omega)D - \omega U]u^{(k-1)} + \omega v\}$. Optimal values of ω can be computed and depend upon the properties of A (Ames, 1993). With optimal values of ω, the convergence rate of this method is $2R\sqrt{2}$ which is much larger than that for Gauss-Seidel (R is usually much less than one).

For other acceleration techniques, see the literature (Ames, 1993).

12.3 Nonlinear Equations in One Variable

Special Methods for Polynomials

The polynomial $P(x) = a_0x^n + a_1x^{n-1} + \cdots + a_{n-1}x + a_n = 0$, with real coefficients a_j, $j = 0, \ldots, n$, has exactly n roots which may be real or complex.

If all the coefficients of $P(x)$ are integers, then any rational roots, say r/s (r and s are integers with no common factors), of $P(x) = 0$ must be such that r is an integral divisor of a_n and s is an integral division of a_0. Any polynomial with rational coefficients may be converted into one with integral coefficients by multiplying the polynomial by the lowest common multiple of the denominators of the coefficients.

Example. $x^4 - 5x^2/3 + x/5 + 3 = 0$. The lowest common multiple of the denominators is 15. Multiplying by 15, which does not change the roots, gives $15x^4 - 25x^2 + 3x + 45 = 0$. The only possible rational roots r/s are such that r may have the value ± 45, ± 15, ± 5, ± 3, and ± 1, while s may have the values ± 15, ± 5, ± 3, and ± 1. All possible rational roots, with no common factors, are formed using all possible quotients.

If $a_0 > 0$, the first negative coefficient is preceded by k coefficients which are positive or zero, and G is the largest of the absolute values of the negative coefficients, then each real root is less than $1 + \sqrt[k]{G/a_0}$ (upper bound on the real roots). For a lower bound to the real roots, apply the criterion to $P(-x) = 0$.

Example. $P(x) = x^5 + 3x^4 - 2x^3 - 12x + 2 = 0$. Here $a_0 = 1$, $G = 12$, and $k = 2$. Thus, the upper bound for the real roots is $1 + \sqrt[2]{12} \approx 4.464$. For the lower bound, $P(-x) = -x^5 + 3x^4 + 2x^3 + 12x + 2 = 0$, which is equivalent to $x^5 - 3x^4 - 2x^3 - 12x - 2 = 0$. Here $k = 1$, $G = 12$, and $a_0 = 1$. A lower bound is $-(1 + 12) = 13$. Hence all real roots lie in $-13 < x < 1 + \sqrt[2]{12}$.

A useful *Descartes rule of signs* for the number of positive or negative real roots is available by observation for polynomials with real coefficients. The number of positive real roots is either equal to the number of sign changes, n, or is less than n by a positive *even* integer. The number of negative real roots is either equal to the number of sign changes, n, of $P(-x)$, or is less than n by a positive even integer.

Example. $P(x) = x^5 - 3x^3 - 2x^2 + x - 1 = 0$. There are three sign changes, so $P(x)$ has either three or one positive roots. Since $P(-x) = -x^5 + 3x^3 - 2x^2 - 1 = 0$, there are either two or zero negative roots.

The Graeffe Root-Squaring Technique

This is an iterative method for finding the roots of the algebraic equation

$$f(x) = a_0x^p + a_1x^{p-1} + \cdots + a_{p-1}x + a_p = 0$$

If the roots are $r_1, r_2, r_3, \ldots$, then one can write

$$S_p = r_1^p\left(1 + \frac{r_2^p}{r_1^p} + \frac{r_3^p}{r_1^p} + \cdots\right)$$

and if one root is larger than all the others, say r_1, then for large enough p all terms (other than 1) would become negligible. Thus,

$$S_p \approx r_1^p$$

or

$$\lim_{p \to \infty} S_p^{1/p} = r_1$$

The Graeffe procedure provides an efficient way for computing S_p via a sequence of equations such that the roots of each equation are the squares of the roots of the preceding equations in the sequence. This serves the purpose of ultimately obtaining an equation whose roots are so widely separated in magnitude that they may be read approximately from the equation by inspection. The basic procedure is illustrated for a polynomial of degree 4:

$$f(x) = a_0 x^4 + a_1 x^3 + a_2 x^2 + a_3 x + a_4 = 0$$

Rewrite this as

$$a_0 x^4 + a_2 x^2 + a_4 = -a_1 x^3 - a_3 x$$

and square both sides so that upon grouping

$$a_0^2 x^8 + \left(2a_0 a_2 - a_1^2\right) x^6 + \left(2a_0 a_4 - 2a_1 a_3 + a_2^2\right) x^4 + \left(2a_2 a_4 - a_3^2\right) x^2 + a_4^2 = 0$$

Because this involves only even powers of x, we may set $y = x^2$ and rewrite it as

$$a_0^2 y^4 + \left(2a_0 a_2 - a_1^2\right) y^3 + \left(2a_0 a_4 - 2a_1 a_3 + a_2^2\right) y^2 + \left(2a_2 a_4 - a_3^2\right) y + a_4^2 = 0$$

whose roots are the squares of the original equation. If we repeat this process again, the new equation has roots which are the fourth power, and so on. After p such operations, the roots are 2^p (original roots). If at any stage we write the coefficients of the unknown in sequence

$$a_0^{(p)} \quad a_1^{(p)} \quad a_2^{(p)} \quad a_3^{(p)} \quad a_4^{(p)}$$

then, to get the new sequence $a_i^{(p+1)}$, write $a_i^{(p+1)} = 2a_0^{(p)}$ (times the symmetric coefficient) with respect to $a_i^{(p)} - 2a_1^{(p)}$ (times the symmetric coefficient) $- \cdots (-1)^i a_i^{(p)2}$. Now if the roots are r_1, r_2, r_3, and r_4, then $a_1/a_0 = -\Sigma_{i-1}^4 r_i$, $a_i^{(1)}/a_0^{(1)} = -\Sigma r_i^2, \ldots, a_i^{(p)}/a_0^{(p)} = -\Sigma r_i^{2p}$. If the roots are all distinct and r_1 is the largest in magnitude, then eventually

$$r_1^{2p} \approx -\frac{a_1^{(p)}}{a_0^{(p)}}$$

And if r_2 is the next largest in magnitude, then

$$r_2^{2p} \approx -\frac{a_2^{(p)}}{a_1^{(p)}}$$

And, in general $a_n^{(p)}/a_{n-1}^{(p)} \approx -r_n^{2p}$. This procedure is easily generalized to polynomials of arbitrary degree and specialized to the case of multiple and complex roots.

Other methods include Bernoulli iteration, Bairstow iteration, and Lin iteration. These may be found in the cited literature. In addition, the methods given below may be used for the numerical solution of polynomials.

12.4 General Methods for Nonlinear Equations in One Variable

Successive Substitutions

Let $f(x) = 0$ be the nonlinear equation to be solved. If this is rewritten as $x = F(x)$, then an iterative scheme can be set up in the form $x_{k+1} = F(x_k)$. To start the iteration, an initial guess must be obtained graphically or otherwise. The convergence or divergence of the procedure depends upon the method of writing $x = F(x)$, of which there will usually be several forms. A general rule to ensure convergence cannot be given. However, if a is a root of $f(x) = 0$, a necessary condition for convergence is that $|F'(x)| < 1$ in that interval about a in which the iteration proceeds (this means the iteration cannot converge unless $|F'(x)| < 1$, but it does not ensure convergence). This process is called *first order* because the error in x_{k+1} is proportional to the first power of the error in x_k.

Example. $f(x) = x^3 - x - 1 = 0$. A rough plot shows a real root of approximately 1.3. The equation can be written in the form $x = F(x)$ in several ways, such as $x = x^3 - 1$, $x = 1/(x^2 - 1)$, and $x = (1 + x)^{1/3}$. In the first case, $F'(x) = 3x^2 = 5.07$ at $x = 1.3$; in the second, $F'(1.3) = 5.46$; only in the third case is $F'(1.3) < 1$. Hence, only the third iterative process has a chance to converge. This is illustrated in the iteration table below.

Step k	$x = \dfrac{1}{x^2 - 1}$	$x = x^3 - 1$	$x = (1 + x)^{1/3}$
0	1.3	1.3	1.3
1	1.4493	1.197	1.32
2	0.9087	0.7150	1.3238
3	-5.737	-0.6345	1.3247
4	...	...	1.3247

12.5 Numerical Solution of Simultaneous Nonlinear Equations

The techniques illustrated here will be demonstrated for two simultaneous equations — $f(x, y) = 0$ and $g(x, y) = 0$. They immediately generalize to more than two simultaneous equations.

The Method of Successive Substitutions

The two simultaneous equations can be written in various ways in equivalent forms

$$x = F(x, y)$$

$$y = G(x, y)$$

and the method of successive substitutions can be based on

$$x_{k+1} = F(x_k, y_k)$$

$$y_{k+1} = G(x_k, y_k)$$

Again, the procedure is of the first order and a necessary condition for convergence is

$$\left|\frac{\partial F}{\partial x}\right| + \left|\frac{\partial F}{\partial y}\right| < 1 \qquad \left|\frac{\partial G}{\partial x}\right| + \left|\frac{\partial G}{\partial y}\right| < 1$$

in the iteration neighborhood of the true solution.

The Newton-Raphson Procedure

Using the two simultaneous equation, start from an approximate, say (x_0, y_0), obtained graphically or from a two-way table. Then, solve successively the linear equations

$$\Delta x_k \frac{\partial f}{\partial x}(x_k, y_k) + \Delta y_k \frac{\partial f}{\partial y}(x_k, y_k) = -f(x_k, y_k)$$

$$\Delta x_k \frac{\partial g}{\partial x}(x_k, y_k) + \Delta y_k \frac{\partial g}{\partial y}(x_k, y_k) = -g(x_k, y_k)$$

for Δx_k and Δy_k. Then, the $k+1$ approximation is given from $x_{k+1} = x_k + \Delta x_k, y_{k+1} = y_k + \Delta y_k$. A modification consists in solving the equations with (x_k, y_k) replaced by (x_0, y_0) (or another suitable pair later on in the iteration) in the derivatives. This means the derivatives (and therefore the coefficients of Δx_k, Δy_k) are independent of k. Hence, the results become

$$\Delta x_k = \frac{-f(x_k, y_k)(\partial g/\partial y)(x_0, y_0) + g(x_k, y_k)(\partial f/\partial y)(x_0, y_0)}{(\partial f/\partial x)(x_0, y_0)(\partial g/\partial y)(x_0, y_0) - (\partial f/\partial y)(x_0, y_0)(\partial g/\partial x)(x_0, y_0)}$$

$$\Delta y_k = \frac{-g(x_k, y_k)(\partial f/\partial x)(x_0, y_0) + f(x_k, y_k)(\partial g/\partial x)(x_0, y_0)}{(\partial f/\partial x)(x_0, y_0)(\partial g/\partial y)(x_0, y_0) - (\partial f/\partial y)(x_0, y_0)(\partial g/\partial x)(x_0, y_0)}$$

and $x_{k+1} = \Delta x_k + x_k$, $y_{k+1} = \Delta y_k + y_k$. Such an alteration of the basic technique reduces the rapidity of convergence.

Example

$$f(x, y) = 4x^2 + 6x - 4xy + 2y^2 - 3$$

$$g(x, y) = 2x^2 - 4xy + y^2$$

By plotting, one of the approximate roots is found to be $x_0 = 0.4$, $y_0 = 0.3$. At this point, there results $\partial f/\partial x = 8$, $\partial f/\partial y = -0.4$, $\partial g/\partial x = 0.4$, and $\partial g/\partial y = -1$. Hence,

$$x_{k+1} = x_k + \Delta x_k = x_k + \frac{-f(x_k, y_k) - 0.4g(x_k, y_k)}{8(-1) - (-0.4)(0.4)}$$

$$= x_k - 0.127\,55 f(x_k, y_k) - 0.051\,02 g(x_k, y_k)$$

and

$$y_{k+1} = y_k - 0.051\,02 f(x_k, y_k) + 1.020\,41 g(x_k, y_k)$$

The first few iteration steps are shown in the following table.

Step k	x_k	y_k	$f(x_k, y_k)$	$g(x_k, y_k)$
0	0.4	0.3	-0.26	0.07
1	0.43673	0.24184	0.078	0.0175
2	0.42672	0.25573	-0.0170	-0.007
3	0.42925	0.24943	0.0077	0.0010

Methods of Perturbation

Let $f(x) = 0$ be the equation. In general, the iterative relation is

$$x_{k+1} = x_k - \frac{f(x_k)}{\alpha_k}$$

where the iteration begins with x_0 as an initial approximation and α_k is some functional.

The Newton-Raphson Procedure. This variant chooses $\alpha_k = f'(x_k)$ where $f' = df/dx$ and geometrically consists of replacing the graph of $f(x)$ by the tangent line at $x = x_k$ in each successive step. If $f'(x)$ and $f''(x)$ have the same sign throughout an interval $a \le x \le b$ containing the solution, with $f(a)$ and $f(b)$ of opposite signs, then the process converges starting from any x_0 in the interval $a \le x \le b$. The process is second order.

Example

$$f(x) = x - 1 + \frac{(0.5)^x - 0.5}{0.3}$$

$$f'(x) = 1 - 2.3105[0.5]^x$$

An approximate root (obtained graphically) is 2.

Step k	x_k	$f(x_k)$	$f'(x_k)$
0	2	0.1667	0.4224
1	1.605	-0.002	0.2655
2	1.6125	-0.0005	...

The Method of False Position. This variant is commenced by finding x_0 and x_1 such that $f(x_0)$ and $f(x_1)$ are of opposite signs. Then, α_1 = slope of secant line joining $[x_0, f(x_0)]$ and $[x_1, f(x_1)]$ so that

$$x_2 = x_1 - \frac{x_1 - x_0}{f(x_1) - f(x_0)} f(x_1)$$

In each following step, α_k is the slope of the line joining $[x_k, f(x_k)]$ to the most recently determined point where $f(x_j)$ has the opposite sign from that of $f(x_k)$. This method is of first order.

The Method of Wegstein

This is a variant of the method of successive substitutions which forces or accelerates convergence. The iterative procedure $x_{k+1} = F(x_k)$ is revised by setting $\hat{x}_{k+1} = F(x_k)$ and then taking $x_{k+1} = qx_k + (1 - q)\ \hat{x}_{k+1}$. Wegstein found that suitably chosen qs are related to the basic process as follows:

Behavior of Successive Substitution Process	Range of Optimum q
Oscillatory convergence	$0 < q < 1/2$
Oscillatory divergence	$1/2 < q < 1$
Monotonic convergence	$q < 0$
Monotonic divergence	$1 < q$

At each step, q may be calculated to give a locally optimum value by setting

$$q = \frac{x_{k+1} - x_k}{x_{k+1} - 2x_k + x_{k-1}}$$

The Method of Continuity

In the case of n equations in n unknowns, when n is large, determining the approximate solution may involve considerable effort. In such a case, the method of continuity is admirably suited for use on either digital or analog computers. It consists basically of the introduction of an extra variable into the n equations

$$f_i(x_1, x_2, \ldots, x_n) = 0, \qquad i = 1, \ldots, n$$

and replacing them by

$$f_i(x_1, x_2, \ldots, x_n, \lambda) = 0, \qquad i = 1, \ldots, n$$

where λ is introduced in such a way that the functions depend in a simple way upon λ and reduce to an easily solvable system for $\lambda = 0$ and to the original equations for $\lambda = 1$. A system of ordinary differential equations, with independent variable λ, is then constructed by differentiating with respect to λ. There results

$$\sum_{j=1}^{n} \frac{\partial f_i}{\partial x_j} \frac{dx_j}{d\lambda} + \frac{\partial f_i}{\partial \lambda} = 0$$

where $x_1, \ldots, x_n$ are considered as functions of λ. The equations are integrated, with initial conditions obtained with $\lambda = 0$, from $\lambda = 0$ to $\lambda = 1$. If the solution can be continued to $\lambda = 1$, the values of $x_1, \ldots, x_n$ for $\lambda = 1$ will be a solution of the original equations. If the integration becomes infinite, the parameter λ must be introduced in a different fashion. Integration of the differential equations (which are usually nonlinear in λ) may be accomplished on an analog computer or by digital means using techniques described in a later section entitled "Numerical Solution of Ordinary Differential Equations."

Example

$$f(x, y) = 2 + x + y - x^2 + 8xy + y^3 = 0$$

$$g(x, y) = 1 + 2x + 3y + x^2 + xy - ye^x = 0$$

Introduce λ as

$$f(x, y, \lambda) = (2 + x + y) + \lambda(-x^2 + 8xy + y^3) = 0$$

$$g(x, y, \lambda) = (1 + 2x - 3y) + \lambda(x^2 + xy - ye^x) = 0$$

For $\lambda = 1$, these reduce to the original equations, but, for $\lambda = 0$, they are the linear systems

$$x + y = -2$$
$$2x - 3y = -1$$

which has the unique solution $x = -1.4$, $y = -0.6$. The differential equations in this case become

$$\frac{\partial f}{\partial x}\frac{dx}{d\lambda} + \frac{\partial f}{\partial y}\frac{dy}{d\lambda} = -\frac{\partial f}{\partial \lambda}$$

$$\frac{\partial g}{\partial x}\frac{dx}{d\lambda} + \frac{\partial g}{\partial y}\frac{dy}{d\lambda} = -\frac{\partial g}{\partial \lambda}$$

or

$$\frac{dx}{d\lambda} = \frac{\dfrac{\partial f}{\partial y}\dfrac{\partial g}{\partial \lambda} - \dfrac{\partial f}{\partial \lambda}\dfrac{\partial g}{\partial x}}{\dfrac{\partial f}{\partial x}\dfrac{\partial g}{\partial y} - \dfrac{\partial f}{\partial y}\dfrac{\partial g}{\partial x}}$$

$$\frac{dy}{d\lambda} = \frac{\dfrac{\partial f}{\partial \lambda}\dfrac{\partial g}{\partial x} - \dfrac{\partial f}{\partial x}\dfrac{\partial g}{\partial \lambda}}{\dfrac{\partial f}{\partial x}\dfrac{\partial g}{\partial y} - \dfrac{\partial f}{\partial y}\dfrac{\partial g}{\partial x}}$$

Integrating in λ, with initial values $x = -1.4$ and $y = -0.6$ at $\lambda = 0$, from $\lambda = 0$ to $\lambda = 1$ gives the solution.

12.6 Interpolation and Finite Differences

The practicing engineer constantly finds it necessary to refer to tables as sources of information. Consequently, interpolation, or that procedure of "reading between the lines of the table," is a necessary topic in numerical analysis.

Linear Interpolation

If a function $f(x)$ is approximately linear in a certain range, then the ratio $[f(x_1) - f(x_0)]/(x_1 - x_0) = f[x_0, x_1]$ is approximately independent of x_0 and x_1 in the range. The linear approximation to the function $f(x)$, $x_0 < x < x_1$, then leads to the interpolation formula

$$f(x) \approx f(x_0) + (x - x_0)f[x_0, x_1] \approx f(x_0) + \frac{x - x_0}{x_1 - x_0}\left[f(x_1) - f(x_0)\right]$$

$$\approx \frac{1}{x_1 - x_0}\left[(x_1 - x)f(x_0) - (x_0 - x)f(x_1)\right]$$

Divided Differences of Higher Order and Higher-Order Interpolation

The first-order divided difference $f[x_0, x_1]$ was defined above. Divided differences of second and higher order are defined iteratively by

$$f[x_0,x_1,x_2] = \frac{f[x_1,x_2] - f[x_0,x_1]}{x_2 - x_0}$$

$$\vdots$$

$$f[x_0,x_1,\ldots,x_k] = \frac{f[x_1,\ldots,x_k] - f[x_0,x_1,\ldots,x_{k-1}]}{x_k - x_0}$$

and a convenient form for computational purposes is

$$f[x_0,x_1,\ldots,x_k] = \sum_{i=0}^{k}{}' \frac{f(x_i)}{(x_i - x_0)(x_i - x_1)\cdots(x_i - x_k)}$$

for any $k \geq 0$, where the $'$ means the term $(x_i - x_i)$ is omitted in the denominator. For example,

$$f[x_0,x_1,x_2] = \frac{f(x_0)}{(x_0 - x_1)(x_0 - x_2)} + \frac{f(x_1)}{(x_1 - x_0)(x_1 - x_2)} + \frac{f(x_2)}{(x_2 - x_0)(x_2 - x_1)}$$

If the accuracy afforded by a linear approximation is inadequate, a generally more accurate result may be based upon the assumption that $f(x)$ may be approximated by a polynomial of degree 2 or higher over certain ranges. This assumptions leads to *Newton's fundamental interpolation formula* with divided differences:

$$f(x) \approx f(x_0) + (x - x_0)f[x_0,x_1] + (x - x_0)(x - x_1)f[x_0,x_1,x_2]$$
$$+ (x - x_0)(x - x_1)\cdots(x - x_{n-1})f[x_0,x_1,\ldots,x_n] + E_n(x)$$

where $E_n(x)$ = error = $[1/(n + 1)!]f^{(n-1)}(\varepsilon)\pi(x)$ where $\min(x_0, \ldots, x) < \varepsilon < \max(x_0, x_1, \ldots, x_n, x)$ and $\pi(x) = (x - x_0)(x - x_1)\ldots(x - x_n)$. In order to use this most effectively, one may first form a divided-difference table. For example, for third-order interpolation, the difference table is

x_0	$f(x_0)$			
		$f[x_0,x_1]$		
x_1	$f(x_1)$		$f[x_0,x_1,x_2]$	
		$f[x_1,x_2]$		$f[x_0,x_1,x_2,x_3]$
x_2	$f(x_2)$		$f[x_1,x_2,x_3]$	
		$f[x_2,x_3]$		
x_3	$f(x_3)$			

where each entry is given by taking the difference between diagonally adjacent entries to the left, divided by the abscissas corresponding to the ordinates intercepted by the diagonals passing through the calculated entry.

Example. Calculate by third-order interpolation the value of cosh 0.83 given cosh 0.60, cosh 0.80, cosh 0.90, and cosh 1.10.

$x_0 = 0.60$	1.185 47			
		0.7598		
$x_1 = 0.80$	1.337 43		0.6560	
		0.9566		0.1586
$x_2 = 0.90$	1.433 09		0.7353	
		1.1772		
$x_3 = 1.10$	1.668 52			

With $n = 3$, we have

$$\cosh 0.83 \approx 1.185\,47 + (0.23)(0.7598) + (0.23)(0.03)(0.6560)$$
$$+ (0.23)(0.03)(-0.07)(0.1586) = 1.364\,64$$

which varies from the true value by 0.000 04.

Lagrange Interpolation Formulas

The Newton formulas are expressed in terms of divided differences. It is often useful to have interpolation formulas expressed explicitly in terms of the ordinates involved. This is accomplished by the Lagrange interpolation polynomial of degree n:

$$y(x) = \sum_{i=0}^{n} \frac{\pi(x)}{(x - x_i)\pi'(x_i)} f(x_i)$$

where

$$\pi(x) = (x - x_0)(x - x_1)\cdots(x - x_n)$$
$$\pi'(x_i) = (x_i - x_0)(x_i - x_1)\cdots(x_i - x_n)$$

where $(x_i - x_i)$ is the omitted factor. Thus,

$$f(x) = y(x) + E_n(x)$$
$$E_n(x) = \frac{1}{(n+1)!}\pi(x) f^{(n+1)}(\varepsilon)$$

Example. The interpolation polynomial of degree 3 is

$$y(x) = \frac{(x - x_1)(x - x_2)(x - x_3)}{(x_0 - x_1)(x_0 - x_2)(x_0 - x_3)} f(x_0) + \frac{(x - x_0)(x - x_2)(x - x_3)}{(x_1 - x_0)(x_1 - x_2)(x_1 - x_3)} f(x_1)$$
$$+ \frac{(x - x_0)(x - x_1)(x - x_3)}{(x_2 - x_0)(x_2 - x_1)(x_2 - x_3)} f(x_2) + \frac{(x - x_0)(x - x_1)(x - x_2)}{(x_3 - x_0)(x_3 - x_1)(x_3 - x_2)} f(x_3)$$

Thus, directly from the data

x	0	1	3	4
$f(x)$	1	1	-1	2

we have as an interpolation polynomial $y(x)$ for (x):

$$y(x) = 1 \cdot \frac{(x-1)(x-3)(x-4)}{(0-1)(0-3)(0-4)} + 1 \cdot \frac{x(x-3)(x-4)}{(1-0)(1-3)(1-4)}$$
$$-1 \cdot \frac{x(x-1)(x-4)}{(3-0)(3-1)(3-4)} + 2 \cdot \frac{(x-0)(x-1)(x-3)}{(4-0)(4-1)(4-3)}$$

Other Difference Methods (Equally Spaced Ordinates)

Backward Differences. The backward differences denoted by

$$\nabla f(x) = f(x) - f(x-h)$$

$$\nabla^2 f(x) = \nabla f(x) - \nabla f(x-h)$$

$$\ldots$$

$$\nabla f^n(x) = \nabla^{n-1} f(x) - \nabla^{n-1} f(x-h)$$

are useful for calculation near the end of tabulated data.

Central Differences. The central differences denoted by

$$\delta f(x) = f\left(x+\frac{h}{2}\right) - f\left(x-\frac{h}{2}\right)$$

$$\delta^n f(x) = \delta^{n-1} f\left(x+\frac{h}{2}\right) - \delta^{n-1} f\left(x-\frac{h}{2}\right)$$

are useful for calculating at the interior points of tabulated data.
Also to be found in the literature are Gaussian, Stirling, Bessel, Everett, Comrie differences, and so forth.

Inverse Interpolation

This is the process of finding the value of the independent variable or abscissa corresponding to a given value of the function when the latter is between two tabulated values of the abscissa. One method of accomplishing this is to use Lagrange's interpolation formula in the form

$$x = \psi(y) = \sum_{i=0}^{n} \frac{\pi(y)}{(y-y_i)\pi'(y_i)} x_i$$

where x is expressed as a function of y. Other methods revolve about methods of iteration.

12.7 Numerical Differentiation

Numerical differentiation should be avoided wherever possible, particularly when data are empirical and subject to appreciable observation errors. Errors in data can affect numerical derivatives quite strongly (i.e., differentiation is a roughening process). When such a calculation must be made, it is usually desirable first to *smooth* the data to a certain extent.

The Use of Interpolation Formulas

If the data are given over equidistant values of the independent variable x, an interpolation formula, such as the Newton formula, may be used, and the resulting formula differentiated analytically. If the independent variable is not at equidistant values, then Lagrange's formulas must be used. By differentiating three- and five-point Lagrange interpolation formulas, the following differentiation formulas result for equally spaced tabular points.

Three-point Formulas. Let x_0, x_1, and x_2 be the three points

$$f'(x_0) = \frac{1}{2h}\left[-3f(x_0) + 4f(x_1) - f(x_2)\right] + \frac{h^2}{3}f'''(\varepsilon)$$

$$f'(x_1) = \frac{1}{2h}\left[-f(x_0) + f(x_2)\right] + \frac{h^2}{6}f'''(\varepsilon)$$

$$f'(x_2) = \frac{1}{2h}\left[f(x_0) - 4f(x_1) + 3f(x_2)\right] + \frac{h^2}{3}f'''(\varepsilon)$$

where the last term is an error term and $\min_i x_i < \varepsilon < \max_i x_i$.

Five-point Formulas. Let x_0, x_1, x_2, x_3, and x_4 be the five values of the equally spaced independent variable and $f_i = f(x_i)$.

$$f'(x_0) = \frac{1}{12h}\left[-25f_0 + 48f_1 - 36f_2 + 16f_3 - 3f_4\right] + \frac{h^4}{5}f^{(5)}(\varepsilon)$$

$$f'(x_1) = \frac{1}{12h}\left[-3f_0 - 10f_1 + 18f_2 - 6f_3 + f_4\right] - \frac{h^4}{20}f^{(5)}(\varepsilon)$$

$$f'(x_2) = \frac{1}{12h}\left[f_0 - 8f_1 + 8f_3 - f_4\right] + \frac{h^4}{30}f^{(5)}(\varepsilon)$$

$$f'(x_3) = \frac{1}{12h}\left[-f_0 + 6f_1 - 18f_2 + 10f_3 + 3f_4\right] - \frac{h^4}{20}f^{(5)}(\varepsilon)$$

$$f'(x_4) = \frac{1}{12h}\left[3f_0 - 16f_1 + 36f_2 - 48f_3 + 25f_4\right] + \frac{h^4}{5}f^{(5)}(\varepsilon)$$

and the last term is again an error term.

Smoothing Techniques

These techniques involve the approximation of the tabular data by a least squares fit of the data using some known functional form, usually a polynomial. In place of approximating $f(x)$ by a single least squares polynomial of degree n over the entire range of the tabulation, it is often desirable to replace each tabulated value by the value taken on by a last squares polynomial of degree n relevant to a subrange of $2M + 1$ points centered, where possible, at the point for which the entry is to be modified. Thus, each smoothed value replaces a tabulated value. Let $f_i = f(x_i)$ be the tabular points and y_i = smoothed values. A first-degree least squares with three points would be

$$y_0 = \tfrac{1}{6}\left[5f_0 + 2f_1 - f_2\right]$$

$$y_1 = \tfrac{1}{3}\left[f_0 + f_1 + f_2\right]$$

$$y_2 = \tfrac{1}{6}\left[-f_0 + 2f_1 + 5f_2\right]$$

A first-degree least squares with five points would be

$$y_0 = \tfrac{1}{5}\left[3f_0 + 2f_1 + f_2 - f_4\right]$$

$$y_1 = \tfrac{1}{10}\left[4f_0 + 3f_1 + 2f_2 + f_3\right]$$

$$y_2 = \tfrac{1}{5}\left[f_0 + f_1 + f_2 + f_3 + f_4\right]$$

$$y_3 = \tfrac{1}{10}\left[f_0 + 2f_1 + 3f_2 + 4f_3\right]$$

$$y_4 = \tfrac{1}{5}\left[-f_0 + f_2 + 2f_3 + 3f_4\right]$$

Thus, for example, if first-degree, five-point least squares are used, the central formula is used for all values except the first two and the last two, where the off-center formulas are used. A third-degree least squares with seven points would be

$$y_0 = \tfrac{1}{42}\left[39f_0 + 8f_1 - 4f_2 - 4f_3 + f_4 + 4f_5 - 2f_6\right]$$

$$y_1 = \tfrac{1}{42}\left[8f_0 + 19f_1 + 16f_2 + 6f_3 - 4f_4 - 7f_5 + 4f_6\right]$$

$$y_2 = \tfrac{1}{42}\left[-4f_0 + 16f_1 + 19f_2 + 12f_3 + 2f_4 - 4f_5 + f_6\right]$$

$$y_3 = \tfrac{1}{21}\left[-2f_0 + 3f_1 + 6f_2 + 7f_3 + 6f_4 + 3f_5 - 2f_6\right]$$

$$y_4 = \tfrac{1}{42}\left[f_0 - 4f_1 + 2f_2 + 12f_3 + 19f_4 + 16f_5 - 4f_6\right]$$

$$y_5 = \tfrac{1}{42}\left[4f_0 - 7f_1 - 4f_2 + 6f_3 + 16f_4 + 19f_5 + 8f_6\right]$$

$$y_6 = \tfrac{1}{42}\left[-2f_0 + 4f_1 + f_2 - 4f_3 - 4f_4 + 8f_5 + 39f_6\right]$$

Additional smoothing formulas may be found in the references. After the data are smoothed, any of the interpolation polynomials, or an appropriate least squares polynomial, may be fitted and the results used to obtain the derivative.

Least Squares Method

Parabolic. For five evenly spaced neighboring abscissas labeled x_{-2}, x_{-1}, x_0, x_1, and x_2, and their ordinates f_{-2}, f_{-1}, f_0, f_1, and f_2, assume a parabola is fit by least squares. There results for all interior points, except the first and last two points of the data, the formula for the numerical derivative:

$$f_0' = \frac{1}{10h}\left[-2f_{-2} - f_{-1} + f_1 + 2f_2\right]$$

For the first two data points designated by 0 and h:

$$f'(0) = \frac{1}{20h}\left[-21f(0) + 13f(h) + 17f(2h) - 9f(3h)\right]$$

$$f'(h) = \frac{1}{20h}\left[-11f(0) + 3f(h) + 7f(2h) + f(3h)\right]$$

and for the last two given by $\alpha - h$ and α:

$$f'(\alpha-h)=\frac{1}{20h}\left[-11f(\alpha)+3f(\alpha-h)+7f(\alpha-2h)+f(\alpha-3h)\right]$$

$$f'(\alpha)=\frac{1}{20h}\left[-21f(\alpha)+13f(\alpha-h)+17f(\alpha-2h)-9(\alpha-3h)\right]$$

Quartic (Douglas-Avakian). A fourth-degree polynomial $y = a + bx + cx^2 + dx^3 + ex^4$ is fitted to seven adjacent equidistant points (spacing h) after a translation of coordinates has been made so that $x = 0$ corresponds to the central point of the seven. Thus, these may be called $-3h$, $-2h$, $-h$, 0, h, $2h$, and $3h$. Let k = coefficient h for the seven points. This is, in $-3h$, $k = -3$. Then, the coefficients for the polynomial are

$$a=\frac{524\sum f(kh)-245\sum k^2 f(kh)+21\sum k^4 f(kh)}{924}$$

$$b=\frac{397\sum kf(kh)}{1512h}-\frac{7\sum k^3 f(kh)}{216h}$$

$$c=\frac{-840\sum f(kh)+679\sum k^2 f(kh)-67\sum k^4 f(kh)}{3168h^2}$$

$$d=\frac{-7\sum kf(kh)+\sum k^3 f(kh)}{216h^3}$$

$$e=\frac{72\sum f(kh)-67\sum k^2 f(kh)+7\sum k^4 f(kh)}{3168h^4}$$

where all summations run from $k = -3$ to $k = +3$ and $f(kh)$ = tabular value at kh. The slope of the polynomial at $x = 0$ is $dy/dx = b$.

12.8 Numerical Integration

Numerical evaluation of the finite integral $\int_a^b f(x)\,dx$ is carried out by a variety of methods. A few are given here.

Newton-Cotes Formulas (Equally Spaced Ordinates)

Trapezoidal Rule. This formula consists of subdividing the interval $a \le x \le b$ into n subintervals a to $a + h$, $a + h$ to $a + 2h$, ..., and replacing the graph of $f(x)$ by the result of joining the ends of adjacent ordinates by line segments. If $f_j = f(x_j) = f(a + jh)$, $f_0 = f(a)$, and $f_n = f(b)$, the integration formula is

$$\int_a^b f(x)\,dx=\frac{h}{2}\left[f_0+2f_1+2f_2+\cdots+2f_{n-1}+f_n\right]+E_n$$

where $|E_n| = (nh^3/12)|f''(\varepsilon)| = |(b-a)^3/12n^2||f''(\varepsilon)|$, $a < \varepsilon < b$. This procedure is not of high accuracy. However, if $f''(x)$ is continuous in $a < x < b$, the error goes to zero as $1/n^2$, $n \to \infty$.

Parabolic Rule (Simpson's Rule). This procedure consists of subdividing the interval $a < x < b$ into $n/2$ subintervals, each of length $2h$, where n is an even integer. Using the notation as above the integration formula is

$$\int_a^b f(x)\,dx = \frac{h}{3}\left[f_0 + 4f_1 + 2f_2 + 4f_3 + \cdots + 4f_{n-3} + 2f_{n-2} + 4f_{n-1} + f_n\right] + E_n$$

where

$$|E_n| = \frac{nh^5}{180}\left|f^{(4)}(\varepsilon)\right| = \frac{(b-a)^5}{180n^4}\left|f^{(4)}(\varepsilon)\right| \qquad a < \varepsilon < b$$

This method approximates $f(x)$ by a parabola on each subinterval. This rule is generally more accurate than the trapezoidal rule. It is the most widely used integration formula.

Weddle's Rule. This procedure consists of subdividing the integral $a < x < b$ into $n/6$ subintervals, each of length $6h$, where n is a multiple of 6. Using the notation from the trapezoidal rule, there results

$$\int_a^b f(x)\,dx = \frac{3h}{10}\left[f_0 + 5f_1 + f_2 + 6f_3 + f_4 + 5f_5 + 2f_6 + 5f_7 + f_8 + \cdots + 6f_{n-3} + f_{n-2} + 5f_{n-1} + f_n\right] + E_n$$

Note that the coefficients of f_i follow the rule 1, 5, 1, 6, 1, 5, 2, 5, 1, 6, 1, 5, 2, 5, etc.... This procedure consists of approximately $f(x)$ by a polynomial of degree 6 on each subinterval. Here,

$$E_n = \frac{nh^7}{1400}\left[10f^{(6)}(\varepsilon_1) + 9h^2 f^{(8)}(\varepsilon_2)\right]$$

Gaussian Integration Formulas (Unequally Spaced Abscissas)

These formulas are capable of yielding comparable accuracy with fewer ordinates than the equally spaced formulas. The ordinates are obtained by optimizing the distribution of the abscissas rather than by arbitrary choice. For the details of these formulas, Hildebrand (1956) is an excellent reference.

Two-Dimensional Formula

Formulas for two-way integration over a rectangle, circle, ellipse, and so forth, may be developed by a double application of one-dimensional integration formulas. The two-dimensional generalization of the parabolic rule is given here. Consider the iterated integral $\int_a^b \int_c^d f(x, y)\,dx\,dy$. Subdivide $c < x < d$ into m (even) subintervals of length $h = (d - c)/m$, and $a < y < b$ into n (even) subintervals of length $k = (b - a)/n$. This gives a subdivision of the rectangle $a \le y \le b$ and $c \le x \le d$ into subrectangles. Let $x_j = c + jh$, $y_j = a + jk$, and $f_{i,j} = f(x_i, y_j)$. Then,

$$\int_a^b \int_c^d f(x,y)\,dx\,dy = \frac{hk}{9}\Big[\left(f_{0,0} + 4f_{1,0} + 2f_{2,0} + \cdots + f_{m,0}\right) + 4\left(f_{0,1} + 4f_{1,1} + 2f_{2,1} + \cdots + f_{m,1}\right)$$

$$+ 2\left(f_{0,2} + 4f_{1,2} + 2f_{2,2} + \cdots + f_{m,2}\right) + \cdots + \left(f_{0,n} + 4f_{1,n} + 2f_{2,n} + \cdots + f_{m,n}\right)\Big] + E_{m,n}$$

where

$$E_{m,n} = -\frac{hk}{90}\left[mh^4 \frac{\partial^4 f(\varepsilon_1, \eta_1)}{\partial x^4} + nk^4 \frac{\partial^4 f(\varepsilon_2, \eta_2)}{\partial y^4}\right]$$

where ε_1 and ε_2 lie in $c < x < d$, and η_1 and η_2 lie in $a < y < b$.

12.9 Numerical Solution of Ordinary Differential Equations

A number of methods have been devised to solve ordinary differential equations numerically. The general references contain some information. A numerical solution of a differential equation means a table of values of the function y and its derivatives over only a limited part of the range of the independent variable. Every differential equation of order n can be rewritten as n first-order differential equations. Therefore, the methods given below will be for first-order equations, and the generalization to simultaneous systems will be developed later.

The Modified Euler Method

This method is simple and yields modest accuracy. If extreme accuracy is desired, a more sophisticated method should be selected. Let the first-order differential equation be $dy/dx = f(x, y)$ with the initial condition (x_0, y_0) (i.e., $y = y_0$ when $x = x_0$). The procedure is as follows.

Step 1. From the given initial conditions (x_0, y_0) compute $y_0' = f(x_0, y_0)$ and $y_0'' = [\partial f(x_0, y_0)/\partial x] + [\partial f(x_0, y_0)/\partial y]\, y_0'$. Then, determine $y_1 = y_0 + h\, y_0' + (h^2/2)\, y_0''$, where h = subdivision of the independent variable.

Step 2. Determine $y_1' = f(x_1, y_1)$ where $x_1 = x_0 + h$. These prepare us for the following.

Predictor Steps.

Step 3. For $n \geq 1$, calculate $(y_{n+1})_1 = y_{n-1} + 2h\, y_n'$.

Step 4. Calculate $(y_{n+1}')_1 = f[x_{n+1}, (y_{n+1})_1]$.

Corrector Steps.

Step 5. Calculate $(y_{n+1})_2 = y_n + (h/2)\,[(y_{n+1}')_1 + y_n']$, where y_n and y_n' without the subscripts are the previous values obtained by this process (or by steps 1 and 2).

Step 6. $(y_{n+1}')_2 = f[x_{n+1}, (y_{n+1})_2]$.

Step 7. Repeat the corrector steps 5 and 6 if necessary until the desired accuracy is produced in y_{n+1}, y_{n+1}'.

Example. Consider the equation $y' = 2y^2 + x$ with the initial conditions $y_0 = 1$ when $x_0 = 0$. Let $h = 0.1$. A few steps of the computation are illustrated.

Step	
1	$y_0' = 2y_0^2 + x_0 = 2$
	$y_0'' = 1 + 4y_0 y_0' = 1 + 8 = 9$
	$y_1 = 1 + (0.1)(2) + [(0.1)^2/2]9 = 1.245$
2	$y_1' = 2y_1^2 + x_1 = 3.100 + 0.1 = 3.200$
3	$(y_2)_1 = y_0 + 2hy_1' = 1 + 2(0.1)3.200 = 1.640$
4	$(y_2')_1 = 2(y_2)_1^2 + x_2 = 5.592$
5	$(y_2)_2 = y_1 + (0.1/2)[(y_2')_1 + y_1'] = 1.685$
6	$(y_2')_2 = 2(y_2)_2^2 + x_2 = 5.878$
5 (repeat)	$(y_2)_3 = y_1 + (0.05)[(y_2')_2 + y_1'] = 1.699$
6 (repeat)	$(y_2')_3 = 2(y_2)_3^2 + x_2 = 5.974$

and so forth. This procedure. may be programmed for a computer. A discussion of the truncation error of this process may be found in Milne (1953).

Modified Adams' Method

The procedure given here was developed retaining third differences. It can then be considered as a more exact predictor-corrector method than the Euler method. The procedure is as follows for $dy/dx = f(x, y)$ and h = interval size.

Steps 1 and 2 are the same as in Euler method.

Predictor Steps.

Step 3. $(y_{n+1})_1 = y_n + (h/24)[55y'_n - 59y'_{n-1} + 37y'_{n-2} - 9y'_{n-3}]$, where y'_n, y'_{n-1}, etc...., are calculated in step 1.
Step 4. $(y'_{n+1})_1 = f[x_{n+1}, (y_{n+1})_1]$.

Corrector Steps.

Step 5. $(y_{n+1})_2 = y_n + (h/24)[9(y'_{n+1})_1 + 19y'_n - 5y'_{n-1} + y'_{n-2}]$.
Step 6. $(y'_{n+1})_2 = f[x_{n+1}, (y_{n+1})_2]$.
Step 7. Iterate steps 5 and 6 if necessary.

Runge-Kutta Methods

These methods are self-starting and are inherently stable. Kopal (1955) is a good reference for their derivation and discussion. Third- and fourth-order procedures are given below for $dy/dx = f(x, y)$ and h = interval size.

For third-order (error $\approx h^4$).

$$k_0 = hf(x_n, y_n)$$

$$k_1 = hf(x_n + \tfrac{1}{2}h, y_n + \tfrac{1}{2}k_0)$$

$$k_2 = hf(x_n + h, y_n + 2k_1 - k_0)$$

and

$$y_{n+1} = y_n + \tfrac{1}{6}(k_0 + 4k_1 + k_2)$$

for all $n \geq 0$, with initial condition (x_0, y_0).

For fourth-order (error $\approx h^5$),

$$k_0 = hf(x_n, y_n)$$

$$k_1 = hf(x_n + \tfrac{1}{2}h, y_n + \tfrac{1}{2}k_0)$$

$$k_2 = hf(x_n + \tfrac{1}{2}h, y_n + \tfrac{1}{2}k_1)$$

$$k_3 = hf(x_n + h, y_n + k_2)$$

and

$$y_{n+1} = y_n + \tfrac{1}{6}(k_0 + 2k_1 + 2k_2 + k_3)$$

Example. (Third-order) Let $dy/dx = x - 2y$, with initial condition $y_0 = 1$ when $x_0 = 0$, and let $h = 0.1$. Clearly, $x_n = nh$. To calculate y_1, proceed as follows:

$$k_0 = 0.1[x_0 - 2y_0] = -0.2$$

$$k_1 = 0.1[0.05 - 2(1 - 0.1)] = -0.175$$

$$k_2 = 0.1[0.1 - 2(1 - 0.35 + 0.2)] = -0.16$$

$$y_1 = 1 + \tfrac{1}{6}(-0.2 - 0.7 - 0.16) = 0.8234$$

Equations of Higher Order and Simultaneous Differential Equations

Any differential equation of second- or higher order can be reduced to a simultaneous system of first-order equations by the introduction of auxiliary variables. Consider the following equations:

$$\frac{d^2x}{dt^2} + xy\frac{dx}{dt} + z = e^t$$

$$\frac{d^2y}{dt^2} + xy\frac{dy}{dt} = 7 + t^2$$

$$\frac{d^2z}{dt^2} + xz\frac{dz}{dt} + x = e^t$$

In the new variables $x_1 = x$, $x_2 = y$, $x_3 = z$, $x_4 = dx_1/dt$, $x_5 = dx_2/dt$, and $x_6 = dx_3/dt$, the equations become

$$\frac{dx_1}{dt} = x_4$$

$$\frac{dx_2}{dt} = x_5$$

$$\frac{dx_3}{dt} = x_6$$

$$\frac{dx_4}{dt} = -x_1x_2x_4 - x_3 + e^t$$

$$\frac{dx_5}{dt} = -x_3x_2x_5 + 7 + t^2$$

$$\frac{dx_6}{dt} = -x_1x_3x_6 - x_1 + e^t$$

which is a system of the general form

$$\frac{dx_i}{dt} = f_i(t, x_1, x_2, x_3, \ldots, x_n)$$

where $i = 1, 2, \ldots, n$. Such systems may be solved by simultaneous application of any of the above numerical techniques. A Runge-Kutta method for

$$\frac{dx}{dt} = f(t,x,y)$$

$$\frac{dy}{dt} = g(t,x,y)$$

is given below. The fourth-order procedure is shown.

Starting at the initial conditions x_0, y_0, and t_0, the next values x_1 and y_1 are computed via the equations below (where $\Delta t = h$, $t_i = h + t_{i-1}$):

$$k_0 = hf(t_0, x_0, y_0) \qquad l_0 = hg(t_0, x_0, y_0)$$

$$k_1 = hf\left(t_0 + \frac{h}{2}, x_0 + \frac{k_0}{2}, y_0 + \frac{l_0}{2}\right) \qquad l_1 = hg\left(t_0 + \frac{h}{2}, x_0 + \frac{k_0}{2}, y_0 + \frac{l_0}{2}\right)$$

$$k_2 = hf\left(t_0 + \frac{h}{2}, x_0 + \frac{k_1}{2}, y_0 + \frac{l_1}{2}\right) \qquad l_2 = hg\left(t_0 + \frac{h}{2}, x_0 + \frac{k_1}{2}, y_0 + \frac{l_1}{2}\right)$$

$$k_3 = hf(t_0 + h, x_0 + k_2, y_0 + l_2) \qquad l_3 = hg(t_0 + h, x_0 + k_2, y_0 + l_2)$$

and

$$x_1 = x_0 + \tfrac{1}{6}(k_0 + 2k_1 + 2k_2 + k_3)$$

$$y_1 = y_0 + \tfrac{1}{6}(l_0 + 2l_1 + 2l_2 + l_3)$$

To continue the computation, replace t_0, x_0, and y_0 in the above formulas by $t_1 = t_0 + h$, x_1, and y_1 just calculated. Extension of this method to more than two equations follows precisely this same pattern.

12.10 Numerical Solution of Integral Equations

This section considers a method of numerically solving the Fredholm integral equation of the second kind:

$$u(x) = f(x) + \lambda \int_a^b k(x,t)u(t)\,dt \qquad \text{for } u(x)$$

The method discussed arises because a definite integral can be closely approximated by any of several numerical integration formulas (each of which arises by approximating the function by some polynomial over an interval). Thus, the definite integral can be replaced by an integration formula which becomes

$$u(x) = f(x) + \lambda(b-a)\left[\sum_{i=1}^{n} c_i k(x,t_i)u(t_i)\right]$$

where $t_1, \ldots, t_n$ are points of subdivision of the t axis, $a \le t \le b$, and the cs are coefficients whose values depend upon the type of numerical integration formula used. Now, this must hold for all values of x, where $a \le x \le b$; so it must hold for $x = t_1$, $x = t_2, \ldots, x = t_n$. Substituting for x successively $t_1, t_2, \ldots, t_n$, and setting $u(t_i) = u_i$ and $f(t_i) = f_i$, we get n linear algebraic equations for the n unknowns $u_1, \ldots, u_n$. That is,

$$u_i = f_i + (b-a)\left[c_1 k(t_i,t_1)u_1 + c_2 k(t_i,t_2)u_2 + \cdots + c_n k(t_i,t_n)u_n\right], \qquad i = 1,2,\ldots,n$$

These u_i may be solved for by the methods under the section entitled "Numerical Solution of Linear Equations."

12.11 Numerical Methods for Partial Differential Equations

The ultimate goal of numerical (discrete) methods for partial differential equations (PDEs) is the reduction of continuous systems (projections) to discrete systems that are suitable for high-speed computer solutions. The user must be cautioned that the seeming elementary nature of the techniques holds pitfalls that can be seriously misleading. These approximations often lead to difficult mathematical questions of adequacy, accuracy, convergence, stability, and consistency. Convergence is concerned with the approach of the approximate numerical solution to the exact solution as the number of mesh units increase indefinitely in some sense. Unless the numerical method can be shown to converge to the exact solution, the chosen method is unsatisfactory.

Stability deals in general with error growth in the calculation. As stated before, any numerical method involves truncation and round-off errors. These errors are not serious unless they grow as the computation proceeds (i.e., the method is unstable).

Finite Difference Methods

In these methods, the derivatives are replaced by various finite differences. The methods will be illustrated for problems in two space dimensions (x, y) or (x, t) where t is timelike. Using subdivisions $\Delta x = h$ and $\Delta y = k$ with $u(i\,h, jk) = u_{i,j}$, approximate $u_x|_{i,j} = [(u_{i+1,j} - u_{i,j})/h] + O(h)$ (forward difference), a first-order $[O(h)]$ method, or $u_x|_{i,j} = [(u_{i+1,j} - u_{i-1,j})/2h] + O(h^2)$ (central difference), a second-order method. The second derivative is usually approximated with the second-order method $[u_{xx}|_{i,j} = [(u_{i+1,j} - 2u_{i,j} + u_{i-1,j})/h^2] + O(h^2)]$.

Example. Using second-order differences for u_{xx} and u_{yy}, the five-point difference equation (with $h = k$) for Laplace's equation $u_{xx} + u_{yy} = 0$ is $u_{i,j} = 1/4[u_{i+1,j} + u_{i-1,j} + u_{i,j+1} + u_{i,j-1}]$. The accuracy is $O(h^2)$. This model is called *implicit* because one must solve for the total number of unknowns at the unknown grid points (i, j) in terms of the given boundary data. In this case, the system of equations is a linear system.

Example. Using a forward-difference approximation for u_t and a second-order approximation for u_{xx}, the diffusion equation $u_t = u_{xx}$ is approximated by the *explicit* formula $u_{i,j+1} = ru_{i-1,j} + (1 - 2r)u_{i,j} + ru_{i+1,j}$. This classic result permits step-by-step advancement in the t direction beginning with the initial data at $t = 0$ ($j = 0$) and guided by the boundary data. Here, the term $r = \Delta t/(\Delta x)^2 = k/h^2$ is restricted to be less than or equal to 1/2 for stability and the truncation error is $O(k^2 + kh^2)$.

The Crank-Nicolson implicit formula which approximates the diffusion equation $u_t = u_{xx}$ is

$$-r\lambda u_{i-1,j+1} + (1 + 2r\lambda)u_{i,j+1} - r\lambda u_{i+1,j+1} = r(1-\lambda)u_{i-1,j} + \left[1 - 2r(1-\lambda)\right]u_{i,j} + r(1-\lambda)u_{i+1,j}$$

The stability of this numerical method was analyzed by Crandall (Ames, 1993) where the λ, r stability diagram is given.

Approximation of the time derivative in $u_t = u_{xx}$ by a central difference leads to an always unstable approximation — the useless approximation

$$u_{i,j+1} = u_{i,j-1} + 2r\left(u_{i+1,j} - 2u_{i,j} + u_{i-1,j}\right)$$

which is a warning to be careful.

The foregoing method is *symmetric* with respect to the point (i, j), where the method is centered. Asymmetric methods have some computational advantages, so the Saul'yev method is described (Ames, 1993). The algorithms ($r = k/h^2$)

$$(1+r)u_{i,j+1} = u_{i,j} + r\left(u_{i-1,j+1} - u_{i,j} + u_{i+1,j}\right) \qquad \text{(Saul'yev A)}$$

$$(1+r)u_{i,j+1} = u_{i,j} + r\left(u_{i+1,j+1} - u_{i,j} + u_{i-1,j}\right) \qquad \text{(Saul'yev B)}$$

are used as in any one of the following options:

1. Use Saul'yev A only and proceed line-by-line in the $t(j)$ direction, but *always* from the left boundary on a line.
2. Use Saul'yev B only and proceed line-by-line in the $t(j)$ direction, but *always* from the right boundary to the left on a line.
3. Alternate from line to line by first using Saul'yev A and then B, or the reverse. This is related to *alternating direction methods.*
4. Use Saul'yev A and Saul'yev B on the same line and average the results for the final answer (A first, and then B). This is equivalent to introducing the dummy variables $P_{i,j}$ and $Q_{i,j}$ such that

$$(1+r)P_{i,j+1} = U_{i,j} + r\left(P_{i-1,j+1} - U_{i,j} + U_{i+1,j}\right)$$

$$(1+r)Q_{i,j+1} = U_{i,j} + r\left(Q_{i+1,j+1} - U_{i,j} + U_{i-1,j}\right)$$

and

$$U_{i,j+1} = \tfrac{1}{2}\left(P_{i,j+1} + Q_{i,j+1}\right)$$

This averaging method has some computational advantage because of the possibility of truncation error cancellation. As an alternative, one can retain the $P_{i,j}$ and $Q_{i,j}$ from the previous step and replace $U_{i,j}$ and $U_{i-1,j}$ by $P_{i,j}$ and $P_{i-1,j}$, respectively, and $U_{i,j}$ and $U_{i+1,j}$ by $Q_{i,j}$ and $Q_{i+1,j}$, respectively.

Weighted Residual Methods (WRMs)

To set the stage for the method of finite elements, we briefly describe the WRMs, which have several variations — the interior, boundary, and mixed methods. Suppose the equation is $Lu = f$, where L is the partial differential operator and f is a known function, of say x and y. The first step in WRM is to select a class of known basis functions b_i (e.g., trigonometric, Bessel, Legendre) to approximate $u(x, y)$ as ~ $\Sigma\, a_i\, b_i\,(x, y) = U(x, y; a)$. Often, the b_i are selected so that $U(x, y; a)$ satisfy the boundary conditions. This is essentially the *interior method.* If the b_i, in $U(x, y; a)$ are selected to satisfy the differential equations, but not the boundary conditions, the variant is called the *boundary method.* When neither the equation nor the boundary conditions are satisfied, the method is said to be *mixed.* The least ingenuity is required here. The usual method of choice is the interior method.

The second step is to select an optimal set of constants a_i, $i = 1, 2, \ldots, n$, by using the residual $R_I(U) = LU - f$. This is done here for the interior method. In the boundary method, there are a set of boundary residual R_B, and, in the mixed method, Both R_I and R_B. Using the spatial average $(w, v) = \int_V wv dV$, the criterion for selecting the values of a_i is the requirement that the n spatial averages

$$\left(b_i, R_L(U)\right) = 0, \qquad i = 1, 2, \ldots, n$$

These represent n equations (linear if the operator L is linear and nonlinear otherwise) for the a_i.

Particular WRMs differ because of the choice of the b_is. The most common follow.

1. *Subdomain* The domain V is divided into n smaller, not necessarily disjoint, subdomains V_i with $w_i(x, y) = 1$ if (x, y) is in V_i, and 0 if (x, y) is not in V_i.

2. *Collocation* Select n points $P_i = (x_i, y_i)$ in V with $w_i(P_i) = \delta(P - P_i)$, where $\int_V \phi(P)\delta(P - P_i)dP = \phi(P_i)$ for all test functions $\phi(P)$ which vanish outside the compact set V. Thus, $(w_i, R_E) = \int_V \delta(P - P_i)R_E dV = R_E[U(P_i) \equiv 0$ (i.e., the residual is set equal to zero at the n points P_i).
3. *Least squares* Here, the functional $I(a) = \int_V R_E^2\, dV$, where $a = (a_1, \ldots, a_n)$, is to be made stationary with respect to the a_j. Thus, $0 = \partial I/\partial a_j = 2\int_V R_E(\partial R_E/\partial a_j)dV$, with $j = 1, 2, \ldots, n$. The w_j in this case are $\partial R_E/\partial a_j$.
4. *Bubnov-Galerkin* Choose $w_j(P) = b_j(P)$. This is perhaps the best-known method.
5. *Stationary Functional (Variational) Method* With ϕ a variational integral (or other functional), set $\partial\phi[U]/\partial a_j = 0$, where $j = 1, \ldots, n$, to generate the n algebraic equations.

Example. $u_{xx} + u_{yy} = -2$, with $u = 0$ on the boundaries of the square $x = \pm 1$, $y = \pm 1$. Select an interior method with $U = a_1(1 - x^2)(1 - y^2) + a_2x^2y^2(1 - x^2)(1 - y^2)$, whereupon the residual $R_E(U) = 2a_1(2 - x^2 - y^2) + 2a_2[(1 - 6x^2)y^2(1 - y^2) + (1 - 6y^2) - (1 - x^2)] + 2$. Collocating at (1/3, 1/3) and (2/3, 2/3) gives the two linear equations $-32a_1/9 + 32a_2/243x^2 + 2 = 0$ and $-20a_1/9 - 400a_2/243 + 2 = 0$ for a_1 and a_2.

WRM methods can obviously be used as approximate methods. We have now set the stage for *finite elements*.

Finite Elements

The WRM methods are more general than the *finite elements* (FE) methods. FE methods require, in addition, that the basis functions be finite elements (i.e., functions that are zero except on a small part of the domain under consideration). A typical example of an often used basis is that of triangular elements. For a triangular element with Cartesian coordinates (x_1, y_1), (x_2, y_2), and (x_3, y_3), define natural coordinates L_1, L_2, and L_3 $(L_i \leftrightarrow (x_i, y_i))$ so that $L_i = A_i/A$ where

$$A = \tfrac{1}{2}\det\begin{bmatrix} 1 & x_1 & y_1 \\ 1 & x_2 & y_2 \\ 1 & x_3 & y_3 \end{bmatrix}$$

is the area of the triangle and

$$A_1 = \tfrac{1}{2}\det\begin{bmatrix} 1 & x & y \\ 1 & x_2 & y_2 \\ 1 & x_3 & y_3 \end{bmatrix}$$

$$A_2 = \tfrac{1}{2}\det\begin{bmatrix} 1 & x_1 & y_1 \\ 1 & x & y \\ 1 & x_3 & y_3 \end{bmatrix}$$

$$A_3 = \tfrac{1}{2}\det\begin{bmatrix} 1 & x_1 & y_1 \\ 1 & x_2 & y_2 \\ 1 & x & y \end{bmatrix}$$

Clearly $L_1 + L_2 + L_3 = 1$, and the L_i are one at node i and zero at the other nodes. In terms of the Cartesian coordinates,

$$\begin{bmatrix} L_1 \\ L_2 \\ L_3 \end{bmatrix} = \frac{1}{2A}\begin{bmatrix} x_2y_3 - x_3y_2, & y_2 - y_3, & x_3 - x_2 \\ x_3y_1 - x_1y_3, & y_3 - y_1, & x_1 - x_3 \\ x_1y_2 - x_2y_1, & y_1 - y_2, & x_2 - x_1 \end{bmatrix}\begin{bmatrix} 1 \\ x \\ y \end{bmatrix}$$

is the linear triangular element relation.

Tables of linear, quadratic, and cubic basis functions are given in the literature. Notice that while the linear basis needs three nodes, the quadratic requires six and the cubic basis ten. Various modifications, such as the Hermite basis, are described in the literature. Triangular elements are useful in approximating irregular domains.

For rectangular elements, the *chapeau* functions are often used. Let us illustrate with an example. Let $u_{xx} + u_{yy} = Q$, $0 < x < 2$, $0 < y < 2$, $u(x, 2) = 1$, $u(0, y) = 1$, $u_y(x, 0) = 0$, $u_x(2, y) = 0$, and $Q(x, y) = Qw\delta(x - 1)\delta(y - 1)$,

$$\delta(x) = \begin{cases} 0 & x \neq 0 \\ 1 & x = 0 \end{cases}$$

Using four equal rectangular elements, map the element I with vertices at (0, 0) (0, 1), (1, 1), and (1, 0) into the local (canonical) coordinates (ξ, η), $-1 \leq \xi \leq 1$, $-1 \leq \eta \leq 1$, by means of $x = 1/2(\xi + 1)$, $y = 1/2(\eta + 1)$. This mapping permits one to develop software that standardizes the treatment of all elements. Converting to (ξ, η) coordinates, our problem becomes $u_{\xi\xi} + u_{\eta\eta} = 1/4Q$, $-1 \leq \xi \leq 1$, $-1 \leq \eta \leq 1$, $Q = Qw\delta(\xi - 1)\delta(\eta - 1)$.

First, a trial function $\bar{u}(\xi, \eta)$ is defined as $u(\xi, \eta) \approx \bar{\mu}(\xi, \eta) = \Sigma_{i=1}^{4} A_i \phi_i(\xi, \eta)$ (in element I) where the ϕ_i are the two-dimensional chapeau functions

$$\phi_1 = \left[\tfrac{1}{2}(1 - \xi)\tfrac{1}{2}(1 - \eta)\right] \qquad \phi_2 = \left[\tfrac{1}{2}(1 + \xi)\tfrac{1}{2}(1 - \eta)\right]$$

$$\phi_3 = \left[\tfrac{1}{2}(1 + \xi)\tfrac{1}{2}(1 + \eta)\right] \qquad \phi_4 = \left[\tfrac{1}{2}(1 - \xi)\tfrac{1}{2}(1 + \eta)\right]$$

Clearly ϕ_i take the value one at node i, provide a bilinear approximation, and are nonzero only over elements adjacent to node i.

Second, the equation residual $R_E = \nabla^2 \bar{u} - 1/4Q$ is formed and a WRM procedure is selected to formulate the algebraic equations for the A_i. This is indicated using the Galerkin method. Thus, for element I, we have

$$\iint_{D_I} \left(\bar{u}_{\xi\xi} + \bar{u}_{\eta\eta} - Q\right)\phi_i(\xi, \eta)\, d\xi\, d\eta = 0. \qquad i = 1, \ldots, 4$$

Applying Green's theorem, this result becomes

$$\iint_{D_I} \left[\bar{u}_\xi (\phi_i)_\xi + \bar{u}_\eta (\phi_i)_\eta + \tfrac{1}{4} Q\phi_i\right] d\xi\, d\eta - \int_{\partial D_I} \left(\bar{u}_\xi c_\xi + \bar{u}_\eta c_\eta\right)\phi_i\, ds = 0. \qquad i = 1, 2, \ldots, 4$$

Using the same procedure in all four elements and recalling the property that the ϕ_i in each element are nonzero only over elements adjacent to node i gives the following nine equations:

$$\sum_{e=1}^{4} \left\{ \iint_{D_e} \sum_{i=1}^{9} A_i \left[(\phi_i)_\xi (\phi_i)_\xi + (\phi_i)_\eta (\phi_i)_\eta \right] + \tfrac{1}{4} Q\phi_i \right\} d\xi\, d\eta$$

$$- \sum_{e=1}^{4} \int_{\partial D_e} \left(\bar{u}_\xi c_\xi + \bar{u}_\eta c_\eta\right) \phi\, ds = 0, \qquad n = 1, 2, \ldots, 9$$

where the c_ξ and c_η are the direction cosines of the appropriate element (e) boundary.

Method of Lines

The *method of lines*, when used on PDEs in two dimensions, reduces the PDE to a system of ordinary differential equations (ODEs), usually by finite difference or finite element techniques. If the original problem is an initial value (boundary value) problem, then the resulting ODEs form an initial value (boundary value) problem. These ODEs are solved by ODE numerical methods.

Example. $u_t = u_{xx} + u^2$, $0 < x < 1$, $0 < t$, with the initial value $u(x, 0) = x$, and boundary data $u(0, t) = 0$, $u(1, t) = \sin t$. A discretization of the space variable (x) is introduced and the time variable is left continuous. The approximation is $\dot{u}_i = (u_{i+1} - 2u_i + u_{i-1})/h^2 + u_i^2$. With $h = 1/5$, the equations become

$$u_0(t) = 0$$

$$\dot{u}_1 = \tfrac{1}{25}\left[u_2 - 2u_1\right] + u_1^2$$

$$\dot{u}_2 = \tfrac{1}{25}\left[u_3 - 2u_2 + u_1\right] + u_2^2$$

$$\dot{u}_3 = \tfrac{1}{25}\left[u_4 - 2u_3 + u_2\right] + u_3^2$$

$$\dot{u}_4 = \tfrac{1}{25}\left[\sin t - 2u_4 + u_3\right] + u_4^2$$

$$u_5 = \sin t$$

and $u_1(0) = 0.2$, $u_2(0) = 0.4$, $u_3(0) = 0.6$, and $u_4(0) = 0.8$.

12.12 Discrete and Fast Fourier Transforms

Let $x(n)$ be a sequence that is nonzero only for a finite number of samples in the interval $0 \le n \le N - 1$. The quantity

$$X(k) = \sum_{n=0}^{N-1} x(n) e^{-i(2\pi/N)nk}, \quad k = 0, 1, \ldots, N-1$$

is called the *discrete Fourier transform* (DFT) of the sequence $x(n)$. Its inverse (IDFT) is given by

$$x(n) = \frac{1}{N} \sum_{k=0}^{N-1} X(k) e^{i(2\pi/N)nk}, \quad n = 0, 1, \ldots, N-1 \quad \left(i^2 = -1\right)$$

Clearly, DFT and IDFT are finite sums and there are N frequency values. Also, $X(k)$ is periodic in k with period N.

Example. $x(0) = 1$, $x(1) = 2$, $x(2) = 3$, $x(3) = 4$

$$X(k) = \sum_{n=0}^{3} x(n) e^{-i(2\pi/4)nk}, \quad k = 0, 1, 2, 3, 4$$

Thus,

$$X(0) = \sum_{n=0}^{3} x(n) = 10$$

and $X(1) = x(0) + x(1)e^{-i\pi/2} + x(2)e^{-i\pi} + x(3)e^{-i3\pi/2} = 1 - 2i - 3 + 4i = -2 + 2i$; $X(2) = -2$; $X(3) = -2 - 2i$.

DFT Properties

1. Linearity: If $x_3(n) = ax_1(n) + bx_2(n)$, then $X_3(k) = aX_1(k) + bX_2(k)$.
2. Symmetry: For $x(n)$ real, $\mathrm{Re}[X(k)] = \mathrm{Re}[X(N - k)]$, $\mathrm{Im}[(k)] = -\mathrm{Im}[X(N - k)]$.
3. Circular shift: By a circular shift of a sequence defined in the interval $0 \le n \le N - 1$, we mean that, as values *fall off* from one end of the sequence, they are appended to the other end. Denoting this by $x(n \oplus m)$, we see that positive m means shift left and negative m means shift right. Thus, $x_2(n) = x_1(n \oplus m) \Leftrightarrow X_2(k) = X_1(k)e^{i(2\pi/N)km}$.
4. Duality: $x(n) \Leftrightarrow X(k)$ implies $(1/N)X(n) \Leftrightarrow x(-k)$.
5. Z-transform relation: $X(k) = X(z)|_{z=e^{i(2\pi k/N)}}$, $k = 0, 1, \ldots, N - 1$.
6. Circular convolution: $x_3(n) = \sum_{m=0}^{N-1} x_1(m)x_2(n \ominus m) = \sum_{\ell=0}^{N-1} x_1(n \ominus \ell)x_2(\ell)$ where $x_2(n \ominus m)$ corresponds to a circular shift to the right for positive m.

One fast algorithm for calculating DFTs is the radix-2 *fast Fourier transform* developed by J. W. Cooley and J. W. Tucker. Consider the two-point DFT $X(k) = \sum_{n=0}^{1} x(n)e^{-i(2\pi/2)nk}$, $k = 0, 1$. Clearly, $X(k) = x(0) + x(1)e^{-i\pi k}$. So, $X(0) = x(0) + x(1)$ and $X(1) = x(0) - x(1)$. This process can be extended to DFTs of length $N = 2^r$, where r is a positive integer. For $N = 2^r$, decompose the N-point DFT into *two* $N/2$-point DFTs. Then, decompose each $N/2$-point DFT into *two* $N/4$-point DFTs, and so on until eventually we have $N/2$ *two*-point DFTs. Computing these as indicated above, we combine them into $N/4$ four-point DFTs and then $N/8$ eight-point DFTs, and so on, until the DFT is computed. The total number of DFT operations (for large N) is $O(N^2)$, and that of the FFT is $O(N \log_2 N)$, quite a saving for large N.

12.13 Software

Some available software is listed here.

General Packages

General software packages include Maple, Mathematica, and Matlab. All contain algorithms for handling a large variety of both numerical and symbolic computations.

Special Packages for Linear Systems

In the IMSL Library, there are three complementary linear system packages of note.

LINPACK is a collection of programs concerned with *direct* methods for general (or full) symmetric, symmetric positive definite, triangular, and tridiagonal matrices. There are also programs for least squares problems, along with the QR algorithm for eigensystems and the singular value decompositions of rectangular matrices. The programs are intended to be completely machine independent, fully portable, and run with good efficiency in most computing environments. The LINPACK User's Guide by Dongarra *et al.* is the basic reference.

ITPACK is a modular set of programs for iterative methods. The package is oriented toward the sparse matrices that arise in the solution of PDEs and other applications. While the programs apply to full matrices, that is rarely profitable. Four basic iteration methods and two convergence acceleration methods are in the package. There is a Jacobi, SOR (with optimum relaxation parameter estimated), symmetric SOR, and reduced system (red-black ordering) iteration, each with semi-iteration and conjugate gradient acceleration. All parameters for these iterations are automatically estimated. The practical and theoretical background for ITPACK is found in Hagemen and Young (1981).

YALEPACK is a substantial collection of programs for sparse matrix computations.

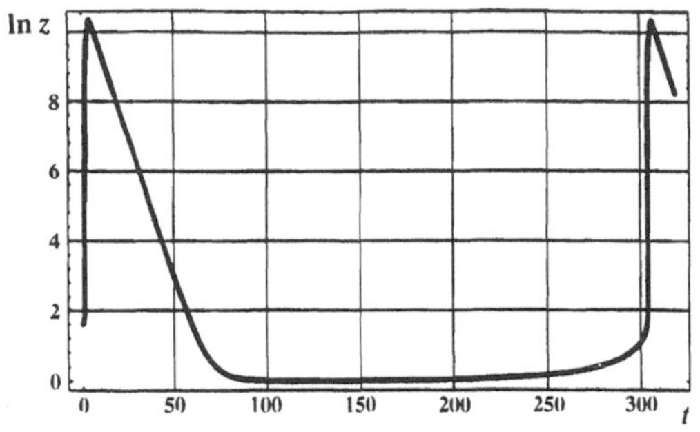

Figure 12.1

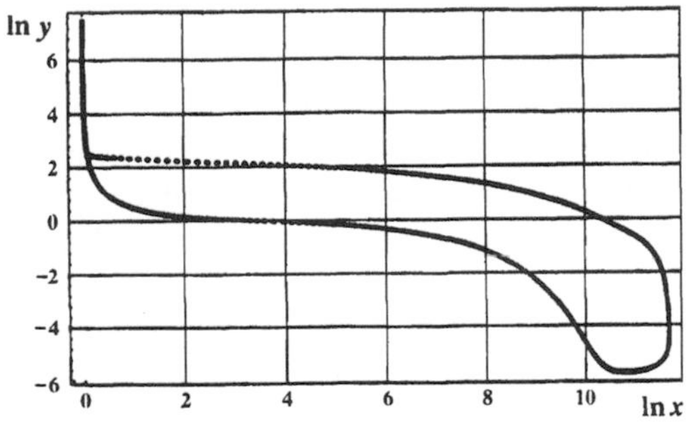

Figure 12.2

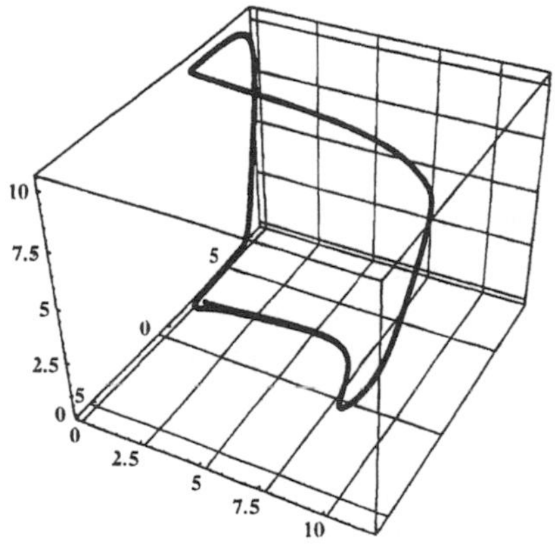

Figure 12.3

FIGURES 12.1 to 12.3 The "Oregonator" is a periodic chemical reaction describable by three nonlinear first-order differential equations. The results (Figure 12.1) illustrate the periodic nature of the major chemical versus time. Figure 12.2 shows the phase diagram of two of the reactants, and Figure 12.3 is the three-dimensional phase diagram of all reactants. The numerical computation was done using a fourth-order Runge-Kuta method on Mathematica by Waltraud Rufeger at the Georgia Institute of Technology.

Ordinary Differential Equations Packages

Also in IMSL, one finds such sophisticated software as DVERK, DGEAR, or DREBS for initial value problems. For two-point boundary value problems, one finds DTPTB (use of DVERK and multiple shooting) or DVCPR.

Partial Differential Equations Packages

DISPL was developed and written at Argonne National Laboratory. DISPL is designed for nonlinear second-order PDEs (parabolic, elliptic, hyperbolic (some cases), and parabolic-elliptic). Boundary conditions of a general nature and material interfaces are allowed. The spatial dimension can be either one or two and in Cartesian, cylindrical, or spherical (one dimension only) geometry. The PDEs are reduced to ordinary DEs by Galerkin discretization of the spatial variables. The resulting ordinary DEs in the timelike variable are then solved by an ODE software package (such as GEAR). Software features include graphics capabilities, printed output, dump/restart/facilities, and free format input. DISPL is intended to be an engineering and scientific tool and is not a finely tuned production code for a small set of problems. DISPL makes no effort to control the spatial discretization errors. It has been used to successfully solve a variety of problems in chemical transport, heat and mass transfer, pipe flow, etc.

PDELIB was developed and written at Los Alamos Scientific Laboratory. PDELIB is a library of subroutines to support the numerical solution of evolution equations with a timelike variable and one or two space variables. The routines are grouped into a dozen independent modules according to their function (i.e., accepting initial data, approximating spatial derivatives, advancing the solution in time). Each task is isolated in a distinct module. Within a module, the basic task is further refined into general-purpose flexible lower-level routines. PDELIB can be understood and used at different levels. Within a small period of time, a large class of problems can be solved by a novice. Moreover, it can provide a wide variety of outputs.

DSS/2 is a differential systems simulator developed at Lehigh University as a transportable numerical method of lines (NMOL) code. See also LEANS.

FORSIM is designed for the automated solution of sets of implicitly coupled PDEs of the form

$$\frac{\partial u_i}{\partial t} = \varphi_i\left(x, t, u_i, u_j, \ldots, (u_i)_x, \ldots, (u_i)_{xx}, (u_j)_{xx}, \ldots\right), \quad \text{for } i = 1, \ldots, N$$

The user specifies the ϕ_i in a simple FORTRAN subroutine. Finite difference formulas of any order may be selected for the spatial discretization and the spatial grid need not be equidistant. The resulting system of time-dependent ODEs is solved by the method of lines.

SLDGL is a program package for the self-adaptive solution of nonlinear systems of elliptic and parabolic PDEs in up to three space dimensions. Variable step size and variable order are permitted. The discretization error is estimated and used for the determination of the optimum grid and optimum orders. This is the most general of the codes described here (not for hyperbolic systems, of course). This package has seen extensive use in Europe.

FIDISOL (finite difference solver) is a program package for nonlinear systems of two-or three-dimensional elliptic and parabolic systems in rectangular domains or in domains that can be transformed analytically to rectangular domains. This package is actually a redesign of parts of SLDGL, primarily for the solution of large problems on vector computers. It has been tested on the CYBER 205, CRAY-IM, CRAY X-MP/22, and VP 200. The program vectorizes very well and uses the vector arithmetic efficiently. In addition to the numerical solution, a reliable error estimate is computed.

CAVE is a program package for conduction analysis via eigenvalues for three-dimensional geometries using the method of lines. In many problems, much time is saved because only a few terms suffice.

Many industrial and university computing services subscribe to the IMSL Software Library. Announcements of new software appear in *Directions*, a publication of IMSL. A brief description of some IMSL packages applicable to PDEs and associated problems is now given. In addition to those packages just described, two additional software packages bear mention. The first of these, the ELLPACK system, solves elliptic problems in two dimensions with general domains and in three dimensions with box-shaped domains. The system contains over 30 numerical methods modules, thereby providing a means of evaluating and comparing different methods for solving elliptic problems. ELLPACK has a special high-level language making it easy to use. New algorithms can be added or deleted from the system with ease.

Second, TWODEPEP is IMSL's general finite element system for two-dimensional elliptic, parabolic, and eigenvalue problems. The Galerkin finite elements available are triangles with quadratic, cubic, or quartic basic functions, with one edge curved when adjacent to a curved boundary, according to the isoparametric method. Nonlinear equatons are solved by Newton's method, with the resulting linear system solved directly by Gauss elimination. PDE/PROTRAN is also available. It uses triangular elements with piecewise polynomials of degree 2, 3, or 4 to solve quite general steady state, time-dependent, and eigenvalue problems in general two-dimensional regions. There is a simple user input. Additional information may be obtained from IMSL. NASTRAN and STRUDL are two advanced finite element computer systems available from a variety of sources. Another, UNAFEM, has been extensively used.

References

General

Adams, E. and Kulisch, U. (Eds.) 1993. *Scientific Computing with Automatic Result Verification*. Academic Press, Boton, MA.

Gerald, C. F. and Wheatley, P. O. 1984. *Applied Numerical Analysis*. Addison-Wesley, Reading, MA.

Hamming, R. W. 1962. *Numerical Methods for Scientists and Engineers*. McGraw-Hill, New York.

Hildebrand, F. B. 1956. *Introduction to Numerical Analysis*. McGraw-Hill, New York.

Isaacson, E. and Keller, H. B. 1966. *Analysis of Numerical Methods*. John Wiley & Sons, New York.

Kopal, Z. 1955. *Numerical Analysis*. John Wiley & Sons, New York.

Rice, J. R. 1993. *Numerical Methods, Software and Analysis*, 2nd ed. Academic Press, Boston, MA.

Stoer, J. and Bulirsch, R. 1976. *Introduction to Numerical Analysis*. Springer, New York.

Linear Equations

Bodewig, E. 1956. *Matrix Calculus*. Wiley (Interscience), New York.

Hageman, L. A. and Young, D. M. 1981. *Applied Iterative Methods*. Academic Press, Boston, MA.

Varga, R. S. 1962. *Matrix Iterative Numerical Analysis*. John Wiley & Sons, New York.

Young, D. M. 1971. *Iterative Solution of Large-Linear Systems*. Academic Press, Boston, MA.

Ordinary Differential Equations

Aiken, R. C. 1985. *Stiff Computation*. Oxford Unitersity Press, New York.

Gear, C. W. 1971. *Numerical Initial Value Problems in Ordinary Differential Equations*. Prentice Hall, Englewood Cliffs, NJ.

Keller, H. B. 1976. *Numerical Solutions of Two Point Boundary Value Problems*. SIAM, Philadelphia, PA.

Lambert, J. D. 1973. *Computational Methods in Ordinary Differential Equations*. Cambridge University Press, New York.

Milne, W.E. 1953. *Numerical Solution of Differential Equations*. John Wiley & Sons, New York.

Rickey, K. C., Evans, H. R., Griffiths, D. W., and Nethercot, D. A. 1983. *The Finite Element Method — A Basic Introduction for Engineers*, 2nd ed. Halstead Press, New York.

Shampine, L. and Gear, C. W. 1979. A User's View of Solving Stiff Ordinary Differential Equatons, *SIAM Rev.* 21:1–17.

Partial Differential Equations

Ames, W. F. 1993. *Numerical Methods for Partial Differential Equations*. 3d ed. Academic Press, Boston, MA.

Brebbia, C. A. 1984. *Boundary Element Techniques in Computer Aided Engineering*. Martinus Nijhoff, Boston, MA.

Burnett, D. S. 1987. *Finite Element Analysis*. Addison-Wesley, Reading, MA.

Lapidus, L. and Pinder, G. F. 1982. *Numerical Solution of Partial Differential Equations in Science and Engineering*. John Wiley & Sons, New York.

Roache, P. 1972. *Computational Fluid Dynamics*, Hermosa, Albuquerque, NM.

13

Experimental Uncertainty Analysis

W.G. Steele
Mississippi State University

H.W. Coleman
University of Alabama

13.1 Introduction

The goal of an experiment is to answer a question by measuring a specific variable, X_i, or by determining a result, r, from a functional relationship among measured variables

$$r = r(X_1, X_2, \ldots, X_i, \ldots, X_J) \tag{13.1}$$

In all experiments there is some error that prevents the measurement of the true value of each variable, and therefore, prevents the determination of r_{true}.

Uncertainty analysis is a technique that is used to estimate the interval about a measured variable or a determined result within which the true value is thought to lie with a certain degree of confidence. As discussed by Coleman and Steele (1989), uncertainty analysis is an extremely useful tool for all phases of an experimental program from initial planning (general uncertainty analysis) to detailed design, debugging, test operation, and data analysis (detailed uncertainty analysis).

The application of uncertainty analysis in engineering has evolved considerably since the classic paper of Kline and McClintock (1953). Developments in the field have been especially rapid and significant over the past decade, with the methods formulated by Abernethy and co-workers (1985) that were incorporated into ANSI/ASME Standards in (1984) and (1986) being superseded by the more rigorous approach presented in the International Organization for Standardization (ISO) *Guide to the Expression of Uncertainty in Measurement* (1993). This guide, published in the name of ISO and six other international organizations, has in everything but name established a new international experimental uncertainty standard.

The approach in the ISO *Guide* deals with "Type A" and "Type B" categories of uncertainties, not the more traditional engineering categories of systematic (bias) and precision (random) uncertainties, and is of sufficient complexity that its application in normal engineering practice is unlikely. This issue has been addressed by AGARD Working Group 15 on Quality Assessment for Wind Tunnel Testing, by the Standards Subcommittee of the AIAA Ground Test Technical Committee, and by the ASME Committee PTC 19.1 that is revising the ANSI/ASME Standard (1986). The documents issued by two of these groups (AGARD-AR-304, 1994) and (AIAA Standard S-071-1995, 1995) and in

0-8493-0056-8/00/$0.00+$.50

preparation by the ASME Committee present and discuss the additional assumptions necessary to achieve a less complex "large sample" methodology that is consistent with the ISO *Guide*, that is applicable to the vast majority of engineering testing, including most single-sample tests, and that retains the use of the traditional engineering concepts of systematic and precision uncertainties. The range of validity of this "large sample" approximation has been presented by Steele et al. (1994) and by Coleman and Steele (1995). The authors of this section are also preparing a second edition of Coleman and Steele (1989), which will incorporate the ISO *Guide* methodology and will illustrate its use in all aspects of engineering experimentation.

In the following, the uncertainties of individual measured variables and of determined results are discussed. This section concludes with an overview of the use of uncertainty analysis in all phases of an experimental program.

13.2 Uncertainty of a Measured Variable

For a measured variable, X_i, the total error is caused by both precision (random) and systematic (bias) errors. This relationship is shown in Figure 13.1. The possible measurement values of the variable are

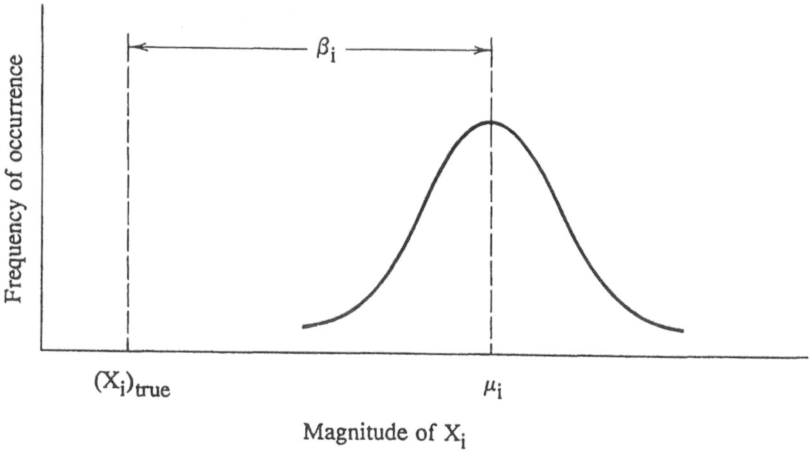

Figure 13.1 Errors in the measurement of variable X_i.

scattered in a distribution (here assumed Gaussian) around the parent population mean, μ_i. The parent population mean differs from $(X_i)_{true}$ by an amount called the systematic (or bias) error, β_i. The quantity β_i is the total fixed error that remains in the measurement process after all calibration corrections have been made. In general, there will be several sources of bias error such as calibration standard errors, data acquisition errors, data reduction errors, and test technique errors. There is usually no direct way to measure these errors, so they must be estimated.

For each bias error source, $(\beta_i)_k$, the experimenter must estimate a systematic uncertainty (or bias limit), $(B_i)_k$, such that there is about a 95% confidence that $(B_i)_k \geq |(\beta_i)_k|$. Systematic uncertainties are usually estimated from previous experience, calibration data, analytical models, and the application of sound engineering judgment. For each variable, there will be a set, K_i, of elemental systematic uncertainties, $(B_i)_k$, for the significant fixed error sources. The overall systematic uncertainty for variable X_i is determined from these estimates as

$$B_i^2 = \sum_{k=1}^{K_i} (B_i)_k^2 \tag{13.2}$$

For a discussion on estimating systematic uncertainties (bias limits), see Coleman and Steele (1989).

The estimate of the precision error for a variable is the sample standard deviation, or the estimate of the error associated with the repeatability of a particular measurement. Unlike the systematic error, the precision error varies from reading to reading. As the number of readings, N_i, of a particular variable tends to infinity, the distribution of these readings becomes Gaussian.

The readings used to calculate the sample standard deviation for each variable must be taken over the time frame and conditions which cover the variation in the variable. For example, taking multiple samples of data as a function of time while holding all other conditions constant will identify the random variation associated with the measurement system and the unsteadiness of the test condition. If the sample standard deviation of the variable being measured is also expected to be representative of other possible variations in the measurement, e.g., repeatability of test conditions, variation in test configuration, then these additional error sources will have to be varied while the multiple data samples are taken to determine the standard deviation.

When the value of a variable is determined as the mean, $\bar{X}_i$, of N_i readings, then the sample standard deviation of the mean is

$$S_{\bar{X}_i} = \left\{ \left[\frac{1}{N_i(N_i - 1)} \right] \sum_{k=1}^{N_i} \left[(X_i)_k - \bar{X}_i \right]^2 \right\}^{1/2} \quad (13.3)$$

where

$$\bar{X}_i = \frac{\sum_{k=1}^{N_i} (X_i)_k}{N_i} \quad (13.4)$$

It must be stressed that these N_i readings have to be taken over the appropriate range of variations for X_i as described above.

When only a single reading of a variable is available so that the value used for the variable is X_i, then N_{P_i} previous readings, $(X_{P_i})_k$, must be used to find the standard deviation for the variable as

$$S_{X_i} = \left\{ \frac{1}{N_{P_i} - 1} \sum_{k=1}^{N_{P_i}} \left[(X_{P_i})_k - \bar{X}_{P_i} \right]^2 \right\}^{1/2} \quad (13.5)$$

where

$$\bar{X}_{P_i} = \frac{1}{N_{P_i}} \sum_{k=1}^{N_{P_i}} (X_{P_i})_k \quad (13.6)$$

Another situation where previous readings of a variable are useful is when a small current sample size, N_i, is used to calculate the mean value, $\bar{X}_i$, of a variable. If a much larger set of previous readings for the same test conditions is available, then it can be used to calculate a more appropriate standard deviation for the variable (Steele et al., 1993) as

$$S_{\bar{X}_i} = \frac{S_{X_i}}{\sqrt{N_i}} \quad (13.7)$$

where N_i is the number of current readings averaged to determine $\bar{X}_i$, and S_{X_i} is computed from N_{P_i} previous readings using Equation (13.5). Typically, these larger data sets are taken in the early "shake-down" or "debugging" phases of an experimental program.

For many engineering applications, the "large sample" approximation applies, and the uncertainty for variable i (X_i or $\bar{X}_i$) is

$$U_i = \sqrt{B_i^2 + (2S_i)^2} \tag{13.8}$$

where S_i is found from the applicable Equation (13.3), (13.5) or (13.7). The interval $X_i \pm U_i$ or $\bar{X}_i$ + U_i, as appropriate, should contain $(X_i)_{true}$ 95 times out of 100. If a small number of samples (N_i or N_{P_i} < 10) is used to determine S_{X_i} or $S_{\bar{X}_i}$, then the "large sample" approximation may not apply and the methods in ISO (1993) or Coleman and Steele (1995) should be used to find U_i.

13.3 Uncertainty of a Result

Consider an experimental result that is determined for J measured variables as

$$r = r(X_1, X_2, \ldots, X_i, \ldots, X_J)$$

where some variables may be single readings and others may be mean values. A typical mechanical engineering experiment would be the determination of the heat transfer in a heat exchanger as

$$q = \dot{m}c_p(T_o - T_i) \tag{13.9}$$

where q is the heat rate, $\dot{m}$ is the flow rate, c_p is the fluid specific heat, and T_o and T_i are the heated fluid outlet and inlet temperatures, respectively. For the "large sample" approximation, U_r is found as

$$U_r = \sqrt{B_r^2 + (2S_r)^2} \tag{13.10}$$

where B_r is the systematic uncertainty of the result

$$B_r^2 = \sum_{i=1}^{J} (\theta_i B_i)^2 + 2\sum_{i=1}^{J-1} \sum_{k=i+1}^{J} \theta_i \theta_k B_{ik} \tag{13.11}$$

with

$$\theta_i = \frac{\partial r}{\partial X_i} \tag{13.12}$$

and S_r is the standard deviation of the result

$$S_r^2 = \sum_{i=1}^{J} (\theta_i S_i)^2 \tag{13.13}$$

The term B_{ik} in Equation (13.11) is the covariance of the systematic uncertainties. When the elemental systematic uncertainties for two separately measured variables are related, for instance when the transducers used to measure different variables are each calibrated against the same standard, the systematic uncertainties are said to be correlated and the covariance of the systematic errors is nonzero. The significance of correlated systematic uncertainties is that they can have the effect of either decreasing or increasing the uncertainty in the result. B_{ik} is determined by summing the products of the elemental systematic uncertainties for variables i and k that arise from the same source and are therefore perfectly correlated (Brown et al., 1996) as

$$B_{ik} = \sum_{\alpha=1}^{L} (B_i)_\alpha (B_k)_\alpha \tag{13.14}$$

where L is the number of elemental systematic error sources that are common for measurements X_i and X_k.

If, for example,

$$r = r(X_1, X_2) \tag{13.15}$$

and it is possible for portions of the systematic uncertainties B_1 and B_2 to arise from the same source(s). Equation (13.11) gives

$$B_r^2 = \theta_1^2 B_1^2 + \theta_2^2 B_2^2 + 2\theta_1\theta_2 B_{12} \tag{13.16}$$

For a case in which the measurements of X_1 and X_2 are each influenced by four elemental systematic error sources and sources two and three are the same for both X_1 and X_2, Equation (13.2) gives

$$B_1^2 = (B_1)_1^2 + (B_1)_2^2 + (B_1)_3^2 + (B_1)_4^2 \tag{13.17}$$

and

$$B_2^2 = (B_2)_1^2 + (B_2)_2^2 + (B_2)_3^2 + (B_2)_4^2 \tag{13.18}$$

while Equation (13.14) gives

$$B_{12} = (B_1)_2 (B_2)_2 + (B_1)_3 (B_2)_3 \tag{13.19}$$

In the general case, there would be additional terms in the expression for the standard deviation of the result, S_r, (Equation 13.13) to take into account the possibility of precision errors in different variables being correlated. These terms have traditionally been neglected, although precision errors in different variables caused by the same uncontrolled factor(s) are certainly possible and can have a substantial impact on the value of S_r (Hudson et al., 1996). In such cases, one would need to acquire sufficient data to allow a valid estimate of the precision covariance terms using standard statistical techniques (ISO, 1993). Note, however, that if multiple test results over an appropriate time period are available, these can be used to directly determine S_r. This value of the standard deviation of the result implicitly includes the correlated error effect.

If a test is performed so that M multiple sets of measurements $(X_1, X_2, \ldots, X_J)_k$ at the same test condition are obtained, then M results can be determined using Equation (13.1) and a mean result, $\bar{r}$, can be determined using

$$\bar{r} = \frac{1}{M} \sum_{k=1}^{M} r_k \tag{13.20}$$

The standard deviation of the sample of M results, S_r, is calculated as

$$S_r = \left[\frac{1}{M-1}\sum_{k=1}^{M}(r_k - \bar{r})^2\right]^{1/2} \tag{13.21}$$

The uncertainty associated with the mean result, $\bar{r}$, for the "large sample" approximation is then

$$U_{\bar{r}} = \sqrt{B_r^2 + (2S_{\bar{r}})^2} \tag{13.22}$$

where

$$S_{\bar{r}} = \frac{S_r}{\sqrt{M}} \tag{13.23}$$

and where B_r is given by Equation (13.11).

The "large sample" approximation for the uncertainty of a determined result (Equations (13.10) or (13.22)) applies for most engineering applications even when some of the variables have fewer than 10 samples. A detailed discussion of the applicability of this approximation is given in Steele et al. (1994) and Coleman and Steele (1995).

The determination of U_r from S_r (or $S_{\bar{r}}$) and B_r using the "large sample" approximation is called detailed uncertainty analysis (Coleman and Steele, 1989). The interval r (or $\bar{r}$) $\pm\, U_r$ (or $U_{\bar{r}}$) should contain r_{true} 95 times out of 100. As discussed in the next section, detailed uncertainty analysis is an extremely useful tool in an experimental program. However, in the early stages of the program, it is also useful to estimate the overall uncertainty for each variable, U_i. The overall uncertainty of the result is then determined as

$$U_r^2 = \sum_{k=1}^{J}(\theta_i U_i)^2 \tag{13.24}$$

This determination of U_r is called general uncertainty analysis.

13.4 Using Uncertainty Analysis in Experimentation

The first item that should be considered in any experimental program is "What question are we trying to answer?" Another key item is how accurately do we need to know the answer, or what "degree of goodness" is required? With these two items specified, general uncertainty analysis can be used in the planning phase of an experiment to evaluate the possible uncertainties from the various approaches that might be used to answer the question being addressed. Critical measurements that will contribute most to the uncertainty of the result can also be identified.

Once past the planning, or preliminary design phase of the experiment, the effects of systematic errors and precision errors are considered separately using the techniques of detailed uncertainty analysis. In the design phase of the experiment, estimates are made of the systematic and precision uncertainties, B_r and $2S_r$, expected in the experimental result. These detailed design considerations guide the decisions made during the construction phase of the experiment.

After the test is constructed, a debugging phase is required before production tests are begun. In the debugging phase, multiple tests are run and the precision uncertainty determined from them is compared with the $2S_r$ value estimated in the design phase. Also, a check is made to see if the test results plus and minus U_r compare favorably with known results for certain ranges of operation. If these checks are not successful, then further test design, construction, and debugging is required.

Once the test operation is fully understood, the execution phase can begin. In this phase, balance checks can be used to monitor the operation of the test apparatus. In a balance check, a quantity, such as flow rate, is determined by different means and the difference in the two determinations, z, is compared to the ideal value of zero. For the balance check to be satisfied, the quantity z must be less than or equal to U_z.

Uncertainty analysis will of course play a key role in the data analysis and reporting phases of an experiment. When the experimental results are reported, the uncertainties should be given along with the systematic uncertainty, B_r, the precision uncertainty, $2S_r$, and the associated confidence level, usually 95%.

References

Abernethy, R.B., Benedict, R.P., and Dowdell, R.B. 1985. ASME Measurement Uncertainty. *J. Fluids Eng.*, 107, 161–164.

AGARD-AR-304. 1994. *Quality Assessment for Wind Tunnel Testing.* AGARD, Neuilly Sur Seine, France.

AIAA Standard S-071-1995. 1995. *Assessment of Wind Tunnel Data Uncertainty.* AIAA, Washington, D.C.

ANSI/ASME MFC-2M-1983. 1984. *Measurement Uncertainty for Fluid Flow in Closed Conduits.* ASME, New York.

ANSI/ASME PTC 19.1-1985, Part 1. 1986. *Measurement Uncertainty.* ASME, New York.

Coleman, H.W. and Steele, W.G. 1989. *Experimentation and Uncertainty Analysis for Engineers.* John Wiley & Sons, New York.

Coleman, H.W. and Steele, W.G. 1995. Engineering Application of Experimental Uncertainty Analysis. *AIAA Journal*, 33(10), 1888–1896.

Brown, K.B., Coleman, H.W., Steele, W.G., and Taylor, R.P. 1996. Evaluation of Correlated Bias Approximations in Experimental Uncertainty Analysis. *AIAA J.*, 34(5), 1013–1018.

Hudson, S.T., Bordelon, Jr., W.J., and Coleman, H.W. 1996. Effect of Correlated Precision Errors on the Uncertainty of a Subsonic Venturi Calibration. *AIAA J.*, 34(9), 1862–1867.

Kline, S.J. and McClintock, F.A. 1953. Describing Uncertainties in Single-Sample Experiments. *Mech. Eng.*, 75, 3–8.

ISO. 1993. *Guide to the Expression of Uncertainty in Measurement.* ISO, Geneva, Switzerland.

Steele, W.G., Taylor, R.P., Burrell, R.E., and Coleman, H.W. 1993. Use of Previous Experience to Estimate Precision Uncertainty of Small Sample Experiments. *AIAA J.*, 31(10), 1891–1896.

Steele, W.G., Ferguson, R.A., Taylor, R.P., and Coleman, H.W. 1994. Comparison of ANSI/ASME and ISO Models for Calculation of Uncertainty. *ISA Trans.*, 33, 339–352.

14

Chaos

R. L. Kautz
National Institute of Standards and Technology

14.1 Introduction

Since the time of Newton, the science of dynamics has provided quantitative descriptions of regular motion, from a pendulum's swing to a planet's orbit, expressed in terms of differential equations. However, the role of Newtonian mechanics has recently expanded with the realization that it can also describe chaotic motion. In elementary terms, **chaos** can be defined as **pseudorandom** behavior observed in the steady-state dynamics of a deterministic **nonlinear system**. How can motion be pseudorandom, or random according to statistical tests and yet be entirely predictable? This is just one of the paradoxes of chaotic motion, which is globally stable but locally unstable, predictable in principle but not in practice, and geometrically complex but derived from simple equations.

The strange nature of chaotic motion was first understood by Henri Poincaré, who established the mathematical foundations of chaos in a treatise published in 1890 (Holmes, 1990). However, the practical importance of chaos began to be widely appreciated only in the 1960s, beginning with the work of Edward Lorenz (1963), a meteorologist who discovered chaos in a simple model for fluid convection. Today, chaos is understood to explain a wide variety of apparently random natural phenomena, ranging from dripping faucets (Martien et al., 1985), to the flutter of a falling leaf (Tanabe and Kaneko, 1994), to the irregular rotation of a moon of Saturn (Wisdom et al., 1984).

Although chaos is used purposely to provide an element of unpredictability in some toys and carnival rides (Kautz and Huggard, 1994), it is important from an engineering point of view primarily as a phenomenon to be avoided. Perhaps the simplest scenario arises when a nonlinear mechanism is used to achieve a desired effect, such as the synchronization of two oscillators. In many such cases, the degree of nonlinearity must be chosen carefully: strong enough to ensure the desired effect but not so strong that chaos results. In another scenario, an engineer might be required to deal with an intrinsically chaotic system. In this case, if the system can be modeled mathematically, then a small feedback signal can often be applied to eliminate the chaos (Ott et al., 1990). For example, low-energy feedback has been used to suppress chaotic behavior in a thermal convection loop (Singer et al., 1991). As such considerations suggest, chaos in rapidly becoming an important topic for engineers.

14.2 Flows, Attractors, and Liapunov Exponents

Dynamic systems can generally be described mathematically in terms of a set of differential equations of the form.

0-8493-0056-8/00/$0.00+$.50

$$d\mathbf{x}(t)/dt = \mathbf{F}[\mathbf{x}(t)] \tag{14.1}$$

where $x = (x_1, \ldots, x_N)$ is an N-dimensional vector called the **state vector** and the vector function $\mathbf{F} = F_1(\mathbf{x}), \ldots, F_N(\mathbf{x}))$ defines how the state vector changes with time. In mechanics, the state variables x_i are typically the positions and velocities associated with the potential and kinetic energies of the system. Because the state vector at times $t > 0$ depends only on the initial state vector $\mathbf{x}(0)$, the system defined by Equation (14.1) is deterministic, and its motion is in principle exactly predictable.

The properties of a dynamic system are often visualized most readily in terms of trajectories $\mathbf{x}(t)$ plotted in **state space**, where points are defined by the coordinates $(x_1, \ldots, x_N)$. As an example, consider the motion of a damped pendulum defined by the normalized equation

$$d^2\theta/dt^2 = -\sin\theta - \rho d\theta/dt \tag{14.2}$$

which expresses the angular acceleration $d^2\theta/dt^2$ in terms of the gravitational torque $-\sin\theta$ and a damping torque $-\rho d\theta/dt$ proportional to the angular velocity $v = d\theta/dt$. If we define the state vector as $x = (x_1, x_2) = (\theta, v)$, then Equation (14.2) can be written in the form of Equation (14.1) with $\mathbf{F} = (x_2, -\sin x_1 - \rho x_2)$. In this case, the state space is two dimensional, and a typical trajectory is a spiral, as shown in Figure 14.1 for the initial condition $x(0) = (0,1)$. If additional trajectories, corresponding to other initial conditions, were plotted in Figure 14.1, we would obtain a set of interleaved spirals, all converging on the point $\mathbf{x} = (0,0)$. Because the direction of a trajectory passing through a given point is uniquely defined by Equation (14.1), state–space trajectories can never cross, and, by analogy with the motion of a fluid, the set of all of trajectories is called a flow.

The tendency of a flow to converge toward a single point or other restricted subset of state space is characteristic of dissipative systems like the damped pendulum. Such an asymptotic set, called an attracting set or **attractor**, can be a fixed point (for which $\mathbf{F}(\mathbf{x}) = 0$) as in Figure 14.1, but might also be a periodic or chaotic trajectory. The convergence of neighboring trajectories is suggested in Figure 14.1 by a series of ellipses spaced at time intervals $\Delta t = 1.5$ that track the flow of all trajectories originating within the circle specified at $t = 0$. In general, the contraction of an infinitesimal state–space volume V as it moves with the flow is given by

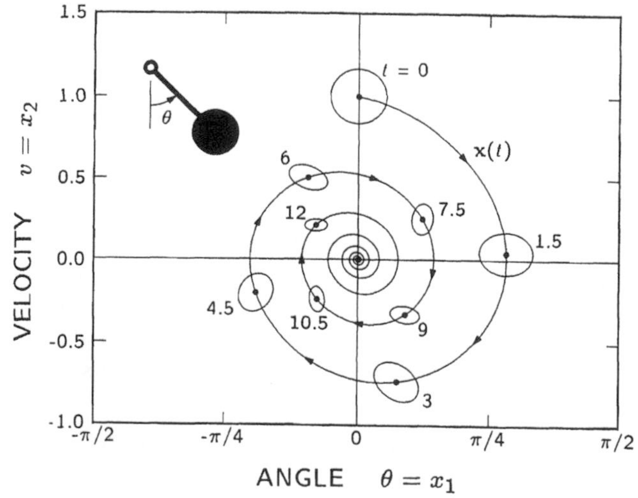

Figure 14.1 The state-space trajectory $\mathbf{x}(t)$ for a pendulum with a damping coefficient $\rho = 0.2$ for the initial condition $\mathbf{x}(0) = (0,1)$. The evolution of trajectories initialized in a small circle surrounding $\mathbf{x} = (0,1)$ is indicated by the ellipses plotted at time intervals of $\Delta t = 1.5$.

$$V^{-1}\partial V/\partial t = \nabla \cdot \mathbf{F} \tag{14.3}$$

where $\nabla \cdot \mathbf{F} = \Sigma_{i=1}^{N} \partial F_i/x_i$ is the divergence of **F**. For the damped pendulum, $\nabla \cdot \mathbf{F} = \rho$, so the area of the ellipse shown in Figure 14.1 shrinks exponentially as $V(t) = V(0)\exp(-\rho t)$. The contraction of state–space volumes explains the existence of attractors in dissipative systems, but in conservative systems such as the pendulum with $\rho = 0$, state–space volumes are preserved, and trajectories are instead confined to constant-energy surfaces.

While the existence of chaotic behavior is generally difficult to predict, two essential conditions are easily stated. First, the complex topology of a chaotic trajectory can exist only in a state–space of dimension $N \geq 3$. Thus, the pendulum defined by Equation (14.2) cannot be chaotic because $N = 2$ for this system. Second, a system must be nonlinear to exhibit chaotic behavior. Linear systems, for which any linear combination $c_1x_a(t) + c_2x_b(t)$ of two solutions $x_a(t)$ and $x_b(t)$ is also a solution, are mathematically simple and amenable to analysis. In contrast, nonlinear systems are noted for their intractability. Thus, chaotic behavior is of necessity explored more frequently by numerical simulation than mathematical analysis, a fact that helps explain why the prevalence of chaos was discovered only after the advent of efficient computation.

A useful criterion for the existence of chaos can be developed from an analysis of a trajectory's local stability. As sketched in Figure 14.2, the local stability of a trajectory x(t) is determined by considering a neighboring trajectory $\tilde{x}(t)$ initiated by an infinitesimal deviation e(t_0) from x(t) at time t_0. The deviation vector e(t) = $\tilde{x}(t)$ – x(t) at times $t_1 > t_0$ can be expressed in terms of the Jacobian matrix

$$J_{ij}(t_1,t_0) = \partial x_i(t_1)/\partial x_j(t_0) \tag{14.4}$$

which measures the change in state variable x_i at time t_1 due to a change in x_j at time t_0. From the Jacobian's definition, we have $\mathbf{e}(t_1) = \mathbf{J}(t_1,t_0)\mathbf{e}(t_0)$. Although the local stability of $\mathbf{x}(t)$ is determined simply by whether deviations grow or decay in time, the analysis is complicated by the fact that deviation vectors can also rotate, as suggested in Figure 14.2. Fortunately, an arbitrary deviation can be written in terms of the eigenvectors $\mathbf{e}^{(i)}$ of the Jacobian, defined by

$$\mathbf{J}(t_1,t_0)\mathbf{e}^{(i)} = \mu_i(t_1,t_0)\mathbf{e}^{(i)} \tag{14.5}$$

which are simply scaled by the eigenvalues $\mu_i(t_1,t_0)$ without rotation. Thus, the N eigenvalues of the Jacobian matrix provide complete information about the growth of deviations. Anticipating that the asymptotic growth will be exponential in time, we define the **Liapunov exponents**,

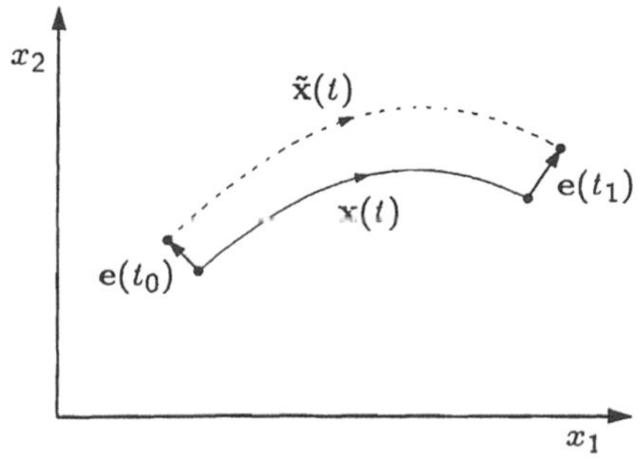

Figure 14.2 A trajectory $\mathbf{x}(t)$ and a neighboring trajectory $\tilde{\mathbf{x}}(t)$ plotted in state space from time t_0 to t_1. The vectors $\mathbf{e}(t_0)$ and $\mathbf{e}(t_1)$ indicate the deviation of $\tilde{\mathbf{x}}(t)$ from $\mathbf{x}(t)$ at times t_0 and t_1.

$$\lambda_i = \lim_{t_1 \to \infty} \frac{\ln|\mu_i(t_1, t_0)|}{t_1 - t_0} \tag{14.6}$$

Because any deviation can be broken into components that grow or decay asymptotically as $\exp(\lambda_i t)$, the N exponents associated with a trajectory determine its local stability.

In dissipative systems, chaos can be defined as motion on an attractor for which one or more Liapunov exponents are positive. Chaotic motion thus combines global stability with local instability in that motion is confined to the attractor, generally a bounded region of state space, but small deviations grow exponentially in time. This mixture of stability and instability in chaotic motion is evident in the behavior of an infinitesimal deviation ellipsoid similar to the finite ellipse shown in Figure 14.1. Because some λ_i are positive, an ellipsoid centered on a chaotic trajectory will expand exponentially in some directions. On the other hand, because state-space volumes always contract in dissipative systems and the asymptotic volume of the ellipsoid scales as $\exp(\Lambda t)$, where $\Lambda = \Sigma_{i=1}^{N} \lambda_i$, the sum of the negative exponents must be greater in magnitude than the sum of the positive exponents. Thus, a deviation ellipsoid tracking a chaotic trajectory expands in some directions while contracting in others. However, because an arbitrary deviation almost always includes a component in a direction of expansion, nearly all trajectories neighboring a chaotic trajectory diverge exponentially.

According to our definition of chaos, neighboring trajectories must diverge exponentially and yet remain on the attractor. How is this possible? Given that the attractor is confined to a bounded region of state space, perpetual divergence can occur only for trajectories that differ infinitesimally. Finite deviations grow exponentially at first but are limited by the bounds of the chaotic attractor and eventually shrink again. The full picture can be understood by following the evolution of a small state-space volume selected in the neighborhood of the chaotic attractor. Initially, the volume expands in some directions and contracts in others. When the expansion becomes too great, however, the volume begins to fold back on itself so that trajectories initially separated by the expansion are brought close together again. As time passes, this stretching and folding is repeated over and over in a process that is often likened to kneading bread or pulling taffy.

Because all neighboring volumes approach the attractor, the stretching and folding process leads to an attracting set that is an infinitely complex filigree of interleaved surfaces. Thus, while the differential equation that defines chaotic motion can be very simple, the resulting attractor is highly complex. Chaotic attractors fall into a class of geometric objects called **fractals**, which are characterized by the presence of structure at arbitrarily small scales and by a dimension that is generally fractional. While the existence of objects with dimensions falling between those of a point and a line, a line and a surface, or a surface and a volume may seem mysterious, fractional dimensions result when dimension is defined by how much of an object is apparent at various scales of resolution. For the dynamical systems encompassed by Equation (14.1), the fractal dimension D of a chaotic attractor falls in the range of $2 < D < N$ where N is the dimension of the state space. Thus, the dimension of a chaotic attractor is large enough that trajectories can continually explore new territory within a bounded region of state space but small enough that the attractor occupies no volume of the space.

14.3 Synchronous Motor

As an example of a system that exhibits chaos, we consider a simple model for a synchronous motor that might be used in a clock. As shown in Figure 14.3, the motor consists of a permanent-magnet rotor subjected to a uniform oscillatory magnetic field $B \sin t$ provided by the stator. In dimensionless notation, its equation of motion is

$$d^2\theta/dt^2 = -f \sin t \sin\theta - \rho d\theta/dt \tag{14.7}$$

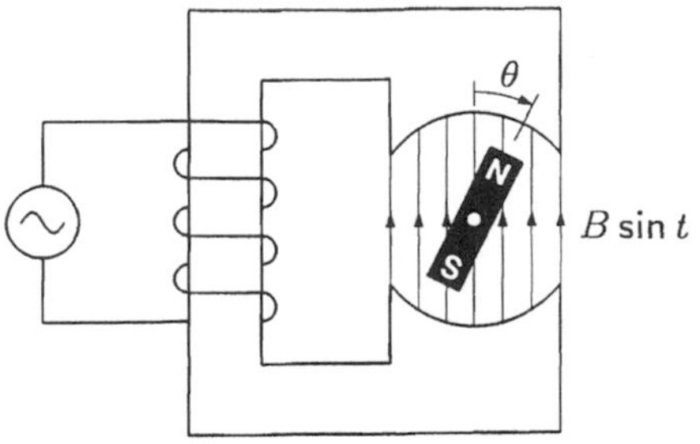

Figure 14.3 A synchronous motor, consisting of a permanent magnet free to rotate in a uniform magnetic field $B \sin t$ with an amplitude that varies sinusoidally in time.

where $d^2\theta/dt^2$ is the angular acceleration of the rotor, $-f \sin t \sin \theta$ is the torque due to the interaction of the rotor's magnetic moment with the stator field, and $-\rho d\theta/dt$ is a viscous damping torque. Although Equation (14.7) is explicitly time dependent, it can be case in the form of Equation (14.1) by defining the state vector as $\mathbf{x} = (x_1, x_2, x_3) = (\theta, \nu, t)$, where $\nu = d\theta/dt$ is the angular velocity, and by defining the flow as $\mathbf{F} = (x_2, -f \sin x_3 \sin x_1 - \rho x_2, 1)$. The state space is thus three dimensional and large enough to allow chaotic motion. Equation (14.7) is also nonlinear due to the term $-f \sin t \sin \theta$, since $\sin(\theta_a + \theta_b)$ is not generally equal to $\sin \theta_a + \sin \theta_b$. Chaos in this system has been investigated by several authors (Ballico et al., 1990).

By intent, the motor uses nonlinearity to synchronize the motion of the rotor with the oscillatory stator field, so it evolves exactly once during each field oscillation. Although synchronization can occur over a range of system parameters, proper operation requires that the drive amplitude f, which measures the strength of the nonlinearity, be chosen large enough to produce the desired rotation but not so large that chaos results. Calculating the motor's dynamics for $\rho = 0.2$, we find that the rotor oscillates without rotating for f less than 0.40 and that the intended rotation is obtained for $0.40 < \rho < 1.87$. The periodic attractor corresponding to synchronized rotation is shown for $f = 1$ in Figure 14.4(a). Here the three-dimensional state-space trajectory is projected onto the (x_1, x_2) or (θ, ν) plane, and a dot marks the point in the rotation cycle at which $t = 0$ modulo 2π. As Figure 14.4(a) indicates, the rotor advances by exactly 2π during each drive cycle.

The utility of the motor hinges on the stability of the synchronous rotation pattern shown in Figure 14.4(a). This periodic pattern is the steady–state motion that develops after initial transients decay and represents the final asymptotic trajectory resulting for initial conditions chosen from a wide area of state space. Because the flow approaches this attracting set from all neighboring points, the effect of a perturbation that displaces the system from the attractor is short lived. This stability is reflected in the Liapunov exponents of the attractor: $\lambda_1 = 0$ and $\lambda_2 = \lambda_3 = -0.100$. The zero exponent is associated with deviations coincident with the direction of the trajectory and is a feature common to all bounded attractors other than fixed points. The zero exponent results because the system is neutrally stable with respect to offsets in the time coordinate. The exponents of -0.100 are associated with deviations transverse to the trajectory and indicate that these deviations decay exponentially with a characteristic time of 1.6 drive cycles. The negative exponents imply that the synchrony between the rotor and the field is maintained in spite of noise or small variations in system parameters, as required of a clock motor.

For drive amplitudes greater than $f = 1.87$, the rotor generally does not advance by precisely 2π during every drive cycle, and its motion is commonly chaotic. An example of chaotic behavior is illustrated for $f = 3$ by the trajectory plotted in Figure 14.4(b) over an interval of 10 drive cycles. In this figure, sequentially numbered dots mark he beginning of each drive cycle. When considered cycle by cycle, the trajectory proves to be a haphazard sequence of oscillations, forward rotations, and reverse rotations. Although we might suppose that this motion is just an initial transient, it is instead characteristic of the steady–state behavior of the motor. If extended, the trajectory continued with an apparently random mixture of oscillation and rotation, without approaching a repetitive cycle. The motion is aptly described as chaotic.

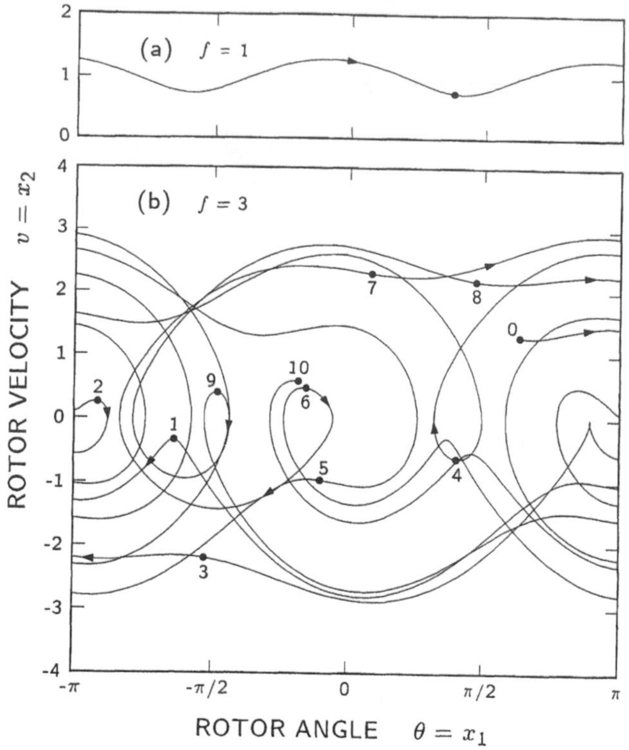

Figure 14.4 State-space trajectories projected onto the (x_1,x_2) or (θ,v) plane, showing attractors of the synchronous motor for $\rho = 0.2$ and two drive amplitudes, $f = 1$ and 3. Dots mark the state of the system at the beginning of each drive cycle ($t = 0$ modulo 2π). The angles $\theta = \pi$ and $-\pi$ are equivalent.

The geometry of the chaotic attractor sampled in Figure 14.4(b) is revealed more fully in Figure 14.5. Here we plot points (θ,v) recording the instantaneous angle and velocity of the rotor at the beginning of each drive cycle for 100,000 successive cycles. Figure 14.5 displays the three-dimensional attractor called a **Poincaré section**, at its intersection with the planes $t = x_3 = 0$ modulo 2π, corresponding to equivalent times in the drive cycle. For the periodic attractor of Figure 14.4(a), the rotor returns to the same position and velocity at the beginning of each drive cycle, so its Poincaré section is a single point, the dot in this figure. For chaotic motion, in contrast, we obtain the complex swirl of points shown in Figure 14.5. If the system is initialized at a point far from the swirl, the motion quickly converges to this attracting set. On succeeding drive cycles, the state of the system jumps from one part of the swirl to another in an apparently random fashion that continues indefinitely. As the number of plotted points approaches infinity, the swirl becomes a cross section of the chaotic attractor. Thus, Figure 14.5 approximates a slice through the infinite filigree of interleaved surfaces that compose the attracting set. In this case, the fractal dimension of the attractor is 2.52 and that of its Poincaré section is 1.52.

The computed Liapunov exponents of the chaotic solution at $\rho = 0.2$ and $f = 3$ are $\lambda_1 = 0$, $\lambda_2 = 0.213$, and $\lambda_3 = -0.413$. As for the periodic solution, the zero exponent implies neutral stability associated with deviations directed along a given trajectory. The positive exponent, which signifies the presence of chaos, is associated with deviations transverse to the given trajectory but tangent to the surface of the attracting set in which it is embedded. The positive exponent implies that such deviations grow exponentially in time and that neighboring trajectories on the chaotic attractor diverge exponentially, a property characteristic of chaotic motion. The negative exponent is associated with deviations transverse to the surface of the attractor and assures the exponential decay of displacements from the attracting set. Thus, the Liapunov exponents reflect both the stability of the chaotic attractor and the instability of a given chaotic trajectory with respect to neighboring trajectories.

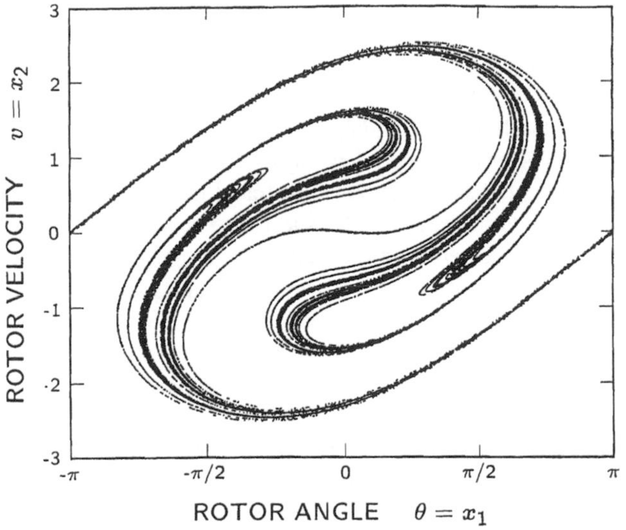

Figure 14.5 Poincaré section of a chaotic attractor of the synchronous motor with $\rho = 0.2$ and $f = 3$, obtained by plotting points $(x_1, x_2) = (\theta, v)$ corresponding to the position and velocity of the rotor at the beginning of 100,000 successive drive cycles.

One sequence of a positive Liapunov exponent is a practical limitation on our ability to predict the future state of a chaotic system. This limitation is illustrated in Figure 14.6, where we plot a given chaotic trajectory (solid line) and three perturbed trajectories (dashed lines) that result by offsetting the initial phase of the given solution by various deviations $e_1(0)$. When the initial angular offset is $e_1(0) = 10^{-3}$ radian, the perturbed trajectory (short dash) closely tracks the given trajectory for about seven drive cycles before the deviation become significant. After seven drive cycles, the perturbed trajectory is virtually independent of the given trajectory, even though it is confined to the same attractor. Similarly, initial offsets of 10^{-6} and 10^{-9} radian lead to perturbed trajectories (medium and long dash) that track the given trajectory for about 12 and 17 drive cycles, respectively, before deviations become significant. These results reflect the fact that small deviations grow exponentially and, in the present case, increase on average by a factor of 10 every 1.7 drive cycles. If the position of the rotor is to be predicted with

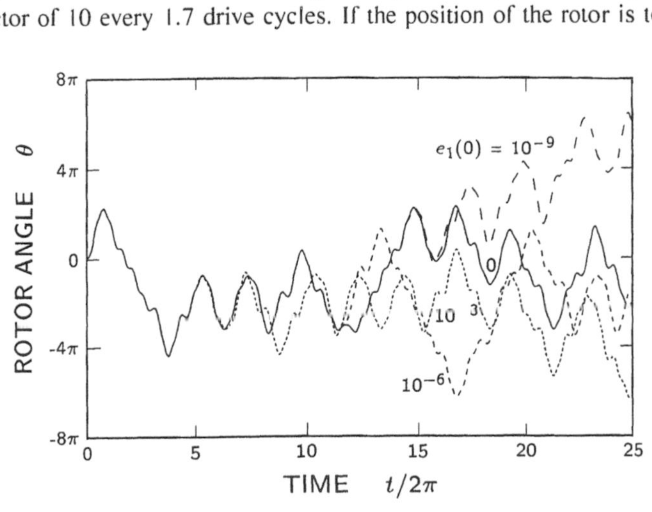

Figure 14.6 Rotor angle as a function of time for chaotic trajectories of the synchronous motor with $\rho = 0.2$ and $f = 3$. Solid line shows a given trajectory and dashed lines show perturbed trajectories resulting from initial angular deviations of $e_1(0) = 10^{-3}$ (short dash), 10^{-6} (medium dash), and 10^{-9} (long dash).

an accuracy of 10^{-1} radian after 20 drive cycles, its initial angle must be known to better than 10^{-13} radian, and the calculation must be carried out with at least 14 significant digits. If a similar prediction is to be made over 40 drive cycles, then 25 significant digits are required. Thus, even though chaotic motion is predictable in principle, the state of a chaotic system can be accurately predicted in practice for only a short time into the future. According to Lorenz (1993), this effect explains why weather forecasts are of limited significance beyond a few days.

This pseudorandom nature of chaotic motion is illustrated in Figure 14.7 for the synchronous motor by a plot of the net rotation during each of 100 successive drive cycles. Although this sequence of rotations results from solving a deterministic equation, it is apparently random, jumping erratically between forward and reverse rotations of various magnitudes up to about 1.3 revolutions. The situation is similar to that of a digital random number generator, in which a deterministic algorithm is used to produce a sequence of pseudorandom numbers. In fact, the similarity is not coincidental since chaotic processes often underlie such algorithms (Li, 1978). For the synchronous motor, statistical analysis reveals almost no correlation between rotations separated by more than a few drive cycles. This statistical independence is a result of the motor's positive Liapunov exponent. Because neighboring trajectories diverge exponentially, a small region of the attractor can quickly expand to cover the entire attractor, and a small range of rotations on one drive cycle can lead to almost any possible rotation a few cycles later. Thus, there is little correlation between rotations separated by a few drive cycles, and on this time scale the motor appears to select randomly between the possible rotations.

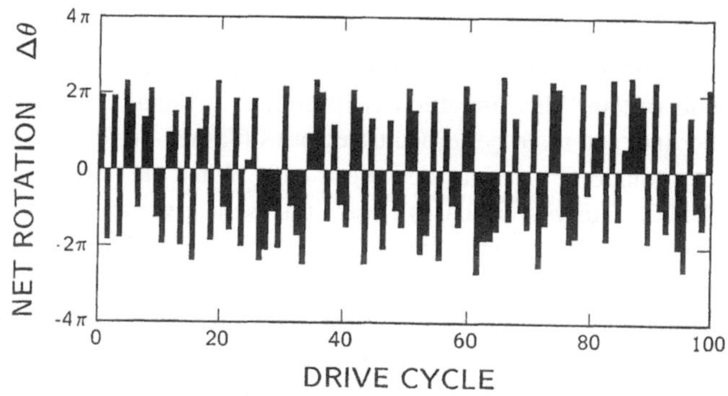

Figure 14.7 Net rotation of a synchronous motor during each of 100 successive drive cycles, illustrating chaotic motion for $\rho = 0.2$ and $f = 3$. By definition, $\Delta\theta = \theta(2\pi n) - \theta(2\pi(n-1))$ on the nth drive cycle.

From an engineering point of view, the problem of chaotic behavior in the synchronous motor can be solved simply by selecting a drive amplitude in the range of $0.40 < f < 1.87$. Within this range, the strength of the nonlinearity is large enough to produce synchronization but not so large as to produce chaos. As this example suggests, it is important to recognize that erratic, apparently random motion can be an intrinsic property of a dynamic system and is not necessarily a product of external noise. Searching a real motor for a source of noise to explain the behavior shown in Figure 14.7 would be wasted effort since the cause is hidden in a noise-free differential equation. Clearly, chaotic motion is a possibility that every engineer should understand.

Defining Terms

Attractor: A set of points in state space to which neighboring trajectories converge in the limit of large time.

Chaos: Pseudorandom behavior observed in the steady-state dynamics of a deterministic nonlinear system.

Fractal: A geometric object characterized by the presence of structure at arbitrarily small scales and by a dimension that is generally fractional.

Liapunov exponent: One of N constants λ_i that characterize the asymptotic exponential growth of infinitesimal deviations from a trajectory in an N-dimensional state space. Various components of a deviation grow or decay on average in proportion to $\exp(\lambda_i t)$.

Nonlinear system: A system of equations for which a linear combination of two solutions is not generally a solution.

Poincaré section: A cross section of a state-space trajectory formed by the intersection of the trajectory with a plane defined by a specified value of one state variable.

Pseudorandom: Random according to statistical tests but derived from a deterministic process.

State space: The space spanned by state vectors.

State vector: A vector **x** whose components are the variables, generally positions and velocities, that define the time evolution of a dynamical system through an equation of the form $x/dt = \mathbf{F}(x)$, where **F** is a vector function.

References

Ballico, M.J., Sawley, M.L., and Skiff, F. 1990. The bipolar motor: A simple demonstration of deterministic chaos. *Am. J. Phys.*, 58, 58–61.

Holmes, P. 1990. Poincaré, celestial mechanics, dynamical-systems theory and "chaos." *Phys. Reports*, 193, 137–163.

Kautz, R.L. and Huggard, B.M. 1994. Chaos at the amusement park: dynamics of the Tilt-A-Whirl. *Am. J. Phys.*, 62, 69–66.

Li, T.Y. and Yorke, J.A. 1978. Ergodic maps on [0,1] and nonlinear pseudo-random number generators. *Nonlinear Anal. Theory Methods Appl.*, 2, 473–481.

Lorenz, E.N. 1963. Deterministic nonperiodic flow. *J. Atmos. Sci.*, 20, 130–141.

Lorenz, E.N. 1993. *The Essence of Chaos*. University of Washington Press, Seattle, WA.

Martien, P., Pope, S.C., Scott, P.L., and Shaw, R.S. 1985. The chaotic behavior of a dripping faucet. *Phys. Lett. A.*, 110, 399–404.

Ott, E., Grebogi, C., and Yorke, J.A. 1990. Controlling chaos. *Phys. Rev. Lett.*, 64, 1196–1199.

Singer, J., Wang, Y.Z., and Bau, H.H. 1991. Controlling a chaotic system. *Phys. Rev. Lett.*, 66, 1123–1125.

Tanabe, Y. and Kaneko, K. 1994. Behavior of falling paper. *Phys. Rev. Lett.*, 73, 1372–1375.

Wisdom, J., Peale, S.J., and Mignard, F. 1984. The chaotic rotation of Hyperion. *Icarus*, 58, 137–152.

For Further Information

A good introduction to deterministic chaos for undergraduates is provided by *Chaotic and Fractal Dynamics: An Introduction for Applied Scientists and Engineers* by Francis C. Moon. This book presents numerous examples drawn from mechanical and electrical engineering.

Chaos in Dynamical Systems by Edward Ott provides a more rigorous introduction to chaotic dynamics at the graduate level.

Practical methods for experimental analysis and control of chaotic systems are presented in *Coping with Chaos: Analysis of Chaotic Data and the Exploitation of Chaotic Systems*, a reprint volume edited by Edward Ott, Tim Sauer, and James A. York.

15

Fuzzy Sets and Fuzzy Logic

Dan M. Frangopol
University of Colorado

15.1 Introduction

In the sixties, Zaheh (1965) introduced the concept of fuzzy sets. Since its inception more than 30 years ago, the theory and methods of fuzzy sets have developed considerably. The demands for treating situations in engineering, social sciences, and medicine, among other applications that are complex and not crisp have been strong driving forces behind these developments.

The concept of the fuzzy set is a generalization of the concept of the ordinary (or crisp) set. It introduces vagueness by eliminating the clear boundary, defined by the ordinary set theory, between full nonmembers (i.e., grade of membership equals zero) and full members (i.e., grade of membership equals one). According to Zaheh (1965) a fuzzy set A, defined as a collection of elements (also called objects) $x \in X$, where X denotes the universal set (also called universe of discourse) and the symbol $\in$ denotes that the element x is a member of X, is characterized by a membership (also called characteristic) function $\mu_A(x)$ which associates each point in X a real member in the unit interval [0,1]. The value of $\mu_A(x)$ at x represents the grade of membership of x in A. Larger values of $\mu_A(x)$ denote higher grades of membership of x in A. For example, a fuzzy set representing the concept of control might assign a degree of membership of 0.0 for no control, 0.1 for weak control, 0.5 for moderate control, 0.9 for strong control, and 1.0 for full control. From this example, it is clear that the two-valued crisp set [i.e., no control (grade of membership 0.0) and full control (grade of membership 1.0)] is a particular case of the general multivalued fuzzy set A in which $\mu_A(x)$ takes its values in the interval [0,1].

Problems in engineering could be very complex and involve various concepts of uncertainty. The use of fuzzy sets in engineering has been quite extensive during this decade. The area of fuzzy control is one of the most developed applications of fuzzy set theory in engineering (Klir and Folger, 1988). Fuzzy controllers have been created for the control of robots, aircraft autopilots, and industrial processes, among others. In Japan, for example, so-called "fuzzy electric appliances," have gained great success from both technological and commercial points of view (Furuta, 1995). Efforts are underway to develop and introduce fuzzy sets as a technical basis for solving various real-world engineering problems in which the underlying information is complex and imprecise. In order to achieve this, a mathematical background in the theory of fuzzy sets is necessary. A brief summary of the fundamental mathematical aspects of the theory of fuzzy sets is presented herein.

0-8493-0056-8/00/$0.00+$.50

15.2 Fundamental Notions

A fuzzy set A is represented by all its elements x_i and associated grades of membership $\mu_A(x_i)$ (Klir and Folger, 1988).

$$A = \{\mu_A(x_1)|x_1, \mu_A(x_2)|x_2, \ldots, \mu_A(x_n)|x_n\} \tag{15.1}$$

where x_i is an element of the fuzzy set, $\mu_A(x_i)$ is its grade of membership in A, and the vertical bar is employed to link the element with their grades of membership in A. Equation (15.1) shows a discrete form of a fuzzy set. For a continuous fuzzy set, the membership function $\mu_A(x)$ is a continuous function of x.

Figure 15.1 illustrates a discrete and a continuous fuzzy set. The larger membership grade *max* $(\mu_A(x_i))$ represents the height of a fuzzy set.

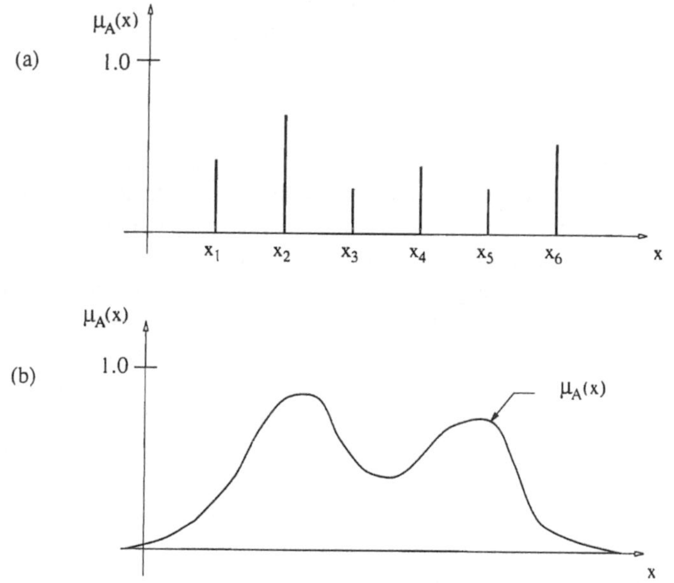

Figure 15.1 (a) Discrete and (b) continuous fuzzy set.

If at least one element of the fuzzy set has a membership grade of 1.0, the fuzzy set is called normalized. Figure 15.2 illustrates both a nonnormalized and a normalized fuzzy set.

The following properties of fuzzy sets, which are obvious extensions of the corresponding definitions for ordinary (crisp) sets, are defined herein according to Zaheh (1965) and Klir and Folger (1988).

Two fuzzy sets A and B are equal, A = B, if and only if $\mu_A(x) = \mu_B(x)$ for every element x in X (see Figure 15.3).

The complement of a fuzzy set A is a fuzzy set $\bar{A}$ defined as

$$\mu_{A}(x) = 1 - \mu_A(x) \tag{15.2}$$

Figure 15.4 shows both discrete and continuous fuzzy sets and their complements.

If the membership grade of each element of the universal set X in fuzzy set B is less than or equal to its membership grade in fuzzy set A, then B is called a subset of A. This is denoted $B \subseteq A$. Figure 15.5 illustrates this situation.

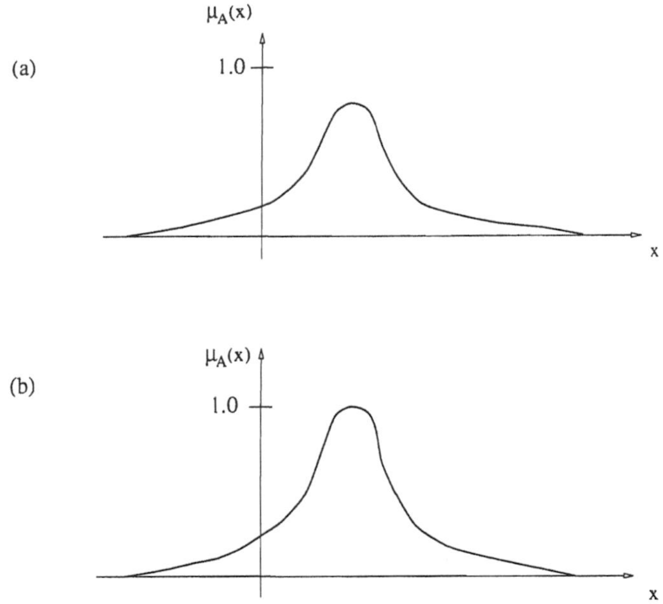

Figure 15.2 (a) Nonnormalized and (b) normalized fuzzy set.

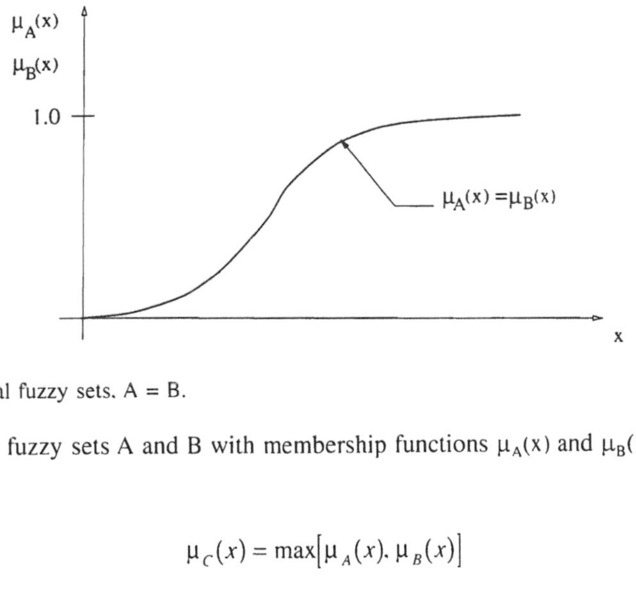

Figure 15.3 Two equal fuzzy sets. A = B.

The union of two fuzzy sets A and B with membership functions $\mu_A(x)$ and $\mu_B(x)$ is a fuzzy set C = A ∪ B such that

$$\mu_C(x) = \max[\mu_A(x), \mu_B(x)] \tag{15.3}$$

for all x in X.

Conversely, the intersection of two fuzzy sets A and B with membership functions $\mu_A(x)$ and $\mu_B(x)$, respectively, is a fuzzy set C = A ∩ B such that

$$\mu_C(x) = \min[\mu_A(x), \mu_B(x)] \tag{15.4}$$

for all x in X.

Figure 15.6 illustrates two fuzzy sets A and B, the union set A ∪ B and the intersection set A ∩ B.

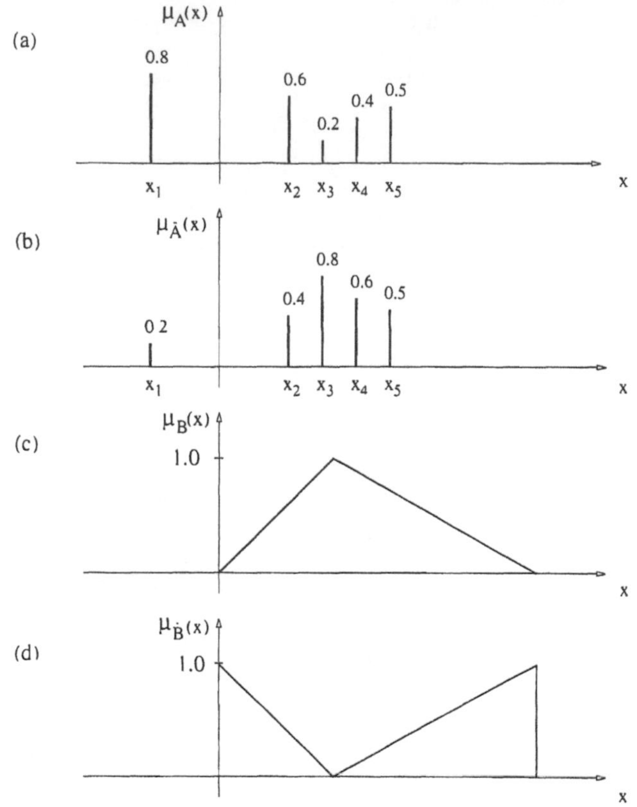

Figure 15.4 (a) Discrete fuzzy set A. (b) complement $\bar{A}$ of fuzzy set A. (c) continuous fuzzy set B. and (d) complement $\bar{B}$ of fuzzy set B

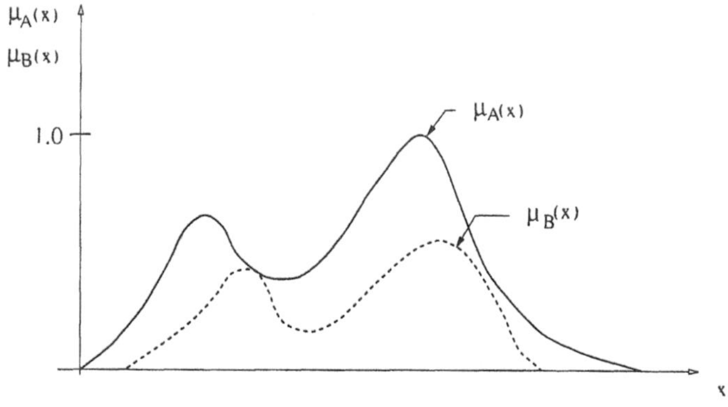

Figure 15.5 Fuzzy set A and its subset B.

An empty fuzzy set A is a fuzzy set with a membership function $\mu_A(x) = 0$ for all elements x in X (see Figure 15.7).

Two fuzzy sets A and B with respective membership function $\mu_A(x)$ and $\mu_B(x)$ are disjoint if their intersection is empty (see Figure 15.8).

An α-cut of a fuzzy set A is an ordinary (crisp) set A_α containing all elements that have a membership grade in A greater or equal to α. Therefore,

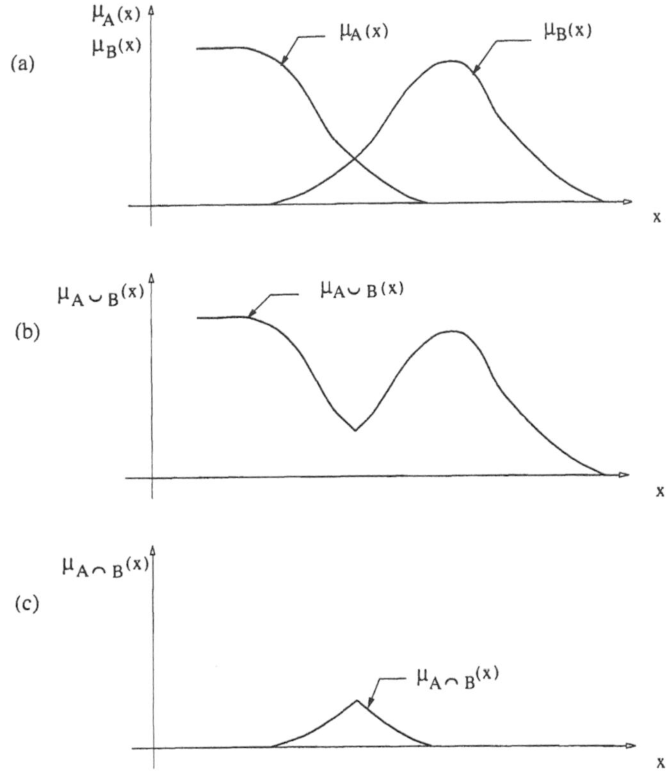

Figure 15.6 (a) Two fuzzy sets, (b) union of fuzzy sets A ∪ B, and (c) intersection of fuzzy sets A ∩ B.

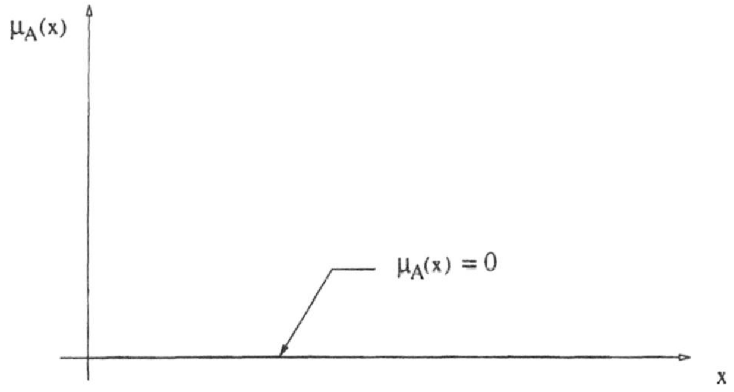

Figure 15.7 Empty fuzzy set.

$$A_\alpha = \{x | \mu_A(x) \geq \alpha\} \tag{15.5}$$

From Figure 15.9, it is clear that $\alpha = 0.5$, the α-cut of the fuzzy set A is the crisp set $A_{0.5} = \{x_5, x_6, x_7, x_8\}$ and for $\alpha = 0.8$, the α-cut of the fuzzy set A is the crisp set $A_{0.8} = \{x_7, x_8\}$.

A fuzzy set is convex if and only if all of its α-cuts are convex for all α in the interval [0,1]. Figure 15.10 shows both a convex and a nonconvex fuzzy set.

A fuzzy number $\tilde{N}$ is a normalized and convex fuzzy set of the real line whose membership function is piecewise continuous and for which it exists exactly one element with $\mu_{\tilde{N}}(x_0) = 1$. As an example, the real numbers close to 50 are shown by four membership functions in Figure 15.11.

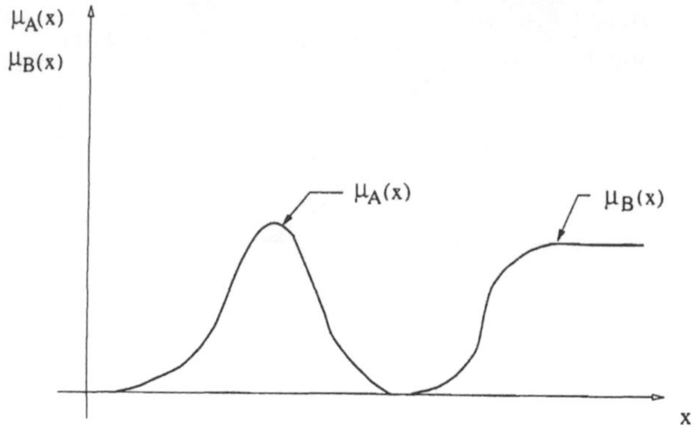

Figure 15.8 Disjoint fuzzy sets.

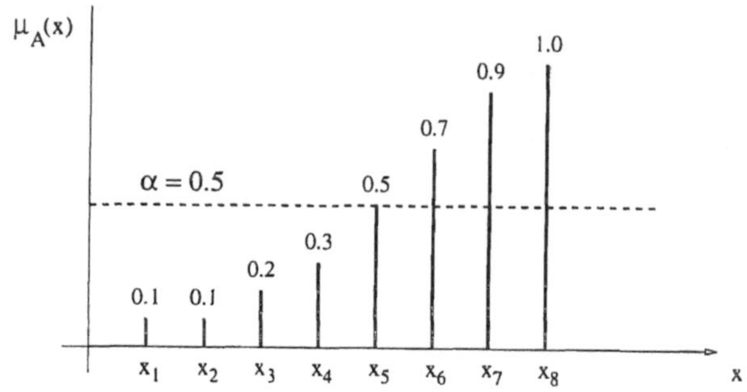

Figure 15.9 α-cut of a fuzzy set.

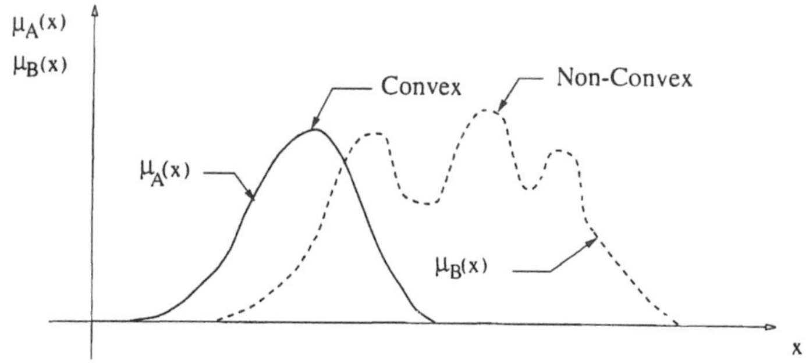

Figure 15.10 Convex and non-convex fuzzy set.

The scalar cardinality of a fuzzy set A is the summation of membership grades of all elements of X in A. Therefore,

$$|A| = \sum_{x} \mu_A(x) \tag{15.6}$$

For example, the scalar cardinality of the fuzzy set A in Figure 15.4(a) is 2.5. Obviously, an empty fuzzy set has a scalar cardinality equal to zero. Also, the scalar cardinality of the fuzzy complement set is equal to scalar cardinality of the original set. Therefore,

$$|A| = |\bar{A}| \tag{15.7}$$

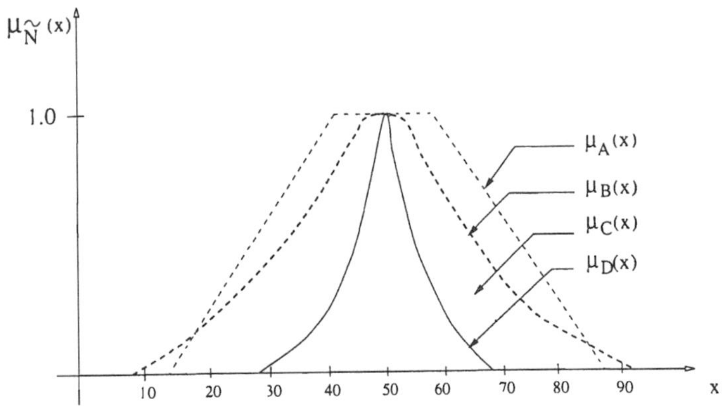

Figure 15.11 Membership functions of fuzzy sets of real numbers close to 50.

One of the basic concepts of fuzzy set theory is the extension principle. According to this principle (Dubois and Prade, 1980), given (a) a function f mapping points in the ordinary set X to points in the ordinary set Y, and (b) any fuzzy set A defined on X,

$$A = \{\mu_A(x_1)|x_1, \mu_A(x_2)|x_2, \dots, \mu_A(x_n)|x_n\}$$

then the fuzzy set B = f(A) is given as

$$B = f(A) = \{\mu_A(x_1)|f(x_1), \mu_A(x_2)|f(x_2), \dots, \mu_A(x_n)|f(x_n)\} \tag{15.8}$$

If more than one element of the ordinary set X is mapped by f to the same element y in Y, then the maximum of the membership grades in the fuzzy set A is considered as the membership grade of y in f(A).

As an example, consider the fuzzy set in Figure 15.4(a), where $x_1 = -2$, $x_2 = 2$, $x_3 = 3$, $x_4 = 4$, and $x_5 = 5$. Therefore, A = {0.8|−2, 0.6|2, 0.2|3, 0.4|4, 0.5|5} and $f(x) = x^4$. By using the extension principle, we obtain

$$f(A) = \{\max(0.8, 0.6)|2^4, 0.2|3^4, 0.4|4^4, 0.5|5^4\}$$

$$= \{0.8|16, 0.2|81, 0.4|256, 0.5|625\}$$

As shown by Klir and Folger (1988), degrees of association can be represented by membership grades in a fuzzy relation. Such a relation can be considered a general case for a crisp relation.

Let P be a binary fuzzy relation between the two crisp sets X = {4, 8, 11} and Y = {4, 7} that represents the relational concept "very close." This relation can be expressed as:

$$P(X, Y) = \{1|(4,4), 0.7|(4,7), 0.6|(8,4), 0.9|(8,7), 0.3|(11,4), 0.6|(11,7)\}$$

or it can be represented by the two dimensional membership matrix

$$\begin{array}{c} \\ x_1 \\ x_2 \\ x_3 \end{array} \begin{array}{c} \begin{array}{cc} y_1 & y_2 \end{array} \\ \begin{bmatrix} 1.0 & 0.7 \\ 0.6 & 0.9 \\ 0.3 & 0.6 \end{bmatrix} \end{array}$$

Fuzzy relations, especially binary relations, are important for many engineering applications.

The concepts of domain, range, and the inverse of a binary fuzzy relation are clearly defined in Zadeh (1971), and Klir and Folger (1988).

The max-min composition operation for fuzzy relations is as follows (Zadeh, 1991; Klir and Folger, 1988):

$$\mu_{P \circ Q}(x,z) = \max_{y \in Y} \min\left[\mu_P(x,y), \mu_Q(y,z)\right] \tag{15.9}$$

for all x in X, y in Y, and z in Z, where the composition of the two binary relations P(X,Y) and Q(Y,Z) is defined as follows:

$$R(X,Z) = P(X,Y) \circ Q(Y,Z) \tag{15.10}$$

As an example, consider the two binary relations

$$P(X,Y) = \{1.0|(4,4), 0.7|(4,7), 0.6|(8,4), 0.9|(8,7), 0.3|(11,4), 0.6|(11,7)\}$$

$$Q(Y,Z) = \{0.8|(4,6), 0.5|(4,9), 0.2|(4,12), 0.0|(4,15), 0.9|(7,6), 0.8|(7,9), 0.5|(7,12), 0.2|(7,15)\}$$

The following matrix equations illustrate the max-min composition for these binary relations

$$\begin{bmatrix} 1.0 & 0.7 \\ 0.6 & 0.9 \\ 0.3 & 0.6 \end{bmatrix} \circ \begin{bmatrix} 0.8 & 0.5 & 0.2 & 0.0 \\ 0.9 & 0.8 & 0.5 & 0.2 \end{bmatrix} = \begin{bmatrix} 0.8 & 0.7 & 0.5 & 0.2 \\ 0.9 & 0.8 & 0.5 & 0.2 \\ 0.6 & 0.6 & 0.5 & 0.2 \end{bmatrix}$$

Zadeh (1971) and Klir and Folger (1988), define also an alternative form of operation on fuzzy relations, called max-product composition. It is denoted as $P(X,Y) \otimes Q(Y,Z)$ and is defined by

$$\mu_{P \otimes Q}(x,z) = \max_{y \in Y}\left[\mu_P(x,y), \mu_Q(y,z)\right] \tag{15.11}$$

for all x in X, y in Y, and z in Z. The matrix equation

$$\begin{bmatrix} 1.0 & 0.7 \\ 0.6 & 0.9 \\ 0.3 & 0.6 \end{bmatrix} \times \begin{bmatrix} 0.8 & 0.5 & 0.2 & 0.0 \\ 0.9 & 0.8 & 0.5 & 0.2 \end{bmatrix} = \begin{bmatrix} 0.8 & 0.7 & 0.5 & 0.2 \\ 0.9 & 0.8 & 0.5 & 0.2 \\ 0.6 & 0.6 & 0.5 & 0.2 \end{bmatrix}$$

illustrates the max product composition for the pair of binary relations P(X,Y) and Q(Y,Z) previously considered.

A crisp binary relation among the elements of a single set can be denoted by R(X,X). If this relation is reflexive, symmetric, and transistive, it is called an equivalence relation (Klir and Folger, 1988).

A fuzzy binary relation S that is reflexive

$$\mu_S(x,x) = 1 \tag{15.12}$$

symmetric

$$\mu_S(x,y) = \mu_S(y,x) \tag{15.13}$$

and transitive

$$\mu_S(x,z) = \max_y \min\left[\mu_S(x,y), \mu_S(y,z)\right] \qquad (15.14)$$

is called a similarity relation (Zadeh, 1971). Equations (15.12), (15.13), and (15.14) are valid for all x,y,z in the domain of S. A similarity relation is a generalization of the notion of equivalence relation.

Fuzzy orderings play a very important role in decision-making in a fuzzy environment. Zadeh (1971) defines fuzzy ordering as a fuzzy relation which is transitive. Fuzzy partial ordering, fuzzy linear ordering, fuzzy preordering, and fuzzy weak ordering are also mathematically defined by Zaheh (1971) and Zimmermann (1991).

The notion of fuzzy relation equation, proposed by Sanchez (1976), is an important notion with various applications. In the context of the max-min composition of two binary relations P(X,Y) and Q(Y,Z), the fuzzy relation equation is as follows

$$\mathbf{P} \circ \mathbf{Q} = \mathbf{R} \qquad (15.15)$$

where **P** and **Q** are matrices of membership functions $\mu_P(x,y)$ and $\mu_Q(y,z)$, respectively, and **R** is a matrix whose elements are determined from Equation (15.9). The solution in this case in unique. However, when **R** and one of the matrices **P**, **Q** are given, the solution is neither guaranteed to exit nor to be unique (Klir and Folger, 1988).

Another important notion is the notion of fuzzy measure. It was introduced by Sugeno (1977). A fuzzy measure is defined by a function which assigns to each crisp subset of X a number in the unit interval [0,1]. This member represents the ambiguity associated with our belief that the crisp subset of X belongs to the subset A. For instance, suppose we are trying to diagnose a mechanical system with a failed component. In other terms, we are trying to assess whether this system belongs to the set of systems with, say, safety problems with regard to failure, serviceability problems with respect to deflections, and serviceability problems with respect to vibrations. Therefore, we might assign a low value, say 0.2 to failure problems, 0.3 to deflection problems, and 0.8 to vibration problems. The collection of these values constitutes a fuzzy measure of the state of the system.

Other measures including plausibility, belief, probability, and possibility measures are also used for defining the ambiguity associated with several crisp defined alternatives. For an excellent treatment of these measures and of the relationship among classes of fuzzy measures see Klir and Folger (1988).

Measures of fuzziness are used to indicate the degree of fuzziness of a fuzzy set (Zimmermann, 1991). One of the most used measures of fuzziness is the entropy. This measure is defined (Zimmermann, 1991) as

$$d(A) = h\sum_{i=1}^{n} S\left(\mu_A(x_i)\right) \qquad (15.16)$$

where h is a positive constant and $S(\alpha)$ is the Shannon function defined as
$S(\alpha) = -\alpha \ln\alpha - (1-\alpha)\ln(1-\alpha)$ for rational α. For the fuzzy set in Figure 15.4(a), defined as

$$A = \{0.8|-2, 0.6|2, 0.2|3, 0.4|4, 0.5|5\}$$

the entropy is

$$d(A) = h(0.5004 + 0.6730 + 0.5004 + 0.6730 + 0.6931)$$
$$= 3.0399\ h$$

Therefore, for h = 1, the entropy of the fuzzy set A is 3.0399.

The notion of linguistic variable, introduced by Zadeh (1973), is a fundamental notion in the development of fuzzy logic and approximate reasoning. According to Zadeh (1973), linguistic variables are "variables whose values are not members but words or sentences in a natural or artificial language. The motivation for the use of words or sentences rather than numbers is that linguistic characterizations are, in general, less specific than numerical ones." The main differences between fuzzy logic and classical two-valued (e.g., true or false) or multivalued (e.g., true, false, and indeterminate) logic are that (a) fuzzy logic can deal with fuzzy quantities (e.g., most, few, quite a few, many, almost all) which are in general represented by fuzzy numbers (see Figure 15.11), fuzzy predicates (e.g., expensive, rare), and fuzzy modifiers (e.g., extremely, unlikely), and (b) the notions of truth and false are both allowed to be fuzzy using fuzzy true/false values (e.g., very true, mostly false). As Klir and Folger (1988) stated, the ultimate goal of fuzzy logic is to provide foundations for approximate reasoning. For a general background on fuzzy logic and approximate reasoning and their applications to expert systems, the reader is referred to Zadeh (1973, 1987), Kaufmann (1975), Negoita (1985), and Zimmermann (1991), among others.

Decision making in a fuzzy environment is an area of continuous growth in engineering and other fields such as economics and medicine. Bellman and Zadeh (1970) define this process as a "decision process in which the goals and/or the constraints, but not necessarily the system under control, are fuzzy in nature."

According to Bellman and Zadeh (1970), a fuzzy goal G associated with a given set of alternatives X = {x} is identified with a given fuzzy set G in X. For example, the goal associated with the statement "x should be in the vicinity of 50" might be represented by a fuzzy set whose membership function is equal to one of the four membership functions shown in Figure 15.11. Similarly, a fuzzy constraint C in X is also a fuzzy set in X, such as "x should be substantially larger than 20."

Bellman and Zadeh (1970) define a fuzzy decision D as the confluence of goals and constraints, assuming, of course, that the goals and constraints conflict with one another. Situations in which the goals and constraints are fuzzy sets in different spaces, multistage decision processes, stochastic systems with implicitly defined termination time, and their associated optimal policies are also studied in Bellman and Zadeh (1970).

References

Bellman, R.E. and Zadeh, L.A. 1970. Decision-making in a fuzzy environment. *Management Science*, 17(4), 141–164.

Dubois, D. and Prade, H. 1980. *Fuzzy Sets and Systems: Theory and Applications*. Academic Press, New York.

Furuta, H. 1995. Fuzzy logic and its contribution to reliability analysis. In *Reliability and Optimization of Structural Systems*, R. Rackwitz, G. Augusti, and A. Borri, Eds., Chapman & Hall, London, pp. 61–76.

Kaufmann, A. 1975. *Introduction to the Theory of Fuzzy Subsets*, Vol. 1, Academic Press, New York.

Klir, G.J. and Folger, T.A. 1988. *Fuzzy Sets, Uncertainty, and Information*. Prentice Hall, Englewood Cliffs, New Jersey.

Negoita, C.V. 1985. *Expert Systems and Fuzzy Systems*, Benjamin/Cummings, Menlo Park, California.

Sanchez, E. 1976. Resolution of composite fuzzy relation equations. *Information and Control*, 30, 38–48.

Sugeno, M. 1977. Fuzzy measures and fuzzy integrals — a survey, in *Fuzzy Automata and Decision Processes*, M.M. Gupta, R.K. Ragade, and R.R. Yager, Eds., North Holland, New York, pp. 89–102.

Zadeh, L.A. 1965. Fuzzy sets. *Information and Control*, 8, 338–353.

Zadeh, L.A. 1971. Similarity relations and fuzzy orderings. *Information Sciences*, 3, 177–200.

Zadeh, L.A. 1973. The concept of a linguistic variable and its applications to approximate reasoning. Memorandum ERL-M 411, Berkeley, California.

Zadeh, L.A. 1987. *Fuzzy Sets and Applications: Selected Papers by L.A. Zadeh*, R.R. Yager, S. Ovchinnikov, R.M. Tong, and H.T. Nguyen, Eds., John Wiley & Sons, New York.

Zimmerman, H.-J. 1991. *Fuzzy Set Theory – and Its Applications*, 2nd ed., Kluwer Academic Publishers, Boston.

Further Information

The more than 5000 publications that exist in the field of fuzzy sets are widely scattered in many books, journals, and conference proceedings. For newcomers, good introductions to the theory and applications of fuzzy sets are presented in (a) *Introduction to the Theory of Fuzzy Sets*, Volume I, Academic Press, New York, 1975, by Arnold Kaufmann; (b) *Fuzzy Sets and Systems: Theory and Applications*, Academic Press, New York, 1980, by Didier Dubois and Henri Prade; (c) *Fuzzy Sets, Uncertainty and Information*, Prentice Hall, Englewood Cliffs, NJ, 1988, by George Klir and Tina Folger, and (d) *Fuzzy Set Theory and Its Applications.*, 2nd ed., Kluwer Academic Publishers, Boston, 1991, by H.-J. Zimmerman, among others.

The eighteen selected papers by Lotfi A. Zadeh grouped in *Fuzzy Sets and Applications*, John Wiley & Sons, New York, 1987, edited by R. Yager, S. Ovchinnikov, R.M. Tong, and H.T. Nguyen are particularly helpful for understanding the developments of issues in fuzzy set and possibility theory. Also, the interview with Professor Zadeh published in this book illustrates the basic philosophy of the founder of fuzzy set theory.

Appendices

Paul Norton
National Renewable Energy Laboratory

0-8493-0056-8/00/$0.00+$.50

Appendix A. Properties of Gases and Vapors

Table A.1 Properties of Dry Air at Atmospheric Pressure

Symbols and Units:

K = absolute temperature, degrees Kelvin
deg C = temperature, degrees Celsius
deg F = temperature, degrees Fahrenheit
ρ = density, kg/m^3
c_p = specific heat capacity, kJ/kg·K
c_p/c_v = specific heat capacity ratio, dimensionless
μ = viscosity, N·s/m^2 × 10^6 (For N·s/m^2 (= kg/m·s) multiply tabulated values by 10^{-6})
k = thermal conductivity, W/m·k × 10^3 (For W/m·K multiply tabulated values by 10^{-3})
Pr = Prandtl number, dimensionless
h = enthalpy, kJ/kg
V_s = sound velocity, m/s

Temperature			Properties							
K	deg C	deg F	ρ	c_p	c_p/c_v	μ	k	Pr	h	V_s
100	- 173 15	- 280	3 598	1 028		6 929	9 248	770	98 42	198 4
110	- 163 15	262	3 256	1 022	1 420 2	7 633	10 15	768	108 7	208 7
120	- 153 15	- 244	2 975	1 017	1 416 6	8 319	11 05	766	118 8	218 4
130	- 143 15	- 226	2 740	1 014	1 413 9	8 990	11 94	763	129 0	227 6
140	- 133 15	- 208	2 540	1 012	1 411 9	9 646	12 84	761	139 1	236 4
150	- 123 15	- 190	2 367	1 010	1 410 2	10 28	13 73	758	149 2	245 0
160	- 113 15	- 172	2 217	1 009	1 408 9	10 91	14 61	754	159 4	253 2
170	- 103 15	- 154	2 085	1 008	1 407 9	11 52	15 49	750	169 4	261 0
180	- 93 15	- 136	1 968	1 007	1 407 1	12 12	16 37	746	179 5	268 7
190	- 83 15	- 118	1 863	1 007	1 406 4	12 71	17 23	743	189 6	276.2
200	- 73 15	- 100	1 769	1 006	1 405 7	13 28	18 09	739	199 7	283 4
205	- 68 15	- 91	1 726	1 006	1 405 5	13 56	18 52	738	204 7	286 9
210	- 63 15	- 82	1 684	1 006	1 405 3	13 85	18 94	736	209 7	290 5
215	- 58 15	- 73	1 646	1 006	1 405 0	14 12	19 36	734	214 8	293 9
220	- 53 15	- 64	1 607	1 006	1 404 8	14 40	19 78	732	219 8	297 4
225	- 48 15	- 55	1 572	1 006	1 404 6	14 67	20 20	731	224 8	300 8
230	- 43 15	- 46	1 537	1 006	1 404 4	14 94	20 62	729	229 8	304 1
235	- 38 15	- 37	1 505	1 006	1 404 2	15 20	21 04	727	234 9	307 4
240	- 33 15	- 28	1 473	1 005	1 404 0	15 47	21 45	725	239 9	310 6
245	- 28 15	- 19	1 443	1 005	1 403 8	15 73	21 86	724	244 9	313 8
250	- 23 15	- 10	1 413	1 005	1 403 6	15 99	22 27	722	250 0	317 1
255	- 18 15	- 1	1 386	1 005	1 403 4	16 25	22 68	721	255 0	320 2
260	- 13 15	8	1 359	1 005	1 403 2	16 50	23 08	719	260 0	323 4
265	- 8 15	17	1 333	1 005	1 403 0	16 75	23 48	717	265 0	326 5
270	- 3 15	26	1 308	1 006	1 402 9	17 00	23 88	716	270 1	329 6
275	+ 1.85	35	1.285	1 006	1 402 6	17 26	24 28	715	275 1	332 6
280	6 85	44	1 261	1 006	1 402 4	17 50	24 67	713	280 1	335 6
285	11 85	53	1 240	1 006	1 402 2	17 74	25 06	711	285 1	338 5
290	16 85	62	1.218	1 006	1 402 0	17 98	25 47	710	290 2	341 5
295	21 85	71	1 197	1 006	1 401 8	18 22	25.85	709	295 2	344 4
300	26 85	80	1 177	1 006	1 401 7	18 46	26 24	708	300 2	347 3
305	31 85	89	1 158	1 006	1 401 5	18.70	26 63	707	305 3	350 2
310	36 85	98	1 139	1 007	1 401 3	18 93	27 01	705	310 3	353 1
315	41 85	107	1 121	1 007	1 401 0	19 15	27 40	704	315 3	355 8
320	46 85	116	1 103	1 007	1 400 8	19 39	27 78	703	320 4	358 7

*Condensed and computed from: "Tables of Thermal Properties of Gases." National Bureau of Standards Circular 564. U.S. Government Printing Office, November 1955.

Table A.1 (continued) Properties of Dry Air at Atmospheric Pressure

Temperature			*Properties*							
K	*deg C*	*deg F*	ρ	c_p	c_p/c_v	μ	k	*Pr*	*h*	V_s
325	51 85	125	1 086	1 008	1 400 6	19 63	28 15	702	325.4	361.4
330	56 85	134	1 070	1 008	1 400 4	19 85	28 53	701	330.4	364.2
335	61 85	143	1 054	1 008	1 400 1	20 08	28 90	700	335.5	366.9
340	66 85	152	1 038	1 008	1 399 9	20 30	29 28	.699	340.5	369 6
345	71 85	161	1 023	1 009	1.399 6	20 52	29 64	.698	345.6	372 3
350	76 85	170	1 008	1 009	1.399 3	20 75	30 03	.697	350.6	375 0
355	81 85	179	0 994 5	1 010	1 399 0	20 97	30 39	.696	355.7	377.6
360	86 85	188	0 980 5	1 010	1.398 7	21 18	30 78	.695	360.7	380.2
365	91 85	197	0 967 2	1 010	1 398 4	21 38	31 14	.694	365.8	382.8
370	96 85	206	0 953 9	1 011	1.398 1	21 60	31 50	.693	370.8	385.4
375	101 85	215	0 941 3	1 011	1.397 8	21 81	31 86	.692	375.9	388.0
380	106 85	224	0 928 8	1 012	1 397 5	22 02	32 23	.691	380.9	390 5
385	111 85	233	0 916 9	1 012	1.397 1	22 24	32 59	.690	386.0	393 0
390	116 85	242	0 905 0	1 013	1 396 8	22 44	32 95	690	391.0	395 5
395	121 85	251	0 893 6	1 014	1 396 4	22 65	33 31	689	396.1	398 0
400	126 85	260	0 882 2	1 014	1.396 1	22 86	33.65	689	401.2	400 4
410	136 85	278	0 860 8	1 015	1 395 3	23 27	34.35	688	411.3	405 3
420	146.85	296	0 840 2	1 017	1.394 6	23 66	35.05	687	421.5	410 2
430	156 85	314	0 820 7	1 018	1 393 8	24 06	35 75	686	431.7	414 9
440	166.85	332	0 802 1	1 020	1.392 9	24 45	36.43	684	441.9	419 6
450	176.85	350	0 784 2	1 021	1 392 0	24 85	37.10	684	452 1	424 2
460	186.85	368	0.767 7	1 023	1 391 1	25 22	37.78	683	462 3	428 7
470	196.85	386	0 750 9	1 024	1 390 1	25 58	38 46	682	472 5	433 2
480	206 85	404	0 735 1	1 026	1 389 2	25 96	39 11	681	482 8	437 6
490	216.85	422	0 720 1	1.028	1 388 1	26 32	39 76	680	493 0	442 0
500	226.85	440	0 705 7	1.030	1 387 1	26 70	40 41	680	503 3	446 4
510	236.85	458	0 691 9	1.032	1 386 1	27 06	41 06	680	513 6	450 6
520	246 85	476	0 678 6	1 034	1 385 1	27 42	41 69	680	524 0	454 9
530	256 85	494	0 665 8	1 036	1 384 0	27 78	42 32	680	534 3	459 0
540	266 85	512	0 653 5	1 038	1 382 9	28 14	42 94	680	544 7	463 2
550	276.85	530	0 641 6	1 040	1 381 8	28 48	43 57	680	555 1	467 3
560	286 85	548	0 630 1	1 042	1 380 6	28 83	44 20	680	565 5	471 3
570	296 85	566	0 619 0	1 044	1 379 5	29 17	44 80	680	575 9	475 3
580	306 85	584	0 608 4	1 047	1 378 3	29 52	45 41	680	586 4	479 2
590	316 85	602	0 598 0	1 049	1 377 2	29 84	46 01	680	596 9	483 2
600	326 85	620	0 588 1	1 051	1 376 0	30 17	46.61	680	607 4	486 9
620	346 85	656	0 569 1	1 056	1 373 7	30 82	47.80	681	628.4	494 5
640	366 85	692	0.551 4	1.061	1 371 4	31 47	48.96	682	649 6	502 1
660	386 85	728	0.534 7	1.065	1 369 1	32 09	50.12	682	670.9	509 4
680	406 85	764	0 518 9	1 070	1 366 8	32 71	51 25	683	692 2	516 7
700	426 85	800	0 504 0	1 075	1 364 6	33 32	52 36	684	713 7	523 7
720	446 85	836	0 490 1	1.080	1 362 3	33.92	53.45	685	735 2	531.0
740	466 85	872	0 476 9	1 085	1 360 1	34 52	54 53	686	756 9	537 6
760	486 85	908	0.464 3	1.089	1 358 0	35 11	55.62	687	778.6	544.6
780	506 85	944	0 452 4	1 094	1 355 9	35 69	56 68	688	800.5	551 2
800	526 85	980	0 441 0	1 099	1 354	36 24	57.74	689	822 4	557 8
850	576 85	1 070	0 415 2	1.110	1 349	37.63	60.30	693	877 5	574 1
900	626 85	1 160	0 392 0	1 121	1 345	38 97	62 76	696	933 4	589 6
950	676 85	1 250	0.371 4	1.132	1 340	40.26	65.20	699	989 7	604 9
1 000	726 85	1 340	0 352 9	1 142	1 336	41.53	67.54	702	1 046	619 5
1 100	826 85	1 520	0 320 8	1 161	1 329	43.96			1 162	648 0
1 200	926 85	1 700	0 294 1	1 179	1 322	46.26			1 279	675 2
1 300	1 026 85	1 880	0 271 4	1 197	1 316	48 46			1 398	701 0
1 400	1 126 85	2 060	0 252 1	1 214	1 310	50.57			1 518	725 9
1 500	1 220 85	2 240	0 235 3	1.231	1 304	52.61			1 640	749 4
1 600	1 326 85	2 420	0 220 6	1 249	1 299	54.57			1 764	772 6
1 800	1 526 85	2 780	0 196 0	1 288	1 288	58 29			2 018	815 7
2 000	1 726 85	3 140	0 176 4	1 338	1 274				2 280	855.5
2 400	2 126 85	3 860	0 146 7	1.574	1 238				2 853	924.4
2 800	2 526 85	4 580	0 124 5	2 259	1 196				3 599	983 1

Table A.2 Ideal Gas Properties of Nitrogen, Oxygen, and Carbon Dioxide

Symbols and Units:

T = absolute temperature, degrees Kelvin
$\bar{h}$ = enthalpy, kJ/kmol
$\bar{u}$ = internal energy, kJ/kmol
$\bar{s}^\circ$ = absolute entropy at standard reference pressure, kJ/kmol K
[$\bar{h}$ = enthalpy of formation per mole at standard state = 0 kJ/kmol]

Part a. Ideal Gas Properties of Nitrogen, N_2

T	$\bar{h}$	$\bar{u}$	$\bar{s}^\circ$
0	0	0	0
220	6,391	4,562	182.639
230	6,683	4,770	183.938
240	6,975	4,979	185.180
250	7,266	5,188	186.370
260	7,558	5,396	187.514
270	7,849	5,604	188.614
280	8,141	5,813	189.673
290	8,432	6,021	190.695
298	8,669	6,190	191.502
300	8,723	6,229	191.682
310	9,014	6,437	192.638
320	9,306	6,645	193.562
330	9,597	6,853	194.459
340	9,888	7,061	195.328
350	10,180	7,270	196.173
360	10,471	7,478	196.995
370	10,763	7,687	197.794
380	11,055	7,895	198.572
390	11,347	8,104	199.331
400	11,640	8,314	200.071
410	11,932	8,523	200.794
420	12,225	8,733	201.499
430	12,518	8,943	202.189
440	12,811	9,153	202.863
450	13,105	9,363	203.523
460	13,399	9,574	204.170
470	13,693	9,786	204.803
480	13,988	9,997	205.424
490	14,285	10,210	206.033
500	14,581	10,423	206.630
510	14,876	10,635	207.216
520	15,172	10,848	207.792
530	15,469	11,062	208.358
540	15,766	11,277	208.914
550	16,064	11,492	209.461
560	16,363	11,707	209.999
570	16,662	11,923	210.528
580	16,962	12,139	211.049
590	17,26[illegible]	12,356	211.562
600	17,563	12,574	212.066
610	17,864	12,792	212.564
620	18,166	13,011	213.055
630	18,468	13,230	213.541
640	18,772	13,450	214.018
650	19,075	13,671	214.489
660	19,380	13,892	214.954
670	19,685	14,114	215.413
680	19,991	14,337	215.866
690	20,297	14,560	216.314
700	20,604	14,784	216.756
710	20,912	15,008	217.192
720	21,220	15,234	217.624
730	21,529	15,460	218.059
740	21,839	15,686	218.472
750	22,149	15,913	218.889
760	22,460	16,141	219.301
770	22,772	16,370	219.709
780	23,085	16,599	220.113
790	23,398	16,830	220.512
800	23,714	17,061	220.907
810	24,027	17,292	221.298
820	24,342	17,524	221.684
830	24,658	17,757	222.067
840	24,974	17,990	222.447
850	25,292	18,224	222.822
860	25,610	18,459	223.194
870	25,928	18,695	223.562
880	26,248	18,931	223.927
890	26,568	19,168	224.288
900	26,890	19,407	224.647
910	27,210	19,644	225.002
920	27,532	19,883	225.353
930	27,854	20,122	225.701
940	28,178	20,362	226.047
950	28,501	20,603	226.389
960	28,826	20,844	226.728
970	29,151	21,086	227.064
980	29,476	21,328	227.398
990	29,803	21,571	227.728

Source: Adapted from M.J. Moran and H.N. Shapiro, *Fundamentals of Engineering Thermodynamics*, 3rd. ed., Wiley, New York, 1995, as presented in K. Wark, *Thermodynamics*, 4th ed., McGraw-Hill, New York, 1983, based on the *JANAF Thermochemical Tables*, NSRDS-NBS-37, 1971.

Table A.2 (continued) Ideal Gas Properties of Nitrogen, Oxygen, and Carbon Dioxide

T	$\bar{h}$	$\bar{u}$	$\bar{s}^o$	T	n	$\bar{u}$	$\bar{s}^o$
1000	30,129	21,815	228.057	1760	56,227	41.594	247.396
1020	30,784	22,304	228.706	1780	56,938	42,139	247.798
1040	31,442	22,795	229.344	1800	57,651	42,685	248.195
1060	32,101	23,288	229.973	1820	58,363	43,231	248.589
1080	32,762	23,782	230.591	1840	59,075	43,777	248.979
1100	33,426	24,280	231.199	1860	59,790	44,324	249.365
1120	34,092	24,780	231.799	1880	60,504	44,873	249.748
1140	34,760	25,282	232.391	1900	61,220	45,423	250.128
1160	35,430	25,786	232.973	1920	61,936	45,973	250.502
1180	36,104	26,291	233.549	1940	62,654	46,524	250.874
1200	36,777	26,799	234.115	1960	63,381	47,075	251.242
1220	37,452	27,308	234.673	1980	64,090	47,627	251.607
1240	38,129	27,819	235.223	2000	64,810	48,181	251.969
1260	38,807	28,331	235.766	2050	66,612	49,567	252.858
1280	39,488	28,845	236.302	2100	68,417	50,957	253.726
1300	40,170	29,361	236.831	2150	70,226	52,351	254.578
1320	40,853	29,878	237.353	2200	72,040	53,749	255.412
1340	41,539	30,398	237.867	2250	73,856	55,149	256.227
1360	42,227	30,919	238.376	2300	75,676	56,553	257.027
1380	42,915	31,441	238.878	2350	77,496	57,958	257.810
1400	43,605	31,964	239.375	2400	79,320	59,366	258.580
1420	44,295	32,489	239.865	2450	81,149	60,779	259.332
1440	44,988	33,014	240.350	2500	82,981	62,195	260.073
1460	45,682	33,543	240.827	2550	84,814	63,613	260.799
1480	46,377	34,071	241.301	2600	86,650	65,033	261.512
1500	47,073	34,601	241.768	2650	88,488	66,455	262.213
1520	47,771	35,133	242.228	2700	90,328	67,880	262.902
1540	48,470	35,665	242.685	2750	92,171	69,306	263.577
1560	49,168	36,197	243.137	2800	94,014	70,734	264.241
1580	49,869	36,732	243.585	2850	95,859	72,163	264.895
1600	50,571	37,268	244.028	2900	97,705	73,593	265.538
1620	51,275	37,806	244.464	2950	99,556	75,028	266.170
1640	51,980	38,344	244.896	3000	101,407	76,464	266.793
1660	52,686	38,884	245.324	3050	103,260	77,902	267.404
1680	53,393	39,424	245.747	3100	105,115	79,341	268.007
1700	54,099	39,965	246.166	3150	106,972	80,782	268.601
1720	54,807	40,507	246.580	3200	108,830	82,224	269.186
1740	55,516	41,049	246.990	3250	110,690	83,668	269.763

Table A.2 (continued) **Ideal Gas Properties of Nitrogen, Oxygen, and Carbon Dioxide**

Part b. Ideal Gas Properties of Oxygen, O_2

T	$\bar{h}$	$\bar{u}$	$\bar{s}°$	T	$\bar{h}$	$\bar{u}$	$\bar{s}°$
0	0	0	0	600	17,929	12,940	226.346
220	6,404	4,575	196.171	610	18,250	13,178	226.877
230	6,694	4,782	197.461	620	18,572	13,417	227.400
240	6,984	4,989	198.696	630	18,895	13,657	227.918
250	7,275	5,197	199.885	640	19,219	13,898	228.429
260	7,566	5,405	201.027	650	19,544	14,140	228.932
270	7,858	5,613	202.128	660	19,870	14,383	229.430
280	8,150	5,822	203.191	670	20,197	14,626	229.920
290	8,443	6,032	204.218	680	20,524	14,871	230.405
298	8,682	6,203	205.033	690	20,854	15,116	230.885
300	8,736	6,242	205.213	700	21,184	15,364	231.358
310	9,030	6,453	206.177	710	21,514	15,611	231.827
320	9,325	6,664	207.112	720	21,845	15,859	232.291
330	9,620	6,877	208.020	730	22,177	16,107	232.748
340	9,916	7,090	208.904	740	22,510	16,357	233.201
350	10,213	7,303	209.765	750	22,844	16,607	233.649
360	10,511	7,518	210.604	760	23,178	16,859	234.091
370	10,809	7,733	211.423	770	23,513	17,111	234.528
380	11,109	7,949	212.222	780	23,850	17,364	234.960
390	11,409	8,166	213.002	790	24,186	17,618	235.387
400	11,711	8,384	213.765	800	24,523	17,872	235.810
410	12,012	8,603	214.510	810	24,861	18,126	236.230
420	12,314	8,822	215.241	820	25,199	18,382	236.644
430	12,618	9,043	215.955	830	25,537	18,637	237.055
440	12,923	9,264	216.656	840	25,877	18,893	237.462
450	13,228	9,487	217.342	850	26,218	19,150	237.864
460	13,535	9,710	218.016	860	26,559	19,408	238.264
470	13,842	9,935	218.676	870	26,899	19,666	238.660
480	14,151	10,160	219.326	880	27,242	19,925	239.051
490	14,460	10,386	219.963	890	27,584	20,185	239.439
500	14,770	10,614	220.589	900	27,928	20,445	239.823
510	15,082	10,842	221.206	910	28,272	20,706	240.203
520	15,395	11,071	221.812	920	28,616	20,967	240.580
530	15,708	11,301	222.409	930	28,960	21,228	240.953
540	16,022	11,533	222.997	940	29,306	21,491	241.323
550	16,338	11,765	223.576	950	29,652	21,754	241.689
560	16,654	11,998	224.146	960	29,999	22,017	242.052
570	16,971	12,232	224.708	970	30,345	22,280	242.411
580	17,290	12,467	225.262	980	30,692	22,544	242.768
590	17,609	12,703	225.808	990	31,041	22,809	243.120

Table A.2 (continued) Ideal Gas Properties of Nitrogen, Oxygen, and Carbon Dioxide

T	$\bar{h}$	$\bar{u}$	$\bar{s}°$	T	$\bar{h}$	$\bar{u}$	$\bar{s}°$
1000	31,389	23,075	243.471	1760	58,880	44,247	263.861
1020	32,088	23,607	244.164	1780	59,624	44,825	264.283
1040	32,789	24,142	244.844	1800	60,371	45,405	264.701
1060	33,490	24,677	245.513	1820	61,118	45,986	265.113
1080	34,194	25,214	246.171	1840	61,866	46,568	265.521
1100	34,899	25,753	246.818	1860	62,616	47,151	265.925
1120	35,606	26,294	247.454	1880	63,365	47,734	266.326
1140	36,314	26,836	248.081	1900	64,116	48,319	266.722
1160	37,023	27,379	248.698	1920	64,868	48,904	267.115
1180	37,734	27,923	249.307	1940	65,620	49,490	267.505
1200	38,447	28,469	249.906	1960	66,374	50,078	267.891
1220	39,162	29,018	250.497	1980	67,127	50,665	268.275
1240	39,877	29,568	251.079	2000	67,881	51,253	268.655
1260	40,594	30,118	251.653	2050	69,772	52,727	269.588
1280	41,312	30,670	252.219	2100	71,668	54,208	270.504
1300	42,033	31,224	252.776	2150	73,573	55,697	271.399
1320	42,753	31,778	253.325	2200	75,484	57,192	272.278
1340	43,475	32,334	253.868	2250	77,397	58,690	273.136
1360	44,198	32,891	254.404	2300	79,316	60,193	273.981
1380	44,923	33,449	254.932	2350	81,243	61,704	274.809
1400	45,648	34,008	255.454	2400	83,174	63,219	275.625
1420	46,374	34,567	255.968	2450	85,112	64,742	276.424
1440	47,102	35,129	256.475	2500	87,057	66,271	277.207
1460	47,831	35,692	256.978	2550	89,004	67,802	277.979
1480	48,561	36,256	257.474	2600	90,956	69,339	278.738
1500	49,292	36,821	257.965	2650	92,916	70,883	279.485
1520	50,024	37,387	258.450	2700	94,881	72,433	280.219
1540	50,756	37,952	258.928	2750	96,852	73,987	280.942
1560	51,490	38,520	259.402	2800	98,826	75,546	281.654
1580	52,224	39,088	259.870	2850	100,808	77,112	282.357
1600	52,961	39,658	260.333	2900	102,793	78,682	283.048
1620	53,696	40,227	260.791	2950	104,785	80,258	283.728
1640	54,434	40,799	261.242	3000	106,780	81,837	284.399
1660	55,172	41,370	261.690	3050	108,778	83,419	285.060
1680	55,912	41,944	262.132	3100	110,784	85,009	285.713
1700	56,652	42,517	262.571	3150	112,795	86,601	286.355
1720	57,394	43,093	263.005	3200	114,809	88,203	286.989
1740	58,136	43,669	263.435	3250	116,827	89,804	287.614

Table A.2 (continued) Ideal Gas Properties of Nitrogen, Oxygen, and Carbon Dioxide

Part c. Ideal Gas Properties of Carbon Dioxide, CO_2

T	$\bar{h}$	$\bar{u}$	$\bar{s}^\circ$	T	$\bar{h}$	$\bar{u}$	$\bar{s}^\circ$
0	0	0	0	600	22,280	17,291	243.199
220	6,601	4,772	202.966	610	22,754	17,683	243.983
230	6,938	5,026	204.464	620	23,231	18,076	244.758
240	7,280	5,285	205.920	630	23,709	18,471	245.524
250	7,627	5,548	207.337	640	24,190	18,869	246.282
260	7,979	5,817	208.717	650	24,674	19,270	247.032
270	8,335	6,091	210.062	660	25,160	19,672	247.773
280	8,697	6,369	211.376	670	25,648	20,078	248.507
290	9,063	6,651	212.660	680	26,138	20,484	249.233
298	9,364	6,885	213.685	690	26,631	20,894	249.952
300	9,431	6,939	213.915	700	27,125	21,305	250.663
310	9,807	7,230	215.146	710	27,622	21,719	251.368
320	10,186	7,526	216.351	720	28,121	22,134	252.065
330	10,570	7,826	217.534	730	28,622	22,552	252.755
340	10,959	8,131	218.694	740	29,124	22,972	253.439
350	11,351	8,439	219.831	750	29,629	23,393	254.117
360	11,748	8,752	220.948	760	30,135	23,817	254.787
370	12,148	9,068	222.044	770	30,644	24,242	255.452
380	12,552	9,392	223.122	780	31,154	24,669	256.110
390	12,960	9,718	224.182	790	31,665	25,097	256.762
400	13,372	10,046	225.225	800	32,179	25,527	257.408
410	13,787	10,378	226.250	810	32,694	25,959	258.048
420	14,206	10,714	227.258	820	33,212	26,394	258.682
430	14,628	11,053	228.252	830	33,730	26,829	259.311
440	15,054	11,393	229.230	840	34,251	27,267	259.934
450	15,483	11,742	230.194	850	34,773	27,706	260.551
460	15,916	12,091	231.144	860	35,296	28,125	261.164
470	16,351	12,444	232.080	870	35,821	28,588	261.770
480	16,791	12,800	233.004	880	36,347	29,031	262.371
490	17,232	13,158	233.916	890	36,876	29,476	262.968
500	17,678	13,521	234.814	900	37,405	29,922	263.559
510	18,126	13,885	235.700	910	37,935	30,369	264.146
520	18,576	14,253	236.575	920	38,467	30,818	264.728
530	19,029	14,622	237.439	930	39,000	31,268	265.304
540	19,485	14,996	238.292	940	39,535	31,719	265.877
550	19,945	15,372	239.135	950	40,070	32,171	266.444
560	20,407	15,751	239.962	960	40,607	32,625	267.007
570	20,870	16,131	240.789	970	41,145	33,081	267.566
580	21,337	16,515	241.602	980	41,685	33,537	268.119
590	21,807	16,902	242.405	990	42,226	33,995	268.670

Table A.2 (continued) Ideal Gas Properties of Nitrogen, Oxygen, and Carbon Dioxide

T	$\bar{h}$	$\bar{u}$	$\bar{s}°$	T	$\bar{h}$	$\bar{u}$	$\bar{s}°$
1000	42,769	34,455	269.215	1760	86,420	71,787	301.543
1020	43,859	35,378	270.293	1780	87,612	72,812	302.271
1040	44,953	36,306	271.354	1800	88,806	73,840	302.884
1060	46,051	37,238	272.400	1820	90,000	74,868	303.544
1080	47,153	38,174	273.430	1840	91,196	75,897	304.198
1100	48,258	39,112	274.445	1860	92,394	76,929	304.845
1120	49,369	40,057	275.444	1880	93,593	77,962	305.487
1140	50,484	41,006	276.430	1900	94,793	78,996	306.122
1160	51,602	41,957	277.403	1920	95,995	80,031	306.751
1180	52,724	42,913	278.362	1940	97,197	81,067	307.374
1200	53,848	43,871	279.307	1960	98,401	82,105	307.992
1220	54,977	44,834	280.238	1980	99,606	83,144	308.604
1240	56,108	45,799	281.158	2000	100,804	84,185	309.210
1260	57,244	46,768	282.066	2050	103,835	86,791	310.701
1280	58,381	47,739	282.962	2100	106,864	89,404	312.160
1300	59,522	48,713	283.847	2150	109,898	92,023	313.589
1320	60,666	49,691	284.722	2200	112,939	94,648	314.988
1340	61,813	50,672	285.586	2250	115,984	97,277	316.356
1360	62,963	51,656	286.439	2300	119,035	99,912	317.695
1380	64,116	52,643	287.283	2350	122,091	102,552	319.011
1400	65,271	53,631	288.106	2400	125,152	105,197	320.302
1420	66,427	54,621	288.934	2450	128,219	107,849	321.566
1440	67,586	55,614	289.743	2500	131,290	110,504	322.808
1460	68,748	56,609	290.542	2550	134,368	113,166	324.026
1480	69,911	57,606	291.333	2600	137,449	115,832	325.222
1500	71,078	58,606	292.114	2650	140,533	118,500	326.396
1520	72,246	59,609	292.888	2700	143,620	121,172	327.549
1540	73,417	60,613	292.654	2750	146,713	123,849	328.684
1560	74,590	61,620	294.411	2800	149,808	126,528	329.800
1580	76,767	62,630	295.161	2850	152,908	129,212	330.896
1600	76,944	63,741	295.901	2900	156,009	131,898	331.975
1620	78,123	64,653	296.632	2950	159,117	134,589	333.037
1640	79,303	65,668	297.356	3000	162,226	137,283	334.084
1660	80,486	66,592	298.072	3050	165,341	139,982	335.114
1680	81,670	67,702	298.781	3100	168,456	142,681	336.126
1700	82,856	68,721	299.482	3150	171,576	145,385	337.124
1720	84,043	69,742	300.177	3200	174,695	148,089	338.109
1740	85,231	70,764	300.863	3250	177,822	150,801	339.069

Table A.3 Psychrometric Table: Properties of Moist Air at 101 325 N/m²

Symbols and Units:

P_s = pressure of water vapor at saturation, N/m²

W_s = humidity ratio at saturation, mass of water vapor associated with unit mass of dry air

V_a = specific volume of dry air, m³/kg

V_s = specific volume of saturated mixture, m³/kg dry air

h_a^a = specific enthalpy of dry air, kJ/kg

h_s = specific enthalpy of saturated mixture, kJ/kg dry air

s_s = specific entropy of saturated mixture, J/K·kg dry air

Temperature			*Properties*						
C	*K*	*F*	P_s	W_s	V_a	V_s	h_a	h_s	s_s
-40	233.15	-40	12.838	0.000 079 25	0.659 61	0.659 68	-22.35	-22.16	-90.659
-30	243.15	-22	37.992	0.000 234 4	0.688 08	0.688 33	-12.29	-11.72	-46.732
-25	248.15	-13	63.248	0.000 390 3	0.702 32	0.702 75	-7.265	-6.306	-24.706
-20	253.15	-4	103.19	0.000 637 1	0.716 49	0.717 24	-2.236	-0.6653	-2.2194
-15	258.15	+5	165.18	0.001 020	0.730 72	0.731 91	+2.794	5.318	21.189
-10	263 15	14	259 72	0.001 606	0.744 95	0.746 83	7.823	11.81	46.104
-5	268.15	23	401.49	0.002 485	0.759 12	0.762 18	12.85	19.04	73.365
0	273.15	32	610 80	0.003 788	0 773 36	0.778 04	17.88	27.35	104.14
5	278.15	41	871 93	0 005 421	0.787 59	0.794 40	22.91	36.52	137.39
10	283.15	50	1 227 2	0 007 658	0.801 76	0.811 63	27.94	47.23	175.54
15	288 15	59	1 704 4	0.010 69	0.816 00	0.829 98	32.97	59.97	220.22
20	293.15	68	2 337.2	0 014 75	0.830 17	0.849 83	38.00	75.42	273.32
25	298.15	77	3 167.0	0 020 16	0.844 34	0.871 62	43.03	94.38	337.39
30	303 15	86	4 242.8	0 027 31	0.858 51	0.896 09	48.07	117.8	415.65
35	308.15	95	5 623.4	0.036 73	0.872 74	0.924 06	53.10	147.3	512.17
40	313 15	104	7 377.6	0.049 11	0.886 92	0.956 65	58.14	184.5	532.31
45	318.15	113	9 584 8	0 065 36	0.901 15	0.995 35	63.17	232.0	783.06
50	323 15	122	12 339	0.086 78	0.915 32	1.042 3	68.21	293.1	975.27
55	328.15	131	15 745	0.115 2	0.929 49	1.100 7	73.25	372.9	1 221.5
60	333.15	140	19 925	0.153 4	0.943 72	1.174 8	78.29	478.5	1 543.5
65	338.15	149	25 014	0.205 5	0.957 90	1.272 1	83.33	621.4	1 973.6
70	343.15	158	31 167	0.278 8	0.972 07	1.404 2	88.38	820.5	2 564.8
75	348.15	167	38 554	0.385 8	0.986 30	1.592 4	93.42	1 110	3 412.8
80	353.15	176	47 365	0.551 9	1.000 5	1.879 1	98.47	1 557	4 710.9
85	358.15	185	57 809	0.836 3	1.014 6	2.363 2	103.5	2 321	6 892.6
90	363 15	194	70 112	1.416	1.028 8	3.340 9	108.6	3 876	11 281

Note: The P_s column in this table gives the vapor pressure of pure water at temperature intervals of five degrees Celsius. For the latest data on vapor pressures at intervals of 0.1 deg C, see "Vapor Pressure Equation for Water," A. Wexler and L. Greenspan, *J. Res. Nat. Bur. Stand.*, 75A(3):213-229, May-June 1971

[r] Fpr very low barometric pressures and high wet-bulb temperatures, the values of h_s in this table are somewhat low; for corrections, see "ASHRAE Handbook of Fundamentals."

[a]Computed from: Psychrometric Tables, in "ASHRAE Handbook of Fundamentals," American Society of Heating, refrigerating and Air-Conditioning Engineers, 1972.

Table A.4 Water Vapor at Low Pressures: Perfect Gas Behavior pv/T = R = 0.461 51 kJ/kg·K

Symbols and Units:

t = thermodynamic temperature, deg C
T = thermodynamic temperature, K
pv = RT, kJ/kg
u_o = specific internal energy at zero pressure, kJ/kg
h_o = specific enthalpy at zero pressure, kJ/kg
s_I = specific entropy of semiperfect vapor at 0.1 MN/m^2, kJ/kg·K
ψ_I = specific Helmholtz free energy of semiperfect vapor at 0.1 MN/m^2, kJ/kg
ψ_I = specific Helmholtz free energy of semiperfect vapor at 0.1 MN/m^2, kJ/kg
ζ_I = specific Gibbs free energy of semiperfect vapor at 0.1 MN/m^2, kJ/kg
p_r = relative pressure, pressure of semiperfect vapor at zero entropy, TN/m^2
v_r = relative specific volume, specific volume of semiperfect vapor at zero entropy, mm^3/kg
c_{po} = specific heat capacity at constant pressure for zero pressure, kJ/kg·K
c_{vo} = specific heat capacity at constant volume for zero pressure, kJ/kg·K
k = c_{po}/c_{vo} = isentropic exponent, $-(\partial \log p/\partial \log v)_s$

t	T	pv	u_o	h_o	s_I	ψ_I	ζ_I	p_r	v_r	c_{po}	c_{vo}	k
0	273.15	126.06	2 375.5	2 501.5	6.804 2	516.9	643.0	.252 9	498.4	1.858 4	1.396 9	1.330 4
10	283.15	130.68	2 389.4	2 520.1	6.871 1	443 9	574.6	292 3	447.0	1.860 1	1.398 6	1.330 0
20	293.15	135.29	2 403.4	2 538 7	6.935 7	370 2	505.5	336 3	402.4	1.862 2	1.400 7	1 329 5
30	303 15	139.91	2 417 5	2 557 4	6 998 2	296 0	435 9	385 0	363.4	1 864 7	1 403 1	1 328 9
40	313.15	144.52	2 431.5	2 576.0	7.058 7	221 1	365.6	439 0	329.2	1 867 4	1.405 9	1 328 3
50	323.15	149.14	2 445.6	2 594.7	7.117 5	145 6	294.7	498 6	299 1	1.870 5	1.409 0	1 327 5
60	333.15	153 75	2 459.7	2 613 4	7.174 5	69 5	223.2	564 2	272 5	1 873 8	1.412 3	1 326 8
70	343.15	158.37	2 473.8	2 632 2	7 230 0	−7 2	151 2	636 3	248 9	1 877 4	1 415 9	1 325 9
80	353.15	162 98	2 488.0	2 651.0	7.284 0	−84 3	78.6	715 2	227 9	1 881 2	1 419 7	1 325 1
90	363.15	167.60	2 502.2	2 669 8	7.336 6	−162 1	5.5	801 5	209 1	1 885 2	1 423 7	1.324 2
100	373.15	172.21	2 516.5	2 688.7	7 387 8	−240 3	−68.1	895 7	192.26	1.889 4	1.427 9	1.323 2
120	393.15	181.44	2 545.1	2 726.6	7 486 7	−398 3	−216.8	1.109 7	163 50	1.898 3	1.436 7	1 321 2
140	413.15	190.67	2 573.9	2 764.6	7 581 1	−558.2	−367.5	1 361 7	140.03	1.907 7	1.446 2	1 319 1
160	433.15	199.90	2 603.0	2 802.9	7.671 5	−720.0	−520.1	1 656 4	120 69	1.917 7	1 456 2	1.316 9
180	453.15	209.13	2 632.2	2 841.3	7 758 3	−883.5	−674.4	1 999 1	104 61	1 928 1	1 466 6	1.314 7
200	473.15	218.4	2 661.6	2 880.0	7.841 8	−1 048.7	−830.4	2 396	91.15	1.938 9	1.477 4	1.312 4
300	573.15	264.5	2 812.3	3 076.8	8.218 9	−1 898 4	−1 633.9	5.423	48 77	1.997 5	1.536 0	1.300 5
400	673.15	310.7	2 969.0	3 279.7	8.545 1	−2 783.1	−2 472.5	10.996	28.25	2.061 4	1.599 9	1.288 5
500	773.15	356.8	3 132.4	3 489.2	8.835 2	−3 699	−3 342	20.61	17.310	2.128 7	1.667 2	1.276 8
600	873.15	403.0	3 302.5	3 705.5	9.098 2	−4 642	−4 239	36.45	11 056	2.198 0	1 736 5	1.265 8
700	973.15	449.1	3 479 7	3 928.8	9.340 3	−5 610	−5 161	61.58	7 293	2 268 3	1 806 8	1 255 4
800	1 073.15	495.3	3 663.9	4 159.2	9.565 5	−6 601	−6 106	100.34	4.936	2.338 7	1.877 1	1.245 9
900	1 173.15	541.4	3 855.1	4 396.5	9.776 9	−7 615	−7 073	158.63	3.413	2.407 8	1.946 2	1.237 1
1 000	1 273.15	587.6	4 053.1	4 640.6	9.976 6	−8 649	−8 061	244.5	2.403	2.474 4	2.012 8	1.299 3
1 100	1 373.15	633.7	4 257.5	4 891.2	10.166 1	−9 702	−9 068	368.6	1.719	2.536 9	2.075 4	1.222 4
1 200	1 473.15	679.9	4 467.9	5 147.8	10.346 4	−10 774	−10 094	544.9	1.248	2.593 8	2 132 3	1.216 4
1 300	1 573.15	726.0	4 683.7	5 409.7	10.518 4	−11 863	−11 137	791 0	.918	2.643 1	2.181 6	1.211 5

· Adapted from: "Steam Tables," J. H. Keenan, F. G. Keyes, P G. Hill, and J. G. Moore, John Wiley & Sons, Inc. 1969 (International Edition · Metric Units).

REFERENCE

For other steam tables in metric units, see "Steam Tables in SI Units," Ministry of Technology, London, 1970.

Table A.5 Properties of Saturated Water and Steam

Part a. Temperature Table

Temp. °C	Press. bars	Specific Volume m³/kg Sat. Liquid $v_f \times 10^3$	Sat. Vapor v_g	Internal Energy kJ/kg Sat. Liquid u_f	Sat. Vapor u_g	Enthalpy kJ/kg Sat. Liquid h_f	Evap. h_{fg}	Sat. Vapor h_g	Entropy kJ/kg · K Sat. Liquid s_f	Sat. Vapor s_g	Temp. °C
.01	0.00611	1.0002	206.136	0.00	2375.3	0.01	2501.3	2501.4	0.0000	9.1562	.01
4	0.00813	1.0001	157.232	16.77	2380.9	16.78	2491.9	2508.7	0.0610	9.0514	4
5	0.00872	1.0001	147.120	20.97	2382.3	20.98	2489.6	2510.6	0.0761	9.0257	5
6	0.00935	1.0001	137.734	25.19	2383.6	25.20	2487.2	2512.4	0.0912	9.0003	6
8	0.01072	1.0002	120.917	33.59	2386.4	33.60	2482.5	2516.1	0.1212	8.9501	8
10	0.01228	1.0004	106.379	42.00	2389.2	42.01	2477.7	2519.8	0.1510	8.9008	10
11	0.01312	1.0004	99.857	46.20	2390.5	46.20	2475.4	2521.6	0.1658	8.8765	11
12	0.01402	1.0005	93.784	50.41	2391.9	50.41	2473.0	2523.4	0.1806	8.8524	12
13	0.01497	1.0007	88.124	54.60	2393.3	54.60	2470.7	2525.3	0.1953	8.8285	13
14	0.01598	1.0008	82.848	58.79	2394.7	58.80	2468.3	2527.1	0.2099	8.8048	14
15	0.01705	1.0009	77.926	62.99	2396.1	62.99	2465.9	2528.9	0.2245	8.7814	15
16	0.01818	1.0011	73.333	67.18	2397.4	67.19	2463.6	2530.8	0.2390	8.7582	16
17	0.01938	1.0012	69.044	71.38	2398.8	71.38	2461.2	2532.6	0.2535	8.7351	17
18	0.02064	1.0014	65.038	75.57	2400.2	75.58	2458.8	2534.4	0.2679	8.7123	18
19	0.02198	1.0016	61.293	79.76	2401.6	79.77	2456.5	2536.2	0.2823	8.6897	19
20	0.02339	1.0018	57.791	83.95	2402.9	83.96	2454.1	2538.1	0.2966	8.6672	20
21	0.02487	1.0020	54.514	88.14	2404.3	88.14	2451.8	2539.9	0.3109	8.6450	21
22	0.02645	1.0022	51.447	92.32	2405.7	92.33	2449.4	2541.7	0.3251	8.6229	22
23	0.02810	1.0024	48.574	96.51	2407.0	96.52	2447.0	2543.5	0.3393	8.6011	23
24	0.02985	1.0027	45.883	100.70	2408.4	100.70	2444.7	2545.4	0.3534	8.5794	24
25	0.03169	1.0029	43.360	104.88	2409.8	104.89	2442.3	2547.2	0.3674	8.5580	25
26	0.03363	1.0032	40.994	109.06	2411.1	109.07	2439.9	2549.0	0.3814	8.5367	26
27	0.03567	1.0035	38.774	113.25	2412.5	113.25	2437.6	2550.8	0.3954	8.5156	27
28	0.03782	1.0037	36.690	117.42	2413.9	117.43	2435.2	2552.6	0.4093	8.4946	28
29	0.04008	1.0040	34.733	121.60	2415.2	121.61	2432.8	2554.5	0.4231	8.4739	29
30	0.04246	1.0043	32.894	125.78	2416.6	125.79	2430.5	2556.3	0.4369	8.4533	30
31	0.04496	1.0046	31.165	129.96	2418.0	129.97	2428.1	2558.1	0.4507	8.4329	31
32	0.04759	1.0050	29.540	134.14	2419.3	134.15	2425.7	2559.9	0.4644	8.4127	32
33	0.05034	1.0053	28.011	138.32	2420.7	138.33	2423.4	2561.7	0.4781	8.3927	33
34	0.05324	1.0056	26.571	142.50	2422.0	142.50	2421.0	2563.5	0.4917	8.3728	34
35	0.05628	1.0060	25.216	146.67	2423.4	146.68	2418.6	2565.3	0.5053	8.3531	35
36	0.05947	1.0063	23.940	150.85	2424.7	150.86	2416.2	2567.1	0.5188	8.3336	36
38	0.06632	1.0071	21.602	159.20	2427.4	159.21	2411.5	2570.7	0.5458	8.2950	38
40	0.07384	1.0078	19.523	167.56	2430.1	167.57	2406.7	2574.3	0.5725	8.2570	40
45	0.09593	1.0099	15.258	188.44	2436.8	188.45	2394.8	2583.2	0.6387	8.1648	45

Table A.5 (continued) Properties of Saturated Water and Steam

Temp. °C	Press. bars	Specific Volume m³/kg Sat. Liquid $v_f \times 10^3$	Sat. Vapor v_g	Internal Energy kJ/kg Sat. Liquid u_f	Sat. Vapor u_g	Enthalpy kJ/kg Sat. Liquid h_f	Evap. h_{fg}	Sat. Vapor h_g	Entropy kJ/kg · K Sat. Liquid s_f	Sat. Vapor s_g	Temp. °C
50	.1235	1.0121	12.032	209.32	2443.5	209.33	2382.7	2592.1	.7038	8.0763	50
55	.1576	1.0146	9.568	230.21	2450.1	230.23	2370.7	2600.9	.7679	7.9913	55
60	.1994	1.0172	7.671	251.11	2456.6	251.13	2358.5	2609.6	.8312	7.9096	60
65	.2503	1.0199	6.197	272.02	2463.1	272.06	2346.2	2618.3	.8935	7.8310	65
70	.3119	1.0228	5.042	292.95	2469.6	292.98	2333.8	2626.8	.9549	7.7553	70
75	.3858	1.0259	4.131	313.90	2475.9	313.93	2321.4	2635.3	1.0155	7.6824	75
80	.4739	1.0291	3.407	334.86	2482.2	334.91	2308.8	2643.7	1.0753	7.6122	80
85	.5783	1.0325	2.828	355.84	2488.4	355.90	2296.0	2651.9	1.1343	7.5445	85
90	.7014	1.0360	2.361	376.85	2494.5	376.92	2283.2	2660.1	1.1925	7.4791	90
95	.8455	1.0397	1.982	397.88	2500.6	397.96	2270.2	2668.1	1.2500	7.4159	95
100	1.014	1.0435	1.673	418.94	2506.5	419.04	2257.0	2676.1	1.3069	7.3549	100
110	1.433	1.0516	1.210	461.14	2518.1	461.30	2230.2	2691.5	1.4185	7.2387	110
120	1.985	1.0603	0.8919	503.50	2529.3	503.71	2202.6	2706.3	1.5276	7.1296	120
130	2.701	1.0697	0.6685	546.02	2539.9	546.31	2174.2	2720.5	1.6344	7.0269	130
140	3.613	1.0797	0.5089	588.74	2550.0	589.13	2144.7	2733.9	1.7391	6.9299	140
150	4.758	1.0905	0.3928	631.68	2559.5	632.20	2114.3	2746.5	1.8418	6.8379	150
160	6.178	1.1020	0.3071	674.86	2568.4	675.55	2082.6	2758.1	1.9427	6.7502	160
170	7.917	1.1143	0.2428	718.33	2576.5	719.21	2049.5	2768.7	2.0419	6.6663	170
180	10.02	1.1274	0.1941	762.09	2583.7	763.22	2015.0	2778.2	2.1396	6.5857	180
190	12.54	1.1414	0.1565	806.19	2590.0	807.62	1978.8	2786.4	2.2359	6.5079	190
200	15.54	1.1565	0.1274	850.65	2595.3	852.45	1940.7	2793.2	2.3309	6.4323	200
210	19.06	1.1726	0.1044	895.53	2599.5	897.76	1900.7	2798.5	2.4248	6.3585	210
220	23.18	1.1900	0.08619	940.87	2602.4	943.62	1858.5	2802.1	2.5178	6.2861	220
230	27.95	1.2088	0.07158	986.74	2603.9	990.12	1813.8	2804.0	2.6099	6.2146	230
240	33.44	1.2291	0.05976	1033.2	2604.0	1037.3	1766.5	2803.8	2.7015	6.1437	240
250	39.73	1.2512	0.05013	1080.4	2602.4	1085.4	1716.2	2801.5	2.7927	6.0730	250
260	46.88	1.2755	0.04221	1128.4	2599.0	1134.4	1662.5	2796.6	2.8838	6.0019	260
270	54.99	1.3023	0.03564	1177.4	2593.7	1184.5	1605.2	2789.7	2.9751	5.9301	270
280	64.12	1.3321	0.03017	1227.5	2586.1	1236.0	1543.6	2779.6	3.0668	5.8571	280
290	74.36	1.3656	0.02557	1278.9	2576.0	1289.1	1477.1	2766.2	3.1594	5.7821	290
300	85.81	1.4036	0.02167	1332.0	2563.0	1344.0	1404.9	2749.0	3.2534	5.7045	300
320	112.7	1.4988	0.01549	1444.6	2525.5	1461.5	1238.6	2700.1	3.4480	5.5362	320
340	145.9	1.6379	0.01080	1570.3	2464.6	1594.2	1027.9	2622.0	3.6594	5.3357	340
360	186.5	1.8925	0.006945	1725.2	2351.5	1760.5	720.5	2481.0	3.9147	5.0526	360
374.14	220.9	3.155	0.003155	2029.6	2029.6	2099.3	0	2099.3	4.4298	4.4298	374.14

Table A.5 (continued) Properties of Saturated Water and Steam

Part b. Pressure Table

		Specific Volume m³/kg		Internal Energy kJ/kg		Enthalpy kJ/kg			Entropy kJ/kg · K		
Press. bars	Temp. °C	Sat. Liquid $v_f \times 10^3$	Sat. Vapor v_g	Sat. Liquid u_f	Sat. Vapor u_g	Sat. Liquid h_f	Evap. h_{fg}	Sat. Vapor h_g	Sat. Liquid s_f	Sat. Vapor s_g	Press. bars
0.04	28.96	1.0040	34.800	121.45	2415.2	121.46	2432.9	2554.4	0.4226	8.4746	0.04
0.06	36.16	1.0064	23.739	151.53	2425.0	151.53	2415.9	2567.4	0.5210	8.3304	0.06
0.08	41.51	1.0084	18.103	173.87	2432.2	173.88	2403.1	2577.0	0.5926	8.2287	0.08
0.10	45.81	1.0102	14.674	191.82	2437.9	191.83	2392.8	2584.7	0.6493	8.1502	0.10
0.20	60.06	1.0172	7.649	251.38	2456.7	251.40	2358.3	2609.7	0.8320	7.9085	0.20
0.30	69.10	1.0223	5.229	289.20	2468.4	289.23	2336.1	2625.3	0.9439	7.7686	0.30
0.40	75.87	1.0265	3.993	317.53	2477.0	317.58	2319.2	2636.8	1.0259	7.6700	0.40
0.50	81.33	1.0300	3.240	340.44	2483.9	340.49	2305.4	2645.9	1.0910	7.5939	0.50
0.60	85.94	1.0331	2.732	359.79	2489.6	359.86	2293.6	2653.5	1.1453	7.5320	0.60
0.70	89.95	1.0360	2.365	376.63	2494.5	376.70	2283.3	2660.0	1.1919	7.4797	0.70
0.80	93.50	1.0380	2.087	391.58	2498.8	391.66	2274.1	2665.8	1.2329	7.4346	0.80
0.90	96.71	1.0410	1.869	405.06	2502.6	405.15	2265.7	2670.9	1.2695	7.3949	0.90
1.00	99.63	1.0432	1.694	417.36	2506.1	417.46	2258.0	2675.5	1.3026	7.3594	1.00
1.50	111.4	1.0528	1.159	466.94	2519.7	467.11	2226.5	2693.6	1.4336	7.2233	1.50
2.00	120.2	1.0605	0.8857	504.49	2529.5	504.70	2201.9	2706.7	1.5301	7.1271	2.00
2.50	127.4	1.0672	0.7187	535.10	2537.2	535.37	2181.5	2716.9	1.6072	7.0527	2.50
3.00	133.6	1.0732	0.6058	561.15	2543.6	561.47	2163.8	2725.3	1.6718	6.9919	3.00
3.50	138.9	1.0786	0.5243	583.95	2546.9	584.33	2148.1	2732.4	1.7275	6.9405	3.50
4.00	143.6	1.0836	0.4625	604.31	2553.6	604.74	2133.8	2738.6	1.7766	6.8959	4.00
4.50	147.9	1.0882	0.4140	622.25	2557.6	623.25	2120.7	2743.9	1.8207	6.8565	4.50
5.00	151.9	1.0926	0.3749	639.68	2561.2	640.23	2108.5	2748.7	1.8607	6.8212	5.00
6.00	158.9	1.1006	0.3157	669.90	2567.4	670.56	2086.3	2756.8	1.9312	6.7600	6.00
7.00	165.0	1.1080	0.2729	696.44	2572.5	697.22	2066.3	2763.5	1.9922	6.7080	7.00
8.00	170.4	1.1148	0.2404	720.22	2576.8	721.11	2048.0	2769.1	2.0462	6.6628	8.00
9.00	175.4	1.1212	0.2150	741.83	2580.5	742.83	2031.1	2773.9	2.0946	6.6226	9.00
10.0	179.9	1.1273	0.1944	761.68	2583.6	762.81	2015.3	2778.1	2.1387	6.5863	10.0
15.0	198.3	1.1539	0.1318	843.16	2594.5	844.84	1947.3	2792.2	2.3150	6.4448	15.0
20.0	212.4	1.1767	0.09963	906.44	2600.3	908.79	1890.7	2799.5	2.4474	6.3409	20.0
25.0	224.0	1.1973	0.07998	959.11	2603.1	962.11	1841.0	2803.1	2.5547	6.2575	25.0
30.0	233.9	1.2165	0.06668	1004.8	2604.1	1008.4	1795.7	2804.2	2.6457	6.1869	30.0
35.0	242.6	1.2347	0.05707	1045.4	2603.7	1049.8	1753.7	2803.4	2.7253	6.1253	35.0
40.0	250.4	1.2522	0.04978	1082.3	2602.3	1087.3	1714.1	2801.4	2.7964	6.0701	40.0
45.0	257.5	1.2692	0.04406	1116.2	2600.1	1121.9	1676.4	2798.3	2.8610	6.0199	45.0
50.0	264.0	1.2859	0.03944	1147.8	2597.1	1154.2	1640.1	2794.3	2.9202	5.9734	50.0
60.0	275.6	1.3187	0.03244	1205.4	2589.7	1213.4	1571.0	2784.3	3.0267	5.8892	60.0
70.0	285.9	1.3513	0.02737	1257.6	2580.5	1267.0	1505.1	2772.1	3.1211	5.8133	70.0
80.0	295.1	1.3842	0.02352	1305.6	2569.8	1316.6	1441.3	2758.0	3.2068	5.7432	80.0
90.0	303.4	1.4178	0.02048	1350.5	2557.8	1363.3	1378.9	2742.1	3.2858	5.6772	90.0
100.	311.1	1.4524	0.01803	1393.0	2544.4	1407.6	1317.1	2724.7	3.3596	5.6141	100.
110.	318.2	1.4886	0.01599	1433.7	2529.8	1450.1	1255.5	2705.6	3.4295	5.5527	110.
120.	324.8	1.5267	0.01426	1473.0	2513.7	1491.3	1193.6	2684.9	3.4962	5.4924	120.
130.	330.9	1.5671	0.01278	1511.1	2496.1	1531.5	1130.7	2662.2	3.5606	5.4323	130.
140.	336.8	1.6107	0.01149	1548.6	2476.8	1571.1	1066.5	2637.6	3.6232	5.3717	140.
150.	342.2	1.6581	0.01034	1585.6	2455.5	1610.5	1000.0	2610.5	3.6848	5.3098	150.
160.	347.4	1.7107	0.009306	1622.7	2431.7	1650.1	930.6	2580.6	3.7461	5.2455	160.
170.	352.4	1.7702	0.008364	1660.2	2405.0	1690.3	856.9	2547.2	3.8079	5.1777	170.
180.	357.1	1.8397	0.007489	1698.9	2374.3	1732.0	777.1	2509.1	3.8715	5.1044	180.
190.	361.5	1.9243	0.006657	1739.9	2338.1	1776.5	688.0	2464.5	3.9388	5.0228	190.
200.	365.8	2.036	0.005834	1785.6	2293.0	1826.3	583.4	2409.7	4.0139	4.9269	200.
220.9	374.1	3.155	0.003155	2029.6	2029.6	2099.3	0	2099.3	4.4298	4.4298	220.9

Source: Adapted from M.J. Moran and H.N. Shapiro, *Fundamentals of Engineering Thermodynamics*, 3rd. ed., Wiley, New York, 1995, as extracted from J.H. Keenan, F.G. Keyes, P.G. Hill, and J.G. Moore, *Steam Tables*, Wiley, New York, 1969.

Table A.6 Properties of Superheated Steam

Symbols and Units:

T = temperature, °C
T_{sat} = Saturation temperature, °C
v = Specific volume, m^3/kg
u = internal energy, kJ/kg
h = enthalpy, kJ/kg
S = entropy, kJ/kg·K
p = pressure, bar and μPa

T °C	v m³/kg	u kJ/kg	h kJ/kg	s kJ/kg · K	v m³/kg	u kJ/kg	h kJ/kg	s kJ/kg · K
	p = 0.06 bar = 0.006 MPa (T_{sat} = 36.16°C)				p = 0.35 bar = 0.035 MPa (T_{sat} = 72.69°C)			
Sat.	23.739	2425.0	2567.4	8.3304	4.526	2473.0	2631.4	7.7158
80	27.132	2487.3	2650.1	8.5804	4.625	2483.7	2645.6	7.7564
120	30.219	2544.7	2726.0	8.7840	5.163	2542.4	2723.1	7.9644
160	33.302	2602.7	2802.5	8.9693	5.696	2601.2	2800.6	8.1519
200	36.383	2661.4	2879.7	9.1398	6.228	2660.4	2878.4	8.3237
240	39.462	2721.0	2957.8	9.2982	6.758	2720.3	2956.8	8.4828
280	42.540	2781.5	3036.8	9.4464	7.287	2780.9	3036.0	8.6314
320	45.618	2843.0	3116.7	9.5859	7.815	2842.5	3116.1	8.7712
360	48.696	2905.5	3197.7	9.7180	8.344	2905.1	3197.1	8.9034
400	51.774	2969.0	3279.6	9.8435	8.872	2968.6	3279.2	9.0291
440	54.851	3033.5	3362.6	9.9633	9.400	3033.2	3362.2	9.1490
500	59.467	3132.3	3489.1	10.1336	10.192	3132.1	3488.8	9.3194

T °C	v m³/kg	u kJ/kg	h kJ/kg	s kJ/kg · K	v m³/kg	u kJ/kg	h kJ/kg	s kJ/kg · K
	p = 0.70 bar = 0.07 MPa (T_{sat} = 89.95°C)				p = 1.0 bar = 0.10 MPa (T_{sat} = 99.63°C)			
Sat.	2.365	2494.5	2660.0	7.4797	1.694	2506.1	2675.5	7.3594
100	2.434	2509.7	2680.0	7.5341	1.696	2506.7	2676.2	7.3614
120	2.571	2539.7	2719.6	7.6375	1.793	2537.3	2716.6	7.4668
160	2.841	2599.4	2798.2	7.8279	1.984	2597.8	2796.2	7.6597
200	3.108	2659.1	2876.7	8.0012	2.172	2658.1	2875.3	7.8343
240	3.374	2719.3	2955.5	8.1611	2.359	2718.5	2954.5	7.9949
280	3.640	2780.2	3035.0	8.3162	2.546	2779.6	3034.2	8.1445
320	3.905	2842.0	3115.3	8.4504	2.732	2841.5	3114.6	8.2849
360	4.170	2904.6	3196.5	8.5828	2.917	2904.2	3195.9	8.4175
400	4.434	2968.2	3278.6	8.7086	3.103	2967.9	3278.2	8.5435
440	4.698	3032.9	3361.8	8.8286	3.288	3032.6	3361.4	8.6636
500	5.095	3131.8	3488.5	8.9991	3.565	3131.6	3488.1	8.8342

T °C	v m³/kg	u kJ/kg	h kJ/kg	s kJ/kg · K	v m³/kg	u kJ/kg	h kJ/kg	s kJ/kg · K
	p = 1.5 bars = 0.15 MPa (T_{sat} = 111.37°C)				p = 3.0 bars = 0.30 MPa (T_{sat} = 133.55°C)			
Sat.	1.159	2519.7	2693.6	7.2233	0.606	2543.6	2725.3	6.9919
120	1.188	2533.3	2711.4	7.2693				
160	1.317	2595.2	2792.8	7.4665	0.651	2587.1	2782.3	7.1276
200	1.444	2656.2	2872.9	7.6433	0.716	2650.7	2865.5	7.3115
240	1.570	2717.2	2952.7	7.8052	0.781	2713.1	2947.3	7.4774
280	1.695	2778.6	3032.8	7.9555	0.844	2775.4	3028.6	7.6299
320	1.819	2840.6	3113.5	8.0964	0.907	2838.1	3110.1	7.7722
360	1.943	2903.5	3195.0	8.2293	0.969	2901.4	3192.2	7.9061
400	2.067	2967.3	3277.4	8.3555	1.032	2965.6	3275.0	8.0330
440	2.191	3032.1	3360.7	8.4757	1.094	3030.6	3358.7	8.1538
500	2.376	3131.2	3487.6	8.6466	1.187	3130.0	3486.0	8.3251
600	2.685	3301.7	3704.3	8.9101	1.341	3300.8	3703.2	8.5892

Table A.6 (continued)   Properties of Superheated Steam

Symbols and Units:

T = temperature, °C

T_{sat} = Saturation temperature, °C

v = Specific volume, m³/kg

u = internal energy, kJ/kg

h = enthalpy, kJ/kg

S = entropy, kJ/kg·K

p = pressure, bar and μPa

T °C	v m³/kg	u kJ/kg	h kJ/kg	s kJ/kg·K	v m³/kg	u kJ/kg	h kJ/kg	s kJ/kg·k
	p = 5.0 bars = 0.50 MPa (T_{sat} = 151.86°C)				p = 7.0 bars = 0.70 MPa (T_{sat} = 164.97°C)			
Sat.	0.3749	2561.2	2748.7	6.8213	0.2729	2572.5	2763.5	6.7080
180	0.4045	2609.7	2812.0	6.9656	0.2847	2599.8	2799.1	6.7880
200	0.4249	2642.9	2855.4	7.0592	0.2999	2634.8	2844.8	6.8865
240	0.4646	2707.6	2939.9	7.2307	0.3292	2701.8	2932.2	7.0641
280	0.5034	2771.2	3022.9	7.3865	0.3574	2766.9	3017.1	7.2233
320	0.5416	2834.7	3105.6	7.5308	0.3852	2831.3	3100.9	7.3697
360	0.5796	2898.7	3188.4	7.6660	0.4126	2895.8	3184.7	7.5063
400	0.6173	2963.2	3271.9	7.7938	0.4397	2960.9	3268.7	7.6350
440	0.6548	3028.6	3356.0	7.9152	0.4667	3026.6	3353.3	7.7571
500	0.7109	3128.4	3483.9	8.0873	0.5070	3126.8	3481.7	7.9299
600	0.8041	3299.6	3701.7	8.3522	0.5738	3298.5	3700.2	8.1956
700	0.8969	3477.5	3925.9	8.5952	0.6403	3476.6	3924.8	8.4391
	p = 10.0 bars = 1.0 MPa (T_{sat} = 179.91°C)				p = 15.0 bars = 1.5 MPa (T_{sat} = 198.32°C)			
Sat.	0.1944	2583.6	2778.1	6.5865	0.1318	2594.5	2792.2	6.4448
200	0.2060	2621.9	2827.9	6.6940	0.1325	2598.1	2796.8	6.4546
240	0.2275	2692.9	2920.4	6.8817	0.1483	2676.9	2899.3	6.6628
280	0.2480	2760.2	3008.2	7.0465	0.1627	2748.6	2992.7	6.8381
320	0.2678	2826.1	3093.9	7.1962	0.1765	2817.1	3081.9	6.9938
360	0.2873	2891.6	3178.9	7.3349	0.1899	2884.4	3169.2	7.1363
400	0.3066	2957.3	3263.9	7.4651	0.2030	2951.3	3255.8	7.2690
440	0.3257	3023.6	3349.3	7.5883	0.2160	3018.5	3342.5	7.3940
500	0.3541	3124.4	3478.5	7.7622	0.2352	3120.3	3473.1	7.5698
540	0.3729	3192.6	3565.6	7.8720	0.2478	3189.1	3560.9	7.6805
600	0.4011	3296.8	3697.9	8.0290	0.2668	3293.9	3694.0	7.8385
640	0.4198	3367.4	3787.2	8.1290	0.2793	3364.8	3783.8	7.9391
	p = 20.0 bars = 2.0 MPa (T_{sat} = 212.42°C)				p = 30.0 bars = 3.0 MPa (T_{sat} = 233.90°C)			
Sat.	0.0996	2600.3	2799.5	6.3409	0.0667	2604.1	2804.2	6.1869
240	0.1085	2659.6	2876.5	6.4952	0.0682	2619.7	2824.3	6.2265
280	0.1200	2736.4	2976.4	6.6828	0.0771	2709.9	2941.3	6.4462
320	0.1308	2807.9	3069.5	6.8452	0.0850	2788.4	3043.4	6.6245
360	0.1411	2877.0	3159.3	6.9917	0.0923	2861.7	3138.7	6.7801
400	0.1512	2945.2	3247.6	7.1271	0.0994	2932.8	3230.9	6.9212
440	0.1611	3013.4	3335.5	7.2540	0.1062	3002.9	3321.5	7.0520
500	0.1757	3116.2	3467.6	7.4317	0.1162	3108.0	3456.5	7.2338
540	0.1853	3185.6	3556.1	7.5434	0.1227	3178.4	3546.6	7.3474
600	0.1996	3290.9	3690.1	7.7024	0.1324	3285.0	3682.3	7.5085
640	0.2091	3362.2	3780.4	7.8035	0.1388	3357.0	3773.5	7.6106
700	0.2232	3470.9	3917.4	7.9487	0.1484	3466.5	3911.7	7.7571

Table A.6 (continued) Properties of Superheated Steam

Symbols and Units:

T = temperature, °C
T_{sat} = Saturation temperature, °C
v = Specific volume, m³/kg
u = internal energy, kJ/kg
h = enthalpy, kJ/kg
S = entropy, kJ/kg·K
p = pressure, bar and μPa

T °C	v m³/kg	u kJ/kg	h kJ/kg	s kJ/kg · K	v m³/kg	u kJ/kg	h kJ/kg	s kJ/kg · K
	p = 40 bars = 4.0 MPa (T_{sat} = 250.4°C)				p = 60 bars = 6.0 MPa (T_{sat} = 275.64°C)			
Sat.	0.04978	2602.3	2801.4	6.0701	0.03244	2589.7	2784.3	5.8892
280	0.05546	2680.0	2901.8	6.2568	0.03317	2605.2	2804.2	5.9252
320	0.06199	2767.4	3015.4	6.4553	0.03876	2720.0	2952.6	6.1846
360	0.06788	2845.7	3117.2	6.6215	0.04331	2811.2	3071.1	6.3782
400	0.07341	2919.9	3213.6	6.7690	0.04739	2892.9	3177.2	6.5408
440	0.07872	2992.2	3307.1	6.9041	0.05122	2970.0	3277.3	6.6853
500	0.08643	3099.5	3445.3	7.0901	0.05665	3082.2	3422.2	6.8803
540	0.09145	3171.1	3536.9	7.2056	0.06015	3156.1	3517.0	6.9999
600	0.09885	3279.1	3674.4	7.3688	0.06525	3266.9	3658.4	7.1677
640	0.1037	3351.8	3766.6	7.4720	0.06859	3341.0	3752.6	7.2731
700	0.1110	3462.1	3905.9	7.6198	0.07352	3453.1	3894.1	7.4234
740	0.1157	3536.6	3999.6	7.7141	0.07677	3528.3	3989.2	7.5190
	p = 80 bars = 8.0 MPa (T_{sat} = 295.06°C)				p = 100 bars = 10.0 MPa (T_{sat} = 311.06°C)			
Sat.	0.02352	2569.8	2758.0	5.7432	0.01803	2544.4	2724.7	5.6141
320	0.02682	2662.7	2877.2	5.9489	0.01925	2588.8	2781.3	5.7103
360	0.03089	2772.7	3019.8	6.1819	0.02331	2729.1	2962.1	6.0060
400	0.03432	2863.8	3138.3	6.3634	0.02641	2832.4	3096.5	6.2120
440	0.03742	2946.7	3246.1	6.5190	0.02911	2922.1	3213.2	6.3805
480	0.04034	3025.7	3348.4	6.6586	0.03160	3005.4	3321.4	6.5282
520	0.04313	3102.7	3447.7	6.7871	0.03394	3085.6	3425.1	6.6622
560	0.04582	3178.7	3545.3	6.9072	0.03619	3164.1	3526.0	6.7864
600	0.04845	3254.4	3642.0	7.0206	0.03837	3241.7	3625.3	6.9029
640	0.05102	3330.1	3738.3	7.1283	0.04048	3318.9	3723.7	7.0131
700	0.05481	3443.9	3882.4	7.2812	0.04358	3434.7	3870.5	7.1687
740	0.05729	3520.4	3978.7	7.3782	0.04560	3512.1	3968.1	7.2670
	p = 120 bars = 12.0 MPa (T_{sat} = 324.75°C)				p = 140 bars = 14.0 MPa (T_{sat} = 336.75°C)			
Sat.	0.01426	2513.7	2684.9	5.4924	0.01149	2476.8	2637.6	5.3717
360	0.01811	2678.4	2895.7	5.8361	0.01422	2617.4	2816.5	5.6602
400	0.02108	2798.3	3051.3	6.0747	0.01722	2760.9	3001.9	5.9448
440	0.02355	2896.1	3178.7	6.2586	0.01954	2868.6	3142.2	6.1474
480	0.02576	2984.4	3293.5	6.4154	0.02157	2962.5	3264.5	6.3143
520	0.02781	3068.0	3401.8	6.5555	0.02343	3049.8	3377.8	6.4610
560	0.02977	3149.0	3506.2	6.6840	0.02517	3133.6	3486.0	6.5941
600	0.03164	3228.7	3608.3	6.8037	0.02683	3215.4	3591.1	6.7172
640	0.03345	3307.5	3709.0	6.9164	0.02843	3296.0	3694.1	6.8326
700	0.03610	3425.2	3858.4	7.0749	0.03075	3415.7	3846.2	6.9939
740	0.03781	3503.7	3957.4	7.1746	0.03225	3495.2	3946.7	7.0952

Table A.7 Chemical, Physical, and Thermal Properties of Gases: Gases and Vapors, Including Fuels and Refrigerants, English and Metric Units

Common name(s)	*Acetylene (Ethyne)*	*Air [mixture]*	*Ammonia, anhyd.*	*Argon*
Chemical formula	C_2H_2		NH_3	*Ar*
Refrigerant number	–	729	717	740
CHEMICAL AND PHYSICAL PROPERTIES				
Molecular weight	26.04	28.966	17 02	39 948
Specific gravity, air = 1	0 90	1 00	0 59	1 38
Specific volume, ft³/lb	14 9	13 5	23 0	9 80
Specific volume, m³/kg	0.93	0.842	1.43	0 622
Density of liquid (at atm bp), lb/ft³	43 0	54 6	42 6	87 0
Density of liquid (at atm bp), kg/m³	693	879	686	1 400
Vapor pressure at 25 deg C, psia			145 4	
Vapor pressure at 25 deg C, MN/m²			1 00	
Viscosity (abs), lbm/ft sec	$6\ 72 \times 10^{-6}$	$12\ 1 \times 10^{-6}$	$6\ 72 \times 10^{-6}$	$13\ 4 \times 10^{-6}$
Viscosity (abs), centipoises[a]	0 01	0 018	0 010	0 02
Sound velocity in gas, m/sec	343	346	415	322
THERMAL AND THERMODYNAMIC PROPERTIES				
Specific heat, c_p, Btu/lb deg F or cal/g deg C	0.40	0 240 3	0 52	0 125
Specific heat, c_p, J/kg K	1 674	1 005	2 175	523
Specific heat ratio, c_p/c_v	1 25	1 40	1 3	1 67
Gas constant R, ft-lb/lb deg R	59 3	53 3	90 8	38 7
Gas constant R, J/kg deg C	319	286 8	488	208
Thermal conductivity, Btu/hr ft deg F	0 014	0 015 1	0 015	0 010 2
Thermal conductivity, W/m deg C	0 024	0 026	0 026	0 017 2
Boiling point (sat 14.7 psia), deg F	-103	-320	-28	-303
Boiling point (sat 760 mm), deg C	-75	-195	-33 3	-186
Latent heat of evap (at bp), Btu/lb	264	88 2	589 3	70
Latent heat of evap (at bp), J/kg	614 000	205 000	1 373 000	163 000
Freezing (melting) point, deg F (1 atm)	-116	-357 2	-107 9	-308 5
Freezing (melting) point, deg C (1 atm)	-82 2	-216 2	-77 7	-189 2
Latent heat of fusion, Btu/lb	23	10 0	143 0	
Latent heat of fusion, J/kg	53 500	23 200	332 300	
Critical temperature, deg F	97 1	-220 5	271 4	-187 6
Critical temperature, deg C	36 2	-140 3	132 5	-122
Critical pressure, psia	907	550	1 650	707
Critical pressure, MN/m²	6 25	3 8	11 4	4 87
Critical volume, ft³/lb		0 050	0.068	0 029 9
Critical volume, m³/kg		0 003	0 004 24	0 001 86
Flammable (yes or no)	Yes	No	No	No
Heat of combustion, Btu/ft³	1 450	–	–	–
Heat of combustion, Btu/lb	21 600			
Heat of combustion, kJ/kg	50 200		–	

[a]For N sec/m² divide by 1 000

Note. The properties of pure gases are given at 25°C (77°F, 298 K) and atmospheric pressure (except as stated).

Table A.7 (continued) Chemical, Physical, and Thermal Properties of Gases: Gases and Vapors, Including Fuels and Refrigerants, English and Metric Units

Common name(s)	*Butadiene*	*n-Butane*	*Isobutane (2-Methyl propane)*	*1-Butene (Butylene)*
Chemical formula	C_4H_6	C_4H_{10}	C_4H_{10}	C_4H_8
Refrigerant number	–	600	600a	–
CHEMICAL AND PHYSICAL PROPERTIES				
Molecular weight	54 09	58 12	58 12	56 108
Specific gravity, air = 1	1 87	2 07	2 07	1 94
Specific volume, ft^3/lb	7 1	6 5	6 5	6 7
Specific volume, m^3/kg	0 44	0 405	0 418	0.42
Density of liquid (at atm bp), lb/ft^3		37 5	37 2	
Density of liquid (at atm bp), kg/m^3		604	599	
Vapor pressure at 25 deg C, psia		35 4	50 4	
Vapor pressure at 25 deg C, MN/m^2		0 024 4	0 347	
Viscosity (abs), lbm/ft sec		$4\ 8 \times 10^{-6}$		
Viscosity (abs), centipoises[a]		0 007		
Sound velocity in gas, m/sec	226	216	216	222
THERMAL AND THERMODYNAMIC PROPERTIES				
Specific heat, c_p, Btu/lb deg F or cal/g·deg C	0 341	0 39	0 39	0 36
Specific heat, c_p, J/kg K	1 427.	1 675	1 630	1 505.
Specific heat ratio, c_p/c_v	1.12	1 096	1 10	1 112
Gas constant R, ft-lb/lb·deg F	28 55	26 56	26 56	27.52
Gas constant R, J/kg deg C	154	143	143	148.
Thermal conductivity, Btu/hr ft deg F		0 01	0 01	
Thermal conductivity, W/m deg C		0 017	0 017	
Boiling point (sat 14 7 psia), deg F	24 1	31 2	10 8	20 6
Boiling point (sat 760 mm), deg C	- 4 5	- 0 4	- 11 8	- 6.3
Latent heat of evap (at bp), Btu/lb		165 6	157 5	167.9
Latent heat of evap (at bp), J/kg		386 000	366 000	391 000
Freezing (melting) point, deg F (1 atm)	- 164	- 217	- 229	- 301 6
Freezing (melting) point, deg C (1 atm)	- 109	- 138	- 145	185 3
Latent heat of fusion, Btu/lb		19 2		16 4
Latent heat of fusion, J/kg		44 700		38 100
Critical temperature, deg F		306	273	291
Critical temperature, deg C	171	152	134	144
Critical pressure, psia	652	550	537	621
Critical pressure, MN/m^2		3 8	3 7	4 28
Critical volume, ft^3/lb		0 070		0 068
Critical volume, m^3/kg		0 004 3		0 004 2
Flammable (yes or no)	Yes	Yes	Yes	Yes
Heat of combustion, Btu/ft^3	2 950	3 300	3 300	3 150
Heat of combustion, Btu/lb	20 900	21 400	21 400	21 000
Heat of combustion, kJ/kg	48 600	49 700	49 700	48 800

[a] For N sec/m^2 divide by 1 000

Table A.7 (continued) Chemical, Physical, and Thermal Properties of Gases: Gases and Vapors, Including Fuels and Refrigerants, English and Metric Units

Common name(s) / *Chemical formula* / *Refrigerant number*	*cis-2-Butene* C_4H_8	*trans-2-Butene* C_4H_8	*Isobutene* C_4H_8	*Carbon dioxide* CO_2
Refrigerant number	·	–	–	744
CHEMICAL AND PHYSICAL PROPERTIES				
Molecular weight	56.108	56.108	56.108	44.01
Specific gravity, air = 1	1.94	1.94	1.94	1.52
Specific volume, ft³/lb	6 7	6.7	6 7	8 8
Specific volume, m³/kg	0.42	0.42	0.42	0.55
Density of liquid (at atm bp), lb/ft³				–
Density of liquid (at atm bp), kg/m³				–
Vapor pressure at 25 deg C, psia				931.
Vapor pressure at 25 deg C, MN/m²				6.42
Viscosity (abs), lbm/ft sec				9.4×10^{-6}
Viscosity (abs), centipoises[a]				0.014
Sound velocity in gas, m/sec	223.	221	221	270.
THERMAL AND THERMODYNAMIC PROPERTIES				
Specific heat, c_p, Btu/lb deg F or cal/g·deg C	0.327	0 365	0.37	0.205
Specific heat, c_p, J/kg·K	1 368	1 527.	1 548	876.
Specific heat ratio, c_p/c_v	1 121	1.107	1 10	1.30
Gas constant *R*, ft-lb/lb deg F				35.1
Gas constant *R*, J/kg·deg C				189.
Thermal conductivity, Btu/hr·ft·deg F				0 01
Thermal conductivity, W/m·deg C				0.017
Boiling point (sat 14.7 psia), deg F	38.6	33 6	19 2	−109 4[b]
Boiling point (sat 760 mm), deg C	3.7	0.9	−7.1	−78.5
Latent heat of evap (at bp), Btu/lb	178.9	174.4	169	246.
Latent heat of evap (at bp), J/kg	416 000	406 000.	393 000	572 000.
Freezing (melting) point, deg F (1 atm)	−218.	−158		
Freezing (melting) point, deg C (1 atm)	−138.9	−105.5		
Latent heat of fusion, Btu/lb	31 2	41.6	25.3	–
Latent heat of fusion, J/kg	72 600	96 800.	58 800	–
Critical temperature, deg F				88
Critical temperature, deg C	160	155.		31
Critical pressure, psia	595.	610		1 072
Critical pressure, MN/m²	4 10	4 20		7.4
Critical volume, ft³/lb				
Critical volume, m³/kg				
Flammable (yes or no)	Yes	Yes	Yes	No
Heat of combustion, Btu/ft³	3 150.	3 150	3 150.	–
Heat of combustion, Btu/lb	21 000	21 000	21 000.	–
Heat of combustion, kJ/kg	48 800	48 800	48 800.	–

[a]For N sec/m² divide by 1 000
[b]Sublimes

Table A.7 (continued) Chemical, Physical, and Thermal Properties of Gases: Gases and Vapors, Including Fuels and Refrigerants, English and Metric Units

Common name(s)	*Carbon monoxide*	*Chlorine*	*Deuterium*	*Ethane*
Chemical formula	CO	Cl_2	D_2	C_2H_6
Refrigerant number	–	–	–	*170*
CHEMICAL AND PHYSICAL PROPERTIES				
Molecular weight	28.011	70.906	2 014	30.070
Specific gravity, air = 1	0.967	2.45	0 070	1.04
Specific volume, ft^3/lb	14 0	5 52	194 5	13.025
Specific volume, m^3/kg	0.874	0 344	12 12	0 815
Density of liquid (at atm bp), lb/ft^3		97.3		28
Density of liquid (at atm bp), kg/m^3		1 559		449.
Vapor pressure at 25 deg C, psia			0 756	
Vapor pressure at 25 deg C, MN/m^2			0 005 2	
Viscosity (abs), lbm/ft·sec	12.1×10^{-6}	$9\ 4 \times 10^{-6}$	$8\ 75 \times 10^{-6}$	$64. \times 10^{-6}$
Viscosity (abs), centipoises[a]	0.018	0 014	0 013	0.095
Sound velocity in gas, m/sec	352.	215.	930	316.
THERMAL AND THERMODYNAMIC PROPERTIES				
Specific heat, c_p, Btu/lb deg F or cal/g·deg C	0.25	0.114	1 73	0.41
Specific heat, c_p, J/kg K	1 046.	477	7 238	1 715.
Specific heat ratio, c_p/c_v	1.40	1.35	1 40	1.20
Gas constant R, ft-lb/lb deg F	55.2	21 8	384	51.4
Gas constant R, J/kg·deg C	297	117	2 066	276
Thermal conductivity, Btu/hr ft deg F	0 014	0 005	0 081	0 010
Thermal conductivity, W/m deg C	0 024	0 008 7	0 140	0.017
Boiling point (sat 14 7 psia), deg F	- 312 7	- 29 2		– 127
Boiling point (sat 760 mm), deg C	- 191 5	- 34		- 88.3
Latent heat of evap (at bp), Btu/lb	92 8	123 7		210.
Latent heat of evap (at bp), J/kg	216 000.	288 000		488 000.
Freezing (melting) point, deg F (1 atm)	- 337	- 150		- 278.
Freezing (melting) point, deg C (1 atm)	– 205	- 101		- 172.2
Latent heat of fusion, Btu/lb	12 8	41 0		41
Latent heat of fusion, J/kg		95 400		95 300
Critical temperature, deg F	- 220	291	- 390 6	90 1
Critical temperature, deg C	– 140	144	- 234 8	32.2
Critical pressure, psia	507	1 120	241	709.
Critical pressure, MN/m^2	3 49	7 72	1 66	4.89
Critical volume, ft^3/lb	0 053	0 028	0 239	0 076
Critical volume, m^3/kg	0 003 3	0 001 75	0 014 9	0.004 7
Flammable (yes or no)	Yes	No		Yes
Heat of combustion, Btu/ft^3	310	–		
Heat of combustion, Btu/lb	4 340	–		22 300.
Heat of combustion, kJ/kg	10 100	–		51 800.

[a]For N sec/m^2 divide by 1 000

Table A.7 (continued) Chemical, Physical, and Thermal Properties of Gases: Gases and Vapors, Including Fuels and Refrigerants, English and Metric Units

Common name(s)	*Ethyl chloride*	*Ethylene (Ethene)*	*Fluorine*
Chemical formula	C_2H_5Cl	C_2H_4	F_2
Refrigerant number	*160*	*1150*	
CHEMICAL AND PHYSICAL PROPERTIES			
Molecular weight	64.515	28.054	37.996
Specific gravity, air = 1	2 23	0.969	1.31
Specific volume, ft³/lb	6 07	13 9	10 31
Specific volume, m³/kg	0 378	0 87	0.706
Density of liquid (at atm bp), lb/ft³	56.5	35.5	
Density of liquid (at atm bp), kg/m³	905.	569	
Vapor pressure at 25 deg C, psia			
Vapor pressure at 25 deg C, MN/m²			
Viscosity (abs), lbm/ft sec		6.72×10^{-6}	16.1×10^{-6}
Viscosity (abs), centipoises[a]		0 010	0 024
Sound velocity in gas, m/sec	204	331	290.
THERMAL AND THERMODYNAMIC PROPERTIES			
Specific heat, c_p, Btu/lb deg F or cal/g·deg C	0 27	0 37	0.198
Specific heat, c_p, J/kg K	1 130	1 548	828
Specific heat ratio, c_p/c_v	1 13	1 24	1.35
Gas constant R, ft-lb/lb deg F	24.0	55 1	40.7
Gas constant R, J/kg·deg C	129	296	219.
Thermal conductivity, Btu/hr ft·deg F		0 010	0.016
Thermal conductivity, W/m·deg C		0 017	0.028
Boiling point (sat 14.7 psia), deg F	54.	- 155	- 306.4
Boiling point (sat 760 mm), deg C	12 2	− 103 8	− 188.
Latent heat of evap (at bp), Btu/lb	166	208	74.
Latent heat of evap (at bp), J/kg	386 000	484 000	172 000
Freezing (melting) point, deg F (1 atm)	−218	- 272	- 364.
Freezing (melting) point, deg C (1 atm)	- 138 9	− 169	- 220
Latent heat of fusion, Btu/lb	29 3	51 5	11
Latent heat of fusion, J/kg	68 100	120 000	25 600.
Critical temperature, deg F	368 6	49	- 200
Critical temperature, deg C	187	9 5	− 129
Critical pressure, psia	764.	741	810
Critical pressure, MN/m²	5 27	5 11	5.58
Critical volume, ft³/lb	0.049	0 073	
Critical volume, m³/kg	0.003 06	0 004 6	
Flammable (yes or no)	No	Yes	
Heat of combustion, Btu/ft³		1 480.	
Heat of combustion, Btu/lb	—	20 600.	
Heat of combustion, kJ/kg	—	47 800.	

[a] For N·sec/m² divide by 1 000.

Table A.7 (continued) Chemical, Physical, and Thermal Properties of Gases: Gases and Vapors, Including Fuels and Refrigerants, English and Metric Units

Common name(s)	*Fluorocarbons*			
Chemical formula	CCl_3F	CCl_2F_2	$CClF_3$	$CBrF_3$
Refrigerant number	*11*	*12*	*13*	*13B1*
CHEMICAL AND PHYSICAL PROPERTIES				
Molecular weight	137.37	120.91	104 46	148.91
Specific gravity, air = 1	4 74	4.17	3 61	5 14
Specific volume, ft³/lb	2.74	3 12	3.58	2.50
Specific volume, m³/kg	0.171	0.195	0.224	0.975
Density of liquid (at atm bp), lb/ft³	92.1	93.0	95.0	124.4
Density of liquid (at atm bp), kg/m³	1 475.	1 490	1 522.	1 993.
Vapor pressure at 25 deg C, psia		94 51	516.	234.8
Vapor pressure at 25 deg C, MN/m²		0.652	3.56	1.619
Viscosity (abs), lbm/ft·sec	$7\ 39 \times 10^{-6}$	8.74×10^{-6}		
Viscosity (abs), centipoises[a]	0.011	0 013		
Sound velocity in gas, m/sec				
THERMAL AND THERMODYNAMIC PROPERTIES				
Specific heat, c_p, Btu/lb·deg F or cal/g·deg C	0 14	0 146	0.154	
Specific heat, c_p, J/kg·K	586.	611.	644.	
Specific heat ratio, c_p/c_v	1 14	1 14	1.145	
Gas constant R, ft-lb/lb·deg F				
Gas constant R, J/kg·deg C				
Thermal conductivity, Btu/hr·ft·deg F	0.005	0.006		
Thermal conductivity, W/m·deg C	0 008 7	0 010 4		
Boiling point (sat 14.7 psia), deg F	74.9	- 21.8	- 114.6	- 72.
Boiling point (sat 760 mm), deg C	23.8	- 29 9	- 81 4	- 57.8
Latent heat of evap (at bp), Btu/lb	77.5	71 1	63.0	51.1
Latent heat of evap (at bp), J/kg	180 000.	165 000	147 000	119 000.
Freezing (melting) point, deg F (1 atm)	- 168.	- 252.	- 294.	- 270.
Freezing (melting) point, deg C (1 atm)	- 111	- 157.8	- 181.1	- 167.8
Latent heat of fusion, Btu/lb				
Latent heat of fusion, J/kg				
Critical temperature, deg F	388 4	233	83 9	152.
Critical temperature, deg C	198.	111.7	28.8	66.7
Critical pressure, psia	635	582	559	573.
Critical pressure, MN/m²	4.38	4 01	3.85	3.95
Critical volume, ft³/lb	0 028 9	0 287	0 027 7	0 021 5
Critical volume, m³/kg	0.001 80	0 018	0.001 73	0.001 34
Flammable (yes or no)	No	No	No	No
Heat of combustion, Btu/ft³	–	–	.	–
Heat of combustion, Btu/lb	–	–	–	–
Heat of combustion, kJ/kg	–	–	–	–

[a]For N sec/m² divide by 1 000

Table A.7 (continued) Chemical, Physical, and Thermal Properties of Gases: Gases and Vapors, Including Fuels and Refrigerants, English and Metric Units

Common name(s)	*Fluorocarbons*			
Chemical formula	CF_4	$CHCl_2F$	$CHClF_2$	$C_2Cl_2F_4$
Refrigerant number	*14*	*21*	*22*	*114*
CHEMICAL AND PHYSICAL PROPERTIES				
Molecular weight	88 00	102 92	86 468	170 92
Specific gravity, air = 1	3 04	3 55	2 99	5 90
Specific volume, ft^3/lb	4 34	3 7	4 35	2 6
Specific volume, m^3/kg	0 271	0 231	0 271	0 162
Density of liquid (at atm bp), lb/ft^3	102.0	87 7	88 2	94 8
Density of liquid (at atm bp), kg/m^3	1 634	1 405	1 413	1 519
Vapor pressure at 25 deg C, psia		26 4	151 4	30 9
Vapor pressure at 25 deg C, MN/m^2		0 182	1 044	0 213
Viscosity (abs), lbm/ft sec		$8\ 06 \times 10^{-6}$	$8\ 74 \times 10^{-6}$	$8\ 06 \times 10^{-6}$
Viscosity (abs), centipoises[a]		0 012	0 013	0 012
Sound velocity in gas, m/sec				
THERMAL AND THERMODYNAMIC PROPERTIES				
Specific heat, c_p, Btu/lb deg F or cal/g deg C		0 139	0.157	0 158
Specific heat, c_p, J/kg K		582	657	661
Specific heat ratio, c_p/c_v		1 18	1 185	1 09
Gas constant *R*, ft-lb/lb deg F				
Gas constant *R*, J/kg deg C				
Thermal conductivity, Btu/hr ft deg F			0 007	0 006
Thermal conductivity, W/m deg C			0 012	0 010
Boiling point (sat 14 7 psia), deg F	− 198 2	48 1	− 41 3	38 4
Boiling point (sat 760 mm), deg C	− 127 9	9 0	− 40 7	3 55
Latent heat of evap (at bp), Btu/lb	58 5	104 1	100 4	58 4
Latent heat of evap (at bp), J/kg	136 000	242 000	234 000	136 000
Freezing (melting) point, deg F (1 atm)	− 299	− 211	− 256	− 137
Freezing (melting) point, deg C (1 atm)	− 183 8	− 135	− 160	− 93 8
Latent heat of fusion, Btu/lb	2 53			
Latent heat of fusion, J/kg	5 880			
Critical temperature, deg F	− 49 9	353 3	204 8	294
Critical temperature, deg C	− 45 5	178 5	96 5	
Critical pressure, psia	610	750	715	475
Critical pressure, MN/m^2	4 21	5 17	4 93	3 28
Critical volume, ft^3/lb	0 025	0 030 7	0 030 5	0 027 5
Critical volume, m^3/kg	0 001 6	0 001 91	0 001 90	0 001 71
Flammable (yes or no)	No	No	No	No
Heat of combustion, Btu/ft^3	−	−	−	
Heat of combustion, Btu/lb	−		−	−
Heat of combustion, kJ/kg	−	−	−	−

[a] For $N\ sec/m^2$ divide by 1 000

Table A.7 (continued) Chemical, Physical, and Thermal Properties of Gases: Gases and Vapors, Including Fuels and Refrigerants, English and Metric Units

Common name(s)	*Fluorocarbons*			*Helium*
Chemical formula	C_2ClF_5	$C_2H_3ClF_2$	$C_2H_4F_2$	*He*
Refrigerant number	*115*	*142b*	*152a*	*704*
CHEMICAL AND PHYSICAL PROPERTIES				
Molecular weight	154 47	100 50	66.05	4 002 6
Specific gravity, air = 1	5 33	3 47	2 28	0 138
Specific volume, ft^3/lb	2 44	3 7	5 9	97 86
Specific volume, m^3/kg	0 152	0 231	0.368	6.11
Density of liquid (at atm bp), lb/ft^3	96.5	74.6	62.8	7 80
Density of liquid (at atm bp), kg/m^3	1 546.	1 195.	1 006.	125
Vapor pressure at 25 deg C, psia	132.1	49 1	86.8	
Vapor pressure at 25 deg C, MN/m^2	0.911	0 338 5	0.596	
Viscosity (abs), lbm/ft·sec				$13 4 \times 10^{-6}$
Viscosity (abs), centipoises[a]				0.02
Sound velocity in gas, m/sec				1 015.
THERMAL AND THERMODYNAMIC PROPERTIES				
Specific heat, c_p, Btu/lb·deg F or cal/g·deg C	0 161			1.24
Specific heat, c_p, J/kg K	674.			5 188.
Specific heat ratio, c_p/c_v	1.091			1.66
Gas constant *R*, ft-lb/lb deg F				386.
Gas constant *R*, J/kg·deg C				2 077
Thermal conductivity, Btu/hr·ft·deg F				0 086
Thermal conductivity, W/m·deg C				0.149
Boiling point (sat 14.7 psia), deg F	- 38 0	14	- 13	- 452.
Boiling point (sat 760 mm), deg C	- 38.9	- 10.0	-25 0	4.22 K
Latent heat of evap (at bp), Btu/lb	53.4	92.5	137 1	10.0
Latent heat of evap (at bp), J/kg	124 000	215 000	319 000	23 300.
Freezing (melting) point, deg F (1 atm)	- 149			[b]
Freezing (melting) point, deg C (1 atm)	- 100.6			—
Latent heat of fusion, Btu/lb				—
Latent heat of fusion, J/kg				
Critical temperature, deg F	176.		387	-450 3
Critical temperature, deg C				5.2 K
Critical pressure, psia	457.6			33.22
Critical pressure, MN/m^2	3 155			
Critical volume, ft^3/lb	0 026 1			0 231
Critical volume, m^3/kg	0.001 63			0 014 4
Flammable (yes or no)	No	No	No	No
Heat of combustion, Btu/ft^3	—	—	—	—
Heat of combustion, Btu/lb	—	—	—	—
Heat of combustion, kJ/kg	—	—	—	—

[a] For N sec/m^2 divide by 1 000

[b] Helium cannot be solidified at atmospheric pressure

Table A.7 (continued) Chemical, Physical, and Thermal Properties of Gases: Gases and Vapors, Including Fuels and Refrigerants, English and Metric Units

Common name(s)	Hydrogen	Hydrogen chloride	Hydrogen sulfide	Krypton
Chemical formula	H_2	HCl	H_2S	Kr
Refrigerant number	702	—	—	
CHEMICAL AND PHYSICAL PROPERTIES				
Molecular weight	2.016	36.461	34 076	83.80
Specific gravity, air = 1	0.070	1 26	1.18	2.89
Specific volume, ft³/lb	194.	10.74	11.5	4.67
Specific volume, m³/kg	12.1	0.670	0 093 0	0.291
Density of liquid (at atm bp), lb/ft³	4.43	74.4	62.	150.6
Density of liquid (at atm bp), kg/m³	71.0	1 192	993.	2 413
Vapor pressure at 25 deg C, psia				
Vapor pressure at 25 deg C, MN/m²				
Viscosity (abs), lbm/ft·sec	6.05×10^{-6}	10.1×10^{-6}	8.74×10^{-6}	16.8×10^{-6}
Viscosity (abs), centipoises[a]	0.009	0.015	0.013	0.025
Sound velocity in gas, m/sec	1 315	310.	302	223.
THERMAL AND THERMODYNAMIC PROPERTIES				
Specific heat, c_p, Btu/lb deg F or cal/g·deg C	3.42	0.194	0.23	0.059
Specific heat, c_p, J/kg K	14 310	812.	962.	247
Specific heat ratio, c_p/c_v	1 405	1.39	1.33	1.68
Gas constant R, ft-lb/lb·deg F	767	42.4	45 3	18.4
Gas constant R, J/kg·deg C	4 126	228	244	99.0
Thermal conductivity, Btu/hr ft deg F	0 105	0.008	0.008	0.005 4
Thermal conductivity, W/m deg C	0 018 2	0 014	0.014	0.009 3
Boiling point (sat 14.7 psia), deg F	- 423	- 121	- 76.	- 244.
Boiling point (sat 760 mm), deg C	20 4 K	- 85	- 60.	- 153.
Latent heat of evap (at bp), Btu/lb	192	190.5	234.	46 4
Latent heat of evap (at bp), J/kg	447 000	443 000	544 000	108 000.
Freezing (melting) point, deg F (1 atm)	- 434 6	- 169 6	- 119 2	- 272.
Freezing (melting) point, deg C (1 atm)	- 259 1	- 112	- 84.	- 169.
Latent heat of fusion, Btu/lb	25 0	23 4	30.2	4.7
Latent heat of fusion, J/kg	58 000	54 400	70 200	10 900.
Critical temperature, deg F	- 399 8	124	213	
Critical temperature, deg C	- 240 0	51 2	100 4	- 63 8
Critical pressure, psia	189	1 201	1 309	800.
Critical pressure, MN/m²	1 30	8 28	9 02	5.52
Critical volume, ft³/lb	0 53	0 038	0 046	0 017 7
Critical volume, m³/kg	0 033	0 002 4	0 002 9	0.001 1
Flammable (yes or no)	Yes	No	Yes	No
Heat of combustion, Btu/ft³	320	—	700	—
Heat of combustion, Btu/lb	62 050	—	8 000	—
Heat of combustion, kJ/kg	144 000	-	18 600.	—

[a] For N sec/m² divide by 1 000

Table A.7 (continued) Chemical, Physical, and Thermal Properties of Gases: Gases and Vapors, Including Fuels and Refrigerants, English and Metric Units

Common name(s)	*Methane*	*Methyl chloride*	*Neon*	*Nitric oxide*
Chemical formula	CH_4	CH_3Cl	*Ne*	*NO*
Refrigerant number	50	40	720	–
CHEMICAL AND PHYSICAL PROPERTIES				
Molecular weight	16 044	50 488	20.179	30 006
Specific gravity, air = 1	0 554	1 74	0 697	1.04
Specific volume, ft^3/lb	24 2	7 4	19 41	13 05
Specific volume, m^3/kg	1 51	0 462	1 211	0.814
Density of liquid (at atm bp), lb/ft^3	26 3	62 7	75.35	
Density of liquid (at atm bp), kg/m^3	421	1 004	1 207	
Vapor pressure at 25 deg C, psia		82 2		
Vapor pressure at 25 deg C, MN/m^2		0 567		
Viscosity (abs), lbm/ft sec	7.39×10^{-6}	7.39×10^{-6}	21.5×10^{-6}	12.8×10^{-6}
Viscosity (abs), centipoises[a]	0 011	0 011	0.032	0.019
Sound velocity in gas, m/sec	446	251	454	341
THERMAL AND THERMODYNAMIC PROPERTIES				
Specific heat, c_p, Btu/lb·deg F or cal/g·deg C	0 54	0 20	0 246	0 235
Specific heat, c_p, J/kg K	2 260	837	1 030.	983
Specific heat ratio, c_p/c_v	1 31	1 28	1.64	1 40
Gas constant *R*, ft-lb/lb·deg F	96	30 6	76 6	51 5
Gas constant *R*, J/kg·deg C	518	165	412	277
Thermal conductivity, Btu/hr ft deg F	0 02	0 006	0.028	0 015
Thermal conductivity, W/m deg C	0 035	0 010	0.048	0 026
Boiling point (sat 14 7 psia), deg F	- 259	- 10 7	- 410.9	- 240
Boiling point (sat 760 mm), deg C	- 434 2	- 23 7	246	- 151 5
Latent heat of evap (at bp), Btu/lb	219 2	184 1	37	
Latent heat of evap (at bp), J/kg	510 000	428 000	86 100.	
Freezing (melting) point, deg F (1 atm)	- 296 6	- 144	- 415 6	- 258.
Freezing (melting) point, deg C (1 atm)	- 182 6	- 97 8	- 248 7	- 161
Latent heat of fusion, Btu/lb	14	56	6 8	32 9
Latent heat of fusion, J/kg	32 600	130 000	15 800	76 500.
Critical temperature, deg F	- 116	289 4	- 379 8	- 136.
Critical temperature, deg C	- 82 3	143	228 8	- 93 3
Critical pressure, psia	673	968	396	945
Critical pressure, MN/m^2	4 64	6 67	2 73	6 52
Critical volume, ft^3/lb	0 099	0 043	0 033	0 033 2
Critical volume, m^3/kg	0.006 2	0 002 7	0 002 0	0 002 07
Flammable (yes or no)	Yes	Yes	No	No
Heat of combustion, Btu/ft^3	985			-
Heat of combustion, Btu/lb	2 290		–	–
Heat of combustion, kJ/kg			–	-

[a] For N sec/m^2 divide by 1 000

Table A.7 (continued) Chemical, Physical, and Thermal Properties of Gases: Gases and Vapors, Including Fuels and Refrigerants, English and Metric Units

Common name(s)	*Nitrogen*	*Nitrous oxide*	*Oxygen*	*Ozone*
Chemical formula	N_2	N_2O	O_2	O_3
Refrigerant number	728	744A	732	–
CHEMICAL AND PHYSICAL PROPERTIES				
Molecular weight	28 013 4	44.012	31 998 8	47.998
Specific gravity, air = 1	0.967	1.52	1 105	1.66
Specific volume, ft^3/lb	13 98	8 90	12 24	8.16
Specific volume, m^3/kg	0 872	0.555	0 764	0.509
Density of liquid (at atm bp), lb/ft^3	50 46	76 6	71.27	
Density of liquid (at atm bp), kg/m^3	808 4	1 227	1 142.	
Vapor pressure at 25 deg C, psia				
Vapor pressure at 25 deg C, MN/m^2				
Viscosity (abs), lbm/ft·sec	$12\ 1 \times 10^{-6}$	$10\ 1 \times 10^{-6}$	13.4×10^{-6}	$8\ 74 \times 10^{-6}$
Viscosity (abs), centipoises[a]	0 018	0 015	0 020	0.013
Sound velocity in gas, m/sec	353.	268	329.	
THERMAL AND THERMO-DYNAMIC PROPERTIES				
Specific heat, c_p, Btu/lb deg F or cal/g·deg C	0.249	0 21	0 220	0 196
Specific heat, c_p, J/kg·K	1 040.	879.	920.	820.
Specific heat ratio, c_p/c_v	1.40	1 31	1 40	
Gas constant R, ft-lb/lb·deg F	55.2	35 1	48 3	32.2
Gas constant R, J/kg deg C	297	189.	260.	173.
Thermal conductivity, Btu/hr ft deg F	0 015	0 010	0.015	0 019
Thermal conductivity, W/m·deg C	0 026	0 017	0.026	0.033
Boiling point (sat 14.7 psia), deg F	– 320 4	– 127 3	– 297.3	– 170
Boiling point (sat 760 mm), deg C	– 195 8	– 88 5	– 182.97	– 112.
Latent heat of evap (at bp), Btu/lb	85 5	161 8	91.7	
Latent heat of evap (at bp), J/kg	199 000	376 000	213 000.	
Freezing (melting) point, deg F (1 atm)	– 346	– 131 5	– 361.1	– 315 5
Freezing (melting) point, deg C (1 atm)	– 210	– 90.8	– 218.4	– 193.
Latent heat of fusion, Btu/lb	11 1	63 9	5.9	97.2
Latent heat of fusion, J/kg	25 800	149 000	13 700	226 000.
Critical temperature, deg F	– 232 6	97 7	– 181 5	16
Critical temperature, deg C	– 147	36 5	– 118 6	– 9.
Critical pressure, psia	493	1 052	726	800
Critical pressure, MN/m^2	3 40	7 25	5 01	5 52
Critical volume, ft^3/lb	0 051	0 036	0 040	0 029 8
Critical volume, m^3/kg	0 003 18	0 002 2	0 002 5	0 001 86
Flammable (yes or no)	No	No	No	No
Heat of combustion, Btu/ft^3		–	–	–
Heat of combustion, Btu/lb			–	–
Heat of combustion, kJ/kg	–	–		–

[a]For N sec/m^2 divide by 1 000

Table A.7 (continued) Chemical, Physical, and Thermal Properties of Gases: Gases and Vapors, Including Fuels and Refrigerants, English and Metric Units

Common name(s)	Propane	Propylene (Propene)	Sulfur dioxide	Xenon
Chemical formula	C_3H_8	C_3H_6	SO_2	Xe
Refrigerant number	290	1 270	764	–
CHEMICAL AND PHYSICAL PROPERTIES				
Molecular weight	44.097	42 08	64 06	131 30
Specific gravity, air = 1	1 52	1 45	2 21	4.53
Specific volume, ft³/lb	8.84	9.3	6.11	2.98
Specific volume, m³/kg	0 552	0.58		
Density of liquid (at atm bp), lb/ft³	36.2	37 5	42 8	190.8
Density of liquid (at atm bp), kg/m³	580.	601	585.	3 060
Vapor pressure at 25 deg C, psia	135 7	166 4	56 6	
Vapor pressure at 25 deg C, MN/m²	0.936	1.147	0.390	
Viscosity (abs), lbm/ft sec	53.8×10^{-6}	57.1×10^{-6}	$8\ 74 \times 10^{-6}$	$15\ 5 \times 10^{-6}$
Viscosity (abs), centipoises[a]	0 080	0 085	0 013	0.023
Sound velocity in gas, m/sec	253.	261	220.	177
THERMAL AND THERMO-DYNAMIC PROPERTIES				
Specific heat, c_p, Btu/lb deg F or cal/g·deg C	0.39	0 36	0 11	0.115
Specific heat, c_p, J/kg·K	1 630.	1 506	460	481
Specific heat ratio, c_p/c_v	1 2	1 16	1 29	1.67
Gas constant R, ft·lb/lb deg F	35.0	36 7	24.1	11.8
Gas constant R, J/kg deg C	188.	197	130.	63 5
Thermal conductivity, Btu/hr·ft·deg F	0 010	0 010	0 006	0 003
Thermal conductivity, W/m·deg C	0 017	0.017	0 010	0.005 2
Boiling point (sat 14 7 psia), deg F	– 44.	– 54	14 0	– 162 5
Boiling point (sat 760 mm), deg C	– 42 2	– 48 3	– 10	– 108
Latent heat of evap (at bp), Btu/lb	184	188 2	155 5	41 4
Latent heat of evap (at bp), J/kg	428 000.	438 000	362 000	96 000
Freezing (melting) point, deg F (1 atm)	– 309 8	– 301	– 104	220
Freezing (melting) point, deg C (1 atm)	– 189 9	– 185	· 75 5	– 140
Latent heat of fusion, Btu/lb	19 1		58 0	10
Latent heat of fusion, J/kg	44 400		135 000	23 300
Critical temperature, deg F	205	197	315 5	61 9
Critical temperature, deg C	96.	91 7	157 6	16 6
Critical pressure, psia	618	668	1 141	852
Critical pressure, MN/m²	4 26	4.61	7 87	5 87
Critical volume, ft³/lb	0 073	0 069	0 03	0 014 5
Critical volume, m³/kg	0.004 5	0 004 3	0 001 9	0 000 90
Flammable (yes or no)	Yes	Yes	No	No
Heat of combustion, Btu/ft³	2 450	2 310	–	
Heat of combustion, Btu/lb	21 660.	21 500	–	–
Heat of combustion, kJ/kg	50 340.	50 000	–	–

[a]For N sec/m² divide by 1 000

Table A.8 Ideal Gas Properties of Air

Part a. SI Units

T(K), *h* and *u*(kJ/kg), *s*°(kJ/kg·K)

T	h	p_r	u	v_r	s°	T	h	p_r	u	v_r	s°
200	199.97	0.3363	142.56	1707.	1.29559	450	451.80	5.775	322.62	223.6	2.11161
210	209.97	0.3987	149.69	1512.	1.34444	460	462.02	6.245	329.97	211.4	2.13407
220	219.97	0.4690	156.82	1346.	1.39105	470	472.24	6.742	337.32	200.1	2.15604
230	230.02	0.5477	164.00	1205.	1.43557	480	482.49	7.268	344.70	189.5	2.17760
240	240.02	0.6355	171.13	1084.	1.47824	490	492.74	7.824	352.08	179.7	2.19876
250	250.05	0.7329	178.28	979.	1.51917	500	503.02	8.411	359.49	170.6	2.21952
260	260.09	0.8405	185.45	887.8	1.55848	510	513.32	9.031	366.92	162.1	2.23993
270	270.11	0.9590	192.60	808.0	1.59634	520	523.63	9.684	374.36	154.1	2.25997
280	280.13	1.0889	199.75	738.0	1.63279	530	533.98	10.37	381.84	146.7	2.27967
285	285.14	1.1584	203.33	706.1	1.65055	540	544.35	11.10	389.34	139.7	2.29906
290	290.16	1.2311	206.91	676.1	1.66802	550	554.74	11.86	396.86	133.1	2.31809
295	295.17	1.3068	210.49	647.9	1.68515	560	565.17	12.66	404.42	127.0	2.33685
300	300.19	1.3860	214.07	621.2	1.70203	570	575.59	13.50	411.97	121.2	2.35531
305	305.22	1.4686	217.67	596.0	1.71865	580	586.04	14.38	419.55	115.7	2.37348
310	310.24	1.5546	221.25	572.3	1.73498	590	596.52	15.31	427.15	110.6	2.39140
315	315.27	1.6442	224.85	549.8	1.75106	600	607.02	16.28	434.78	105.8	2.40902
320	320.29	1.7375	228.42	528.6	1.76690	610	617.53	17.30	442.42	101.2	2.42644
325	325.31	1.8345	232.02	508.4	1.78249	620	628.07	18.36	450.09	96.92	2.44356
330	330.34	1.9352	235.61	489.4	1.79783	630	638.63	19.84	457.78	92.84	2.46048
340	340.42	2.149	242.82	454.1	1.82790	640	649.22	20.64	465.50	88.99	2.47716
350	350.49	2.379	250.02	422.2	1.85708	650	659.84	21.86	473.25	85.34	2.49364
360	360.58	2.626	257.24	393.4	1.88543	660	670.47	23.13	481.01	81.89	2.50985
370	370.67	2.892	264.46	367.2	1.91313	670	681.14	24.46	488.81	78.61	2.52589
380	380.77	3.176	271.69	343.4	1.94001	680	691.82	25.85	496.62	75.50	2.54175
390	390.88	3.481	278.93	321.5	1.96633	690	702.52	27.29	504.45	72.56	2.55731
400	400.98	3.806	286.16	301.6	1.99194	700	713.27	28.80	512.33	69.76	2.57277
410	411.12	4.153	293.43	283.3	2.01699	710	724.04	30.38	520.23	67.07	2.58810
420	421.26	4.522	300.69	266.6	2.04142	720	734.82	32.02	528.14	64.53	2.60319
430	431.43	4.915	307.99	251.1	2.06533	730	745.62	33.72	536.07	62.13	2.61803
440	441.61	5.332	315.30	236.8	2.08870	740	756.44	35.50	544.02	59.82	2.63280

Table A.8 (continued) Ideal Gas Properties of Air

T(K), *h* and *u*(kJ/kg), s°(kJ/kg·K)

T	h	p_r	u	v_r	s°	T	h	p_r	u	v_r	s°
750	767.29	37.35	551.99	57.63	2.64737	1300	1395.97	330.9	1022.82	11.275	3.27345
760	778.18	39.27	560.01	55.54	2.66176	1320	1419.76	352.5	1040.88	10.747	3.29160
770	789.11	41.31	568.07	53.39	2.67595	1340	1443.60	375.3	1058.94	10.247	3.30959
780	800.03	43.35	576.12	51.64	2.69013	1360	1467.49	399.1	1077.10	9.780	3.32724
790	810.99	45.55	584.21	49.86	2.70400	1380	1491.44	424.2	1095.26	9.337	3.34474
800	821.95	47.75	592.30	48.08	2.71787	1400	1515.42	450.5	1113.52	8.919	3.36200
820	843.98	52.59	608.59	44.84	2.74504	1420	1539.44	478.0	1131.77	8.526	3.37901
840	866.08	57.60	624.95	41.85	2.77170	1440	1563.51	506.9	1150.13	8.153	3.39586
860	888.27	63.09	641.40	39.12	2.79783	1460	1587.63	537.1	1168.49	7.801	3.41247
880	910.56	68.98	657.95	36.61	2.82344	1480	1611.79	568.8	1186.95	7.468	3.42892
900	932.93	75.29	674.58	34.31	2.84856	1500	1635.97	601.9	1205.41	7.152	3.44516
920	955.38	82.05	691.28	32.18	2.87324	1520	1660.23	636.5	1223.87	6.854	3.46120
940	977.92	89.28	708.08	30.22	2.89748	1540	1684.51	672.8	1242.43	6.569	3.47712
960	1000.55	97.00	725.02	28.40	2.92128	1560	1708.82	710.5	1260.99	6.301	3.49276
980	1023.25	105.2	741.98	26.73	2.94468	1580	1733.17	750.0	1279.65	6.046	3.50829
1000	1046.04	114.0	758.94	25.17	2.96770	1600	1757.57	791.2	1298.30	5.804	3.52364
1020	1068.89	123.4	776.10	23.72	2.99034	1620	1782.00	834.1	1316.96	5.574	3.53879
1040	1091.85	133.3	793.36	22.39	3.01260	1640	1806.46	878.9	1335.72	5.355	3.55381
1060	1114.86	143.9	810.62	21.14	3.03449	1660	1830.96	925.6	1354.48	5.147	3.56867
1080	1137.89	155.2	827.88	19.98	3.05608	1680	1855.50	974.2	1373.24	4.949	3.58335
1100	1161.07	167.1	845.33	18.896	3.07732	1700	1880.1	1025	1392.7	4.761	3.5979
1120	1184.28	179.7	862.79	17.886	3.09825	1750	1941.6	1161	1439.8	4.328	3.6336
1140	1207.57	193.1	880.35	16.946	3.11883	1800	2003.3	1310	1487.2	3.944	3.6684
1160	1230.92	207.2	897.91	16.064	3.13916	1850	2065.3	1475	1534.9	3.601	3.7023
1180	1254.34	222.2	915.57	15.241	3.15916	1900	2127.4	1655	1582.6	3.295	3.7354
1200	1277.79	238.0	933.33	14.470	3.17888	1950	2189.7	1852	1630.6	3:022	3.7677
1220	1301.31	254.7	951.09	13.747	3.19834	2000	2252.1	2068	1678.7	2.776	3.7994
1240	1324.93	272.3	968.95	13.069	3.21751	2050	2314.6	2303	1726.8	2.555	3.8303
1260	1348.55	290.8	986.90	12.435	3.23638	2100	2377.4	2559	1775.3	2.356	3.8605
1280	1372.24	310.4	1004.76	11.835	3.25510	2150	2440.3	2837	1823.8	2.175	3.8901
						2200	2503.2	3138	1872.4	2.012	3.9191
						2250	2566.4	3464	1921.3	1.864	3.9474

TABLE A.8 (continued) Ideal Gas Properties of Air

Part b. English Units

T(°R), h and u(Btu/lb), $s°$(Btu/lb · °R)

T	h	p_r	u	v_r	$s°$	T	h	p_r	u	v_r	$s°$
360	85.97	0.3363	61.29	396.6	0.50369	940	226.11	9.834	161.68	35.41	0.73509
380	90.75	0.4061	64.70	346.6	0.51663	960	231.06	10.61	165.26	33.52	0.74030
400	95.53	0.4858	68.11	305.0	0.52890	980	236.02	11.43	168.83	31.76	0.74540
420	100.32	0.5760	71.52	270.1	0.54058	1000	240.98	12.30	172.43	30.12	0.75042
440	105.11	0.6776	74.93	240.6	0.55172	1040	250.95	14.18	179.66	27.17	0.76019
460	109.90	0.7913	78.36	215.33	0.56235	1080	260.97	16.28	186.93	24.58	0.76964
480	114.69	0.9182	81.77	193.65	0.57255	1120	271.03	18.60	194.25	22.30	0.77880
500	119.48	1.0590	85.20	174.90	0.58233	1160	281.14	21.18	201.63	20.29	0.78767
520	124.27	1.2147	88.62	158.58	0.59172	1200	291.30	24.01	209.05	18.51	0.79628
537	128.34	1.3593	91.53	146.34	0.59945	1240	301.52	27.13	216.53	16.93	0.80466
540	129.06	1.3860	92.04	144.32	0.60078	1280	311.79	30.55	224.05	15.52	0.81280
560	133.86	1.5742	95.47	131.78	0.60950	1320	322.11	34.31	231.63	14.25	0.82075
580	138.66	1.7800	98.90	120.70	0.61793	1360	332.48	38.41	239.25	13.12	0.82848
600	143.47	2.005	102.34	110.88	0.62607	1400	342.90	42.88	246.93	12.10	0.83604
620	148.28	2.249	105.78	102.12	0.63395	1440	353.37	47.75	254.66	11.17	0.84341
640	153.09	2.514	109.21	94.30	0.64159	1480	363.89	53.04	262.44	10.34	0.85062
660	157.92	2.801	112.67	87.27	0.64902	1520	374.47	58.78	270.26	9.578	0.85767
680	162.73	3.111	116.12	80.96	0.65621	1560	385.08	65.00	278.13	8.890	0.86456
700	167.56	3.446	119.58	75.25	0.66321	1600	395.74	71.73	286.06	8.263	0.87130
720	172.39	3.806	123.04	70.07	0.67002	1650	409.13	80.89	296.03	7.556	0.87954
740	177.23	4.193	126.51	65.38	0.67665	1700	422.59	90.95	306.06	6.924	0.88758
760	182.08	4.607	129.99	61.10	0.68312	1750	436.12	101.98	316.16	6.357	0.89542
780	186.94	5.051	133.47	57.20	0.68942	1800	449.71	114.0	326.32	5.847	0.90308
800	191.81	5.526	136.97	53.63	0.69558	1850	463.37	127.2	336.55	5.388	0.91056
820	196.69	6.033	140.47	50.35	0.70160	1900	477.09	141.5	346.85	4.974	0.91788
840	201.56	6.573	143.98	47.34	0.70747	1950	490.88	157.1	357.20	4.598	0.92504
860	206.46	7.149	147.50	44.57	0.71323	2000	504.71	174.0	367.61	4.258	0.93205
880	211.35	7.761	151.02	42.01	0.71886	2050	518.61	192.3	378.08	3.949	0.93891
900	216.26	8.411	154.57	39.64	0.72438	2100	532.55	212.1	388.60	3.667	0.94564
920	221.18	9.102	158.12	37.44	0.72979	2150	546.54	233.5	399.17	3.410	0.95222

TABLE A.8 (continued) Ideal Gas Properties of Air

T(°R), h and u(Btu/lb), $s°$(Btu/lb · °R)

T	h	p_r	u	v_r	$s°$	T	h	p_r	u	v_r	$s°$
2200	560.59	256.6	409.78	3.176	0.95868	3700	998.11	2330	744.48	.5882	1.10991
2250	574.69	281.4	420.46	2.961	0.96501	3750	1013.1	2471	756.04	.5621	1.11393
2300	588.82	308.1	431.16	2.765	0.97123	3800	1028.1	2618	767.60	.5376	1.11791
2350	603.00	336.8	441.91	2.585	0.97732	3850	1043.1	2773	779.19	.5143	1.12183
2400	617.22	367.6	452.70	2.419	0.98331	3900	1058.1	2934	790.80	.4923	1.12571
2450	631.48	400.5	463.54	2.266	0.98919	3950	1073.2	3103	802.43	.4715	1.12955
2500	645.78	435.7	474.40	2.125	0.99497	4000	1088.3	3280	814.06	.4518	1.13334
2550	660.12	473.3	485.31	1.996	1.00064	4050	1103.4	3464	825.72	.4331	1.13709
2600	674.49	513.5	496.26	1.876	1.00623	4100	1118.5	3656	837.40	.4154	1.14079
2650	688.90	556.3	507.25	1.765	1.01172	4150	1133.6	3858	849.09	.3985	1.14446
2700	703.35	601.9	518.26	1.662	1.01712	4200	1148.7	4067	860.81	.3826	1.14809
2750	717.83	650.4	529.31	1.566	1.02244	4300	1179.0	4513	884.28	.3529	1.15522
2800	732.33	702.0	540.40	1.478	1.02767	4400	1209.4	4997	907.81	.3262	1.16221
2850	746.88	756.7	551.52	1.395	1.03282	4500	1239.9	5521	931.39	.3019	1.16905
2900	761.45	814.8	562.66	1.318	1.03788	4600	1270.4	6089	955.04	.2799	1.17575
2950	776.05	876.4	573.84	1.247	1.04288	4700	1300.9	6701	978.73	.2598	1.18232
3000	790.68	941.4	585.04	1.180	1.04779	4800	1331.5	7362	1002.5	.2415	1.18876
3050	805.34	1011	596.28	1.118	1.05264	4900	1362.2	8073	1026.3	.2248	1.19508
3100	820.03	1083	607.53	1.060	1.05741	5000	1392.9	8837	1050.1	.2096	1.20129
3150	834.75	1161	618.82	1.006	1.06212	5100	1423.6	9658	1074.0	.1956	1.20738
3200	849.48	1242	630.12	.9546	1.06676	5200	1454.4	10539	1098.0	.1828	1.21336
3250	864.24	1328	641.46	.9069	1.07134	5300	1485.3	11481	1122.0	.1710	1.21923
3300	879.02	1418	652.81	.8621	1.07585						
3350	893.83	1513	664.20	.8202	1.08031						
3400	908.66	1613	675.60	.7807	1.08470						
3450	923.52	1719	687.04	.7436	1.08904						
3500	938.40	1829	698.48	.7087	1.09332						
3550	953.30	1946	709.95	.6759	1.09755						
3600	968.21	2068	721.44	.6449	1.10172						
3650	983.15	2196	732.95	.6157	1.10584						

Source: Adapted from M.J. Moran and H.N. Shapiro, *Fundamentals of Engineering Thermodynamics*, 3rd. ed., Wiley, New York, 1995, as based on J.H. Keenan and J. Kaye, *Gas Tables*, Wiley, New York, 1945.

Table A.9 Equations for Gas Properties

Gas	Molar Mass M kg/kmol	Gas Constant R kJ/kg K	Specific Heats at 25°C c_p kJ/kg K	c_v kJ/kg K	k	Equation Coefficients for $c_p/R = a + bT + cT^2 + dT^3 + eT^4$: Temperature Range	a	$b \times 10^3$ K^{-1}	$c \times 10^6$ K^{-2}	$d \times 10^{10}$ K^{-3}	$e \times 10^{11}$ K^{-4}	Critical State Properties p_c MPa	T_c K	Redlich–Kwong Constants a kPa m^6 K$^{1/2}$/kmol^2	b m^3/kmol	Gas
Acetylene, C_2H_2	26.04	0.319	1.69	1.37	1.232	300–1000K	0.8021	23.51	−35.95	286.1	−87.64	6.14	308	8030	0.0362	Acetylene, C_2H_2
						1000–3000K	3.825	6.767	−3.014	6.931	−0.6469					
Air	28.97	0.287	1.01	0.718	1.400	300–1000K	3.721	−1.874	4.719	−34.45	8.531	3.77	132	1580	0.0253	Air
						1000–3000K	2.786	1.925	−0.9465	2.321	−0.2229					
Argon, Ar	39.95	0.208	0.520	0.312	1.667		2.50	0	0	0	0	4.90	151	1680	0.0222	Argon, Ar
Butane, C_4H_{10}	58.12	0.143	1.67	1.53	1.094	300–1500K	0.4756	44.65	−22.04	42.07	0	3.80	425	29000	0.0806	Butane, C_4H_{10}
Carbon Dioxide CO_2	44.01	0.189	0.844	0.655	1.289	300–1000K	2.227	9.992	−9.802	53.97	−12.81	7.38	304	6450	0.0297	Carbon Dioxide CO_2
						1000–3000K	3.247	5.847	−3.412	9.469	−1.009					
Carbon Monoxide CO	28.01	0.297	1.04	0.744	1.399	300–1000K	3.776	−2.093	4.880	−32.71	6.984	3.50	133	1720	0.0274	Carbon Monoxide, CO
						1000–3000K	2.654	2.226	−1.146	2.851	−0.2762					
Ethane, C_2H_6	30.07	0.276	1.75	1.48	1.187	300–1500K	0.8293	20.75	−7.704	8.756	0	4.88	306	9860	0.0450	Ethane, C_2H_6
Ethylene, C_2H_4	28.05	0.296	1.53	1.23	1.240	300–1000K	1.575	10.19	11.25	−199.1	81.98	5.03	282	7860	0.0404	Ethylene, C_2H_4
						1000–3000K	0.2530	18.67	−9.978	26.03	−2.668					
Helium, He	4.003	2.08	5.19	3.12	1.667		2.50	0	0	0	0	0.228	5.20	8.00	0.0165	Helium, He
Hydrogen H_2	2.016	4.12	14.3	10.2	1.405	300–1000K	2.892	3.884	−8.850	86.94	−29.88	1.31	33.2	143	0.0182	Hydrogen H_2
						1000–3000K	3.717	−0.9220	1.221	−4.328	0.5202					
Hydrogen, H	1.008	8.25	20.6	12.4	1.667	300–1000K	2.496	0.02977	−0.07655	0.8238	−0.3158					Hydrogen, H
						1000–3000K	2.567	−0.1509	0.1219	−0.4184	0.05182					
Hydroxyl, OH	17.01	0.489	1.76	1.27	1.384	300–1000K	3.874	−1.349	1.670	−5.670	0.6189					Hydroxyl, OH
						1000–3000K	3.229	0.2014	0.4357	−2.043	0.2696					
Methane, CH_4	16.04	0.518	2.22	1.70	1.304	300–1000K	4.503	−8.965	37.38	−364.9	122.2	4.60	191	3210	0.0298	Methane, CH_4
						1000–3000K	−0.6992	15.31	−7.695	18.96	−1.849					
Neon, Ne	20.18	0.412	1.03	0.618	1.667		2.50	0	0	0	0	2.65	44.4	146	0.0120	Neon, Ne
Nitric Oxide, NO	30.01	0.277	0.995	0.718	1.386	300–1000K	4.120	−4.225	10.77	−97.64	31.85	6.48	180	1980	0.0200	Nitric Oxide, NO
						1000–3000K	2.730	2.372	−1.338	3.604	−0.3743					
Nitrogen, N_2	28.01	0.297	1.04	0.743	1.400	300–1000K	3.725	−1.562	3.208	−15.54	1.154	3.39	126	1550	0.0267	Nitrogen, N_2
						1000–3000K	2.469	2.467	−1.312	3.401	−0.3454					
Nitrogen, N	14.01	0.594	1.48	0.890	1.667	300–1000K	2.496	0.02977	−0.07655	0.8238	−0.3158					Nitrogen, N
						1000–3000K	2.483	0.03033	−0.01517	0.001879	0.009657					
Oxygen, O_2	32.00	0.260	0.919	0.659	1.395	300–1000K	3.837	−3.420	10.99	−109.6	37.47	5.04	155	1740	0.0221	Oxygen, O_2
						1000–3000K	3.156	1.809	−1.052	3.190	−0.3629					
Oxygen, O	16.00	0.520	1.37	0.850	1.612	300–1000K	3.020	−2.176	3.793	−30.62	9.402					Oxygen, O
						1000–3000K	2.662	−0.3051	0.2250	−0.7447	0.09383					
Propane, C_3H_8	44.10	0.189	1.67	1.48	1.127	300–1500K	−0.4861	36.63	−18.91	38.14	0	4.26	370	18300	0.0626	Propane, C_3H_8
Water, H_2O	18.02	0.462	1.86	1.40	1.329	300–1000K	4.132	−1.559	5.315	−42.09	12.84	22.1	647	14300	0.0211	Water, H_2O
						1000–3000K	2.798	2.693	−0.5392	−0.01783	0.09027					

Source: Adapted from J.B. Jones and R.E. Dugan, *Engineering Thermodynamics*, Prentice-Hall, Englewood Cliffs, NJ 1996 from various sources: *JANAF Thermochemical Tables*, 3rd ed., published by the American Chemical Society and the American Institute of Physics for the National Bureau of Standards, 1986. Data for butane, ethane, and propane from K.A. Kobe and E.G. Long, "Thermochemistry for the Petrochemical Industry, Part II — Paraffinic Hydrocarbons, C_1-C_{16}," *Petroleum Refiner*, Vol. 28, No. 2, 1949, pp 113-116.

Appendix B. Properties of Liquids

Table B.1 Properties of Liquid Water*

Symbols and Units:

ρ = density, lbm/ft³. For g/cm³ multiply by 0.016018. For kg/m³ multiply by 16.018.

c_p = specific heat, Btu/lbm·deg R = cal/g·K. For J/kg·K multiply by 4186.8

μ = viscosity. For lbf·sec/ft² = slugs/sec·ft, multiply by 10^{-7}. For lbm·sec·ft multiply by 10^{-7} and by 32.174. For g/sec·cm (poises) multiply by 10^{-7} and by 478.80. For N·sec/m² multiply by 10^{-7} and by 478.880.

k = thermal conductivity, Btu/hr·ft·deg R. For W/m·K multiply by 1.7307.

Temp. °F	*At 1 atm or 14.7 psia*				*At 1,000 psia*				*At 10,000 psia*			
	ρ	c_p	μ	k	ρ	c_p	μ	k	ρ	c_p	μ	k†
32	62.42	1.007	366	0.3286	62.62	0.999	365	0.3319	64.5	0.937	357	0.3508
40	62.42	1.004	323	0.334	62.62	0.997	323	0.337	64.5	0.945	315	0.356
50	62.42	1.002	272	0.3392	62.62	0.995	272	0.3425	64.5	0.951	267	0.3610
60	62.38	1.000	235	0.345	62.58	0.994	235	0.348	64.1	0.956	233	0.366
70	62.31	0.999	204	0.350	62.50	0.994	204	0.353	64.1	0.960	203	0.371
80	62.23	0.998	177	0.354	62.42	0.994	177	0.358	64.1	0.962	176	0.376
90	62.11	0.998	160	0.359	62.31	0.994	160	0.362	63.7	0.964	159	0.380
100	62.00	0.998	142	0.3633	62.19	0.994	142	0.3666	63.7	0.965	142	0.3841
110	61.88	0.999	126	0.367	62.03	0.994	126	0.371	63.7	0.966	126	0.388
120	61.73	0.999	114	0.371	61.88	0.995	114	0.374	63.3	0.967	114	0.391
130	61.54	0.999	105	0.374	61.73	0.995	105	0.378	63.3	0.968	105	0.395
140	61.39	0.999	96	0.378	61.58	0.996	96	0.381	63.3	0.969	98	0.398
150	61.20	1.000	89	0.3806	61.39	0.996	89	0.3837	63.0	0.970	91	0.4003
160	61.01	1.001	83	0.383	61.20	0.997	83	0.386	62.9	0.971	85	0.403
170	60.79	1.002	77	0.386	60.98	0.998	77	0.389	62.5	0.972	79	0.405
180	60.57	1.003	72	0.388	60.75	0.999	72	0.391	62.5	0.973	74	0.407
190	60.35	1.004	68	0.390	60.53	1.001	68	0.393	62.1	0.974	70	0.409
200	60.10	1.005	62.5	0.3916	60.31	1.002	62.9	0.3944	62.1	0.975	65.4	0.4106
250	boiling point 212°F				59.03	1.001	47.8	0.3994	60.6	0.981	50.6	0.4158
300					57.54	1.024	38.4	0.3993	59.5	0.988	41.3	0.4164
350					55.83	1.044	32.1	0.3944	58.1	0.999	35.1	0.4132
400					53.91	1.072	27.6	0.3849	56.5	1.011	30.6	0.4064
500					49.11	1.181	21.6	0.3508	52.9	1.051	24.8	0.3836
600					boiling point 544.58°F				48.3	1.118	21.0	0.3493

†At 7,500 psia.

*From: "1967 ASME Steam Tables", American Society of Mechanical Engineers, Tables 9, 10, and 11 and Figures 6, 7, 8, and 9.

The ASME compilation is a 330-page book of tables and charts, including a 2½ × 3½-ft Mollier chart. All values have been computed in accordance with the 1967 specifications of the International Formulation Committee (IFC) and are in conformity with the 1963 International Skeleton Tables. This standardization of tables began in 1921 and was extended through the International Conferences in London (1929), Berlin (1930), Washington (1934), Philadelphia (1954), London (1956), New York (1963) and Glasgow (1966). Based on these world-wide standard data, the 1967 ASME volume represents detailed computer output in both tabular and graphic form. Included are density and volume, enthalpy, entropy, specific heat, viscosity, thermal conductivity, Prandtl number, isentropic exponent, choking velocity, p-v product, etc., over the entire range (to 1500 psia 1500°F). English units are used, but all conversion factors are given.

Table B.2 Physical and Thermal Properties of Common Liquids

Part a. SI Units

(At 1.0 Atm Pressure (0.101 325 MN/m²), 300 K, except as noted.)

Common name	Density, kg/m³	Specific heat, kJ/kg K	Viscosity, N s/m²	Thermal conductivity, W/m K	Freezing point, K	Latent heat of fusion, kJ/kg	Boiling point, K	Latent heat of evaporation, kJ/kg	Coefficient of cubical expansion per K
Acetic acid	1 049	2 18	001 155	0 171	290	181	391	402	0 001 1
Acetone	784 6	2 15	000 316	0 161	179 0	98.3	329	518	0 001 5
Alcohol, ethyl	785 1	2 44	001 095	0 171	158 6	108	351.46	846	0 001 1
Alcohol, methyl	786 5	2 54	000 56	0 202	175 5	98 8	337 8	1 100	0 001 4
Alcohol, propyl	800 0	2 37	001 92	0 161	146	86 5	371	779	
Ammonia (aqua)	823 5	4 38		0 353					
Benzene	873 8	1 73	000 601	0 144	278 68	126	353 3	390	0 001 3
Bromine		.473	000 95		245 84	66 7	331 6	193	0 001 2
Carbon disulfide	1 261	.992	000 36	0 161	161 2	57 6	319 40	351	0 001 3
Carbon tetrachloride	1 584	.866	.000 91	0 104	250 35	174	349 6	194	0 001 3
Castor oil	956 1	1 97	650	0 180	263.2				
Chloroform	1 465	1 05	000 53	0 118	209.6	77.0	334 4	247	0 001 3
Decane	726 3	2 21	000 859	0 147	243 5	201	447 2	263	
Dodecane	754 6	2 21	001 374	0 140	247 18	216	489 4	256	
Ether	713 5	2 21	000 223	0 130	157	96 2	307 7	372	0 001 6
Ethylene glycol	1 097	2 36	016 2	0 258	260 2	181	470	800	
Fluorine refrigerant R-11	1 476	870[a]	000 42	0 093[a]	162		297 0	180[b]	
Fluorine refrigerant R-12	1 311	.971[a]		0 071[a]	115	34.4	243 4	165[b]	
Fluorine refrigerant R-22	1 194	1 26[a]		0 086[a]	113	183	232.4	232[b]	
Glycerine	1 259	2 62	950	0 287	264.8	200	563.4	974	0.000 54
Heptane	679 5	2 24	000 376	0 128	182.54	140	371.5	318	
Hexane	654 8	2 26	000 297	0 124	178.0	152	341.84	365	
Iodine		2 15			386 6	62 2	457 5	164	
Kerosene	820 1	2 09	001 64	0 145				251	
Linseed oil	929 1	1 84	033 1		253		560		
Mercury		139	001 53		234 3	11 6	630	295	0 000 18
Octane	698 6	2.15	000 51	0 131	216.4	181	398	298	0 000 72
Phenol	1 072	1.43	008 0	0 190	316.2	121	455		0 000 90
Propane	493.5	2.41[a]	000 11		85.5	79.9	231.08	428[b]	
Propylene	514 4	2.85	000 09		87.9	71.4	225.45	342	
Propylene glycol	965 3	2 50	042		213		460	914	
Sea water	1 025	3.76–4 10			270.6				
Toluene	862 3	1.72	000 550	0 133	178	71.8	383.6	363	
Turpentine	868 2	1.78	001 375	0 121	214		433	293	0.000 99
Water	997 1	4 18	000 89	0 609	273	333	373	2 260	0.000 20

[a]At 297 K, liquid
[b]At 101 325 meganewtons, saturation temperature

TABLE B.2 (continued) Physical and Thermal Properties of Common Liquids

Part b. English Units

(At 1.0 Atm Pressure 77°F (25°C), except as noted.)

For viscosity in N·s/m² (=kg m·s), multiply values in centipoises by 0.001. For surface tension in N/m, multiply values in dyne/cm by 0.001.

Common name	*Density, lb/ft³*	*Specific gravity*	*Viscosity* $lb_m/ft\ sec \times 10^4$	*Viscosity* *cp*	*Sound velocity, meters/sec*	*Dielectric constant*	*Refractive index*
Acetic acid	65.493	1.049	7 76	1.155	1584[50]	6.15	1.37
Acetone	48.98	.787	2.12	0.316	1174	20.7	1.36
Alcohol, ethyl	49.01	787	7.36	1.095	1144	24.3	1.36
Alcohol, methyl	49.10	.789	3.76	0.56	1103	32.6	1.33
Alcohol, propyl	49.94	.802	12.9	1.92	1205	20.1	1.38
Ammonia (aqua)	51.411	.826	—	—	—	16.9	—
Benzene	54.55	.876	4.04	0.601	1298	2.2	1.50
Bromine	—	—	6.38	0.95	—	3.20	—
Carbon disulfide	78.72	1.265	2.42	0.36	1149	2.64	1.63
Carbon tetrachloride	98.91	1.59	6.11	0.91	924	2.23	1.46
Castor oil	59.69	0.960	—	650	1474	4.7	—
Chloroform	91 44	1.47	3.56	0.53	995	4.8	1.44
Decane	45.34	728	5.77	0.859	—	2.0	1.41
Dodecane	47.11	—	9.23	1.374	—	—	1.41
Ether	44.54	0.715	1.50	0.223	985	4.3	1.35
Ethylene glycol	68.47	1.100	109	16.2	1644	37.7	1.43
Fluorine refrigerant R-11	92.14	1.480	2.82	0.42	—	2 0	1.37
Fluorine refrigerant R-12	81.84	1.315	—	—	—	2.0	1 29
Fluorine refrigerant R-22	74.53	1.197	—	—	—	2.0	1 26
Glycerine	78.62	1.263	6380	950	1909	40	1.47
Heptane	42.42	.681	2.53	0.376	1138	1.92	1.38
Hexane	40.88	.657	2.00	0.297	1203	—	1.37
Iodine	—	—	—	—	—	11	—
Kerosene	51.2	0.823	11.0	1.64	1320	—	—
Linseed oil	58.0	0.93	222	33.1	—	3.3	—
Mercury	—	13.633	10.3	1.53	1450	—	—
Octane	43.61	.701	3.43	0.51	1171	—	1.40
Phenol	66.94	1.071	54	8.0	1274[100]	9.8	—
Propane	30.81	.495	0.74	0.11	—	1.27	1.34
Propylene	32.11	.516	0.60	0.09	—	—	1.36
Propylene glycol	60.26	.968	—	42	—	—	1.43
Sea water	64.0	1.03	—	—	1535	—	—
Toluene	53.83	0.865	3.70	0.550	1275[30]	2.4	1.49
Turpentine	54.2	0.87	9.24	1.375	1240	—	1.47
Water	62.247	1.00	6.0	0.89	1498	78.54[a]	1.33

[a] The dielectric constant of water near the freezing point is 87.8; it decreases with increase in temperature to about 55.6 near the boiling point.

Appendix C. Properties of Solids

Table C.1 Properties of Common Solids[a]

Material	Specific gravity	Specific heat: Btu/lbm·deg R	Specific heat: kJ/kg·K	Thermal conductivity: Btu/hr·ft·deg F	Thermal conductivity: W/m·K
Asbestos cement board	1.4	0.2	.837	0.35	0.607
Asbestos millboard	1.0	0.2	.837	0.08	0.14
Asphalt	1.1	0.4	1.67		
Beeswax	0.95	0.82	3.43		
Brick, common	1.75	0.22	.920	0.42	0.71
Brick, hard	2.0	0.24	1.00	0.75	1.3
Chalk	2.0	0.215	.900	0.48	0.84
Charcoal, wood	0.4	0.24	1.00	0.05	0.088
Coal, anthracite	1.5	0.3	1.26		
Coal, bituminous	1.2	0.33	1.38		
Concrete, light	1.4	0.23	.962	0.25	0.42
Concrete, stone	2.2	0.18	.753	1.0	1.7
Corkboard	0.2	0.45	1.88	0.025	0.04
Earth, dry	1.4	0.3	1.26	0.85	1.5
Fiberboard, light	0.24	0.6	2.51	0.035	0.058
Fiber hardboard	1.1	0.5	2.09	0.12	0.2
Firebrick	2.1	0.25	1.05	0.8	1.4
Glass, window	2.5	0.2	.837	0.55	0.96
Gypsum board	0.8	0.26	1.09	0.1	0.17
Hairfelt	0.1	0.5	2.09	0.03	0.050
Ice (32°)	0.9	0.5	2.09	1.25	2.2
Leather, dry	0.9	0.36	1.51	0.09	0.2
Limestone	2.5	0.217	.908	1.1	1.9
Magnesia (85%)	0.25	0.2	.837	0.04	0.071
Marble	2.6	0.21	.879	1.5	2.6
Mica	2.7	0.12	.502	0.4	0.71
Mineral wool blanket	0.1	0.2	.837	0.025	0.04
Paper	0.9	0.33	1.38	0.07	0.1
Paraffin wax	0.9	0.69	2.89	0.15	0.2
Plaster, light	0.7	0.24	1.00	0.15	0.2
Plaster, sand	1.8	0.22	.920	0.42	0.71
Plastics, foamed	0.2	0.3	1.26	0.02	0.03
Plastics, solid	1.2	0.4	1.67	0.11	0.19
Porcelain	2.5	0.22	.920	0.9	1.5
Sandstone	2.3	0.22	.920	1.0	1.7
Sawdust	0.15	0.21	.879	0.05	0.08
Silica aerogel	0.11	0.2	.837	0.015	0.02
Vermiculite	0.13	0.2	.837	0.035	0.058
Wood, balsa	0.16	0.7	2.93	0.03	0.050
Wood, oak	0.7	0.5	2.09	0.10	0.17
Wood, white pine	0.5	0.6	2.51	0.07	0.12
Wool, felt	0.3	0.33	1.38	0.04	0.071
Wool, loose	0.1	0.3	1.26	0.02	0.3

[a]Compiled from several sources.

Table C.2 Density of Various Solids:* Approximate Density of Solids at Ordinary Atmospheric Temperature

Substance	Grams per cu cm	Pounds per cu ft
Agate	2.5–2.7	156–168
Alabaster		
Carbonate	2.69–2.78	168–173
Sulfate	2.26–2.32	141–145
Albite	2.62–2.65	163–165
Amber	1.06–1.11	66–69
Amphiboles	2.9–3.2	180–200
Anorthite	2.74–2.76	171–172
Asbestos	2.0–2.8	125–175
Asbestos slate	1.8	112
Asphalt	1.1–1.5	69–94
Basalt	2.4–3.1	150–190
Beeswax	0.96–0.97	60–61
Beryl	2.69–2.7	168–169
Biotite	2.7–3.1	170–190
Bone	1.7–2.0	106–125
Brick	1.4–2.2	87–137
Butter	0.86–0.87	53–54
Calamine	4.1–4.5	255–280
Calcspar	2.6–2.8	162–175
Camphor	0.99	62
Caoutchouc	0.92–0.99	57–62
Cardboard	0.69	43
Celluloid	1.4	87
Cement, set	2.7–3.0	170–190
Chalk	1.9–2.8	118–175
Charcoal		
Oak	0.57	35
Pine	0.28–0.44	18–28
Cinnabar	8.12	507
Clay	1.8–2.6	112–162
Coal		
Anthracite	1.4–1.8	87–112
Bituminous	1.2–1.5	75–94
Cocoa butter	0.89–0.91	56–57
Coke	1.0–1.7	62–105
Copal	1.04–1.14	65–71
Cork	0.22–0.26	14–16
Cork linoleum	0.54	34
Corundum	3.9–4.0	245–250
Diamond	3.01–3.52	188–220
Dolomite	2.84	177
Ebonite	1.15	72
Emery	4.0	250
Epidote	3.25–3.50	203–218
Feldspar	2.55–2.75	159–172
Flint	2.63	164
Fluorite	3.18	198
Galena	7.3–7.6	460–470
Gamboge	1.2	75
Garnet	3.15–4.3	197–268
Gas carbon	1.88	117
Gelatin	1.27	79

Substance	Grams per cu cm	Pounds per cu ft
Glass		
Common	2.4–2.8	150–175
Flint	2.9–5.9	180–370
Glue	1.27	79
Granite	2.64–2.76	165–172
Graphite†	2.30–2.72	144–170
Gum arabic	1.3–1.4	81–87
Gypsum	2.31–2.33	144–145
Hematite	4.9–5.3	306–330
Hornblende	3.0	187
Ice	0.917	57.2
Ivory	1.83–1.92	114–120
Leather, dry	0.86	54
Lime, slaked	1.3–1.4	81–87
Limestone	2.68–2.76	167–171
Linoleum	1.18	74
Magnetite	4.9–5.2	306–324
Malachite	3.7–4.1	231–256
Marble	2.6–2.84	160–177
Meerschaum	0.99–1.28	62–80
Mica	2.6–3.2	165–200
Muscovite	2.76–3.00	172–187
Ochre	3.5	218
Opal	2.2	137
Paper	0.7–1.15	44–72
Paraffin	0.87–0.91	54–57
Peat blocks	0.84	52
Pitch	1.07	67
Porcelain	2.3–2.5	143–156
Porphyry	2.6–2.9	162–181
Pressed wood pulp board	0.19	12
Pyrite	4.95–5.1	309–318
Quartz	2.65	165
Resin	1.07	67
Rock salt	2.18	136
Rubber, hard	1.19	74
Rubber, soft		
Commercial	1.1	69
Pure gum	0.91–0.93	57–58
Sandstone	2.14–2.36	134–147
Serpentine	2.50–2.65	156–165
Silica		
Fused transparent	2.21	138
Translucent	2.07	129
Slag	2.0–3.9	125–240
Slate	2.6–3.3	162–205
Soapstone	2.6–2.8	162–175
Spermaceti	0.95	59
Starch	1.53	95
Sugar	1.59	99
Talc	2.7–2.8	168–174

Substance	Grams per cu cm	Pounds per cu ft
Tallow		
Beef	0.94	59
Mutton	0.94	59
Tar	1.02	66
Topaz	3.5–3.6	219–223
Tourmaline	3.0–3.2	190–200
Wax, sealing	1.8	112
Wood (seasoned)		
Alder	0.42–0.68	26–42
Apple	0.66–0.84	41–52
Ash	0.65–0.85	40–53
Balsa	0.11–0.14	7–9
Bamboo	0.31–0.40	19–25
Basswood	0.32–0.59	20–37
Beech	0.70–0.90	32–56
Birch	0.51–0.77	32–48
Blue gum	1.00	62
Box	0.95–1.16	59–72
Butternut	0.38	24
Cedar	0.49–0.57	30–35
Cherry	0.70–0.90	43–56
Dogwood	0.76	47
Ebony	1.11–1.33	69–83
Elm	0.54–0.60	34–37
Hickory	0.60–0.93	37–58
Holly	0.76	47
Juniper	0.56	35
Larch	0.50–0.56	31–35
Lignum vitae	1.17–1.33	73–83
Locust	0.67–0.71	42–44
Logwood	0.91	57
Mahogany		
Honduras	0.66	41
Spanish	0.85	53
Maple	0.62–0.75	39–47
Oak	0.60–0.90	37–56
Pear	0.61–0.73	38–45
Pine		
Pitch	0.83–0.85	52–53
White	0.35–0.50	22–31
Yellow	0.37–0.60	23–37
Plum	0.66–0.78	41–49
Poplar	0.35–0.5	22–31
Satinwood	0.95	59
Spruce	0.48–0.70	30–44
Sycamore	0.40–0.60	24–37
Teak		
Indian	0.66–0.88	41–55
African	0.98	61
Walnut	0.64–0.70	40–43
Water gum	1.00	62
Willow	0.40–0.60	24–37

†Some values reported as low as 1.6

*Based largely on "Smithsonian Physical Tables" 9th rev. ed., W. E. Forsythe, Ed., The Smithsonian Institution, 1956, p. 292

Note: In the case of substances with voids, such as paper or leather, the bulk density is indicated rather than the density of the solid portion. For density in kg/m³, multiply values in g/cm³ by 1,000.

Table C.3 Specific Stiffness of Metals, Alloys, and Certain Non-Metallics*

Specific stiffness is usually expressed as the modulus of elasticity (in tension) per unit weight-density, i.e., E/ρ, in units of pounds and inches. While the stiffness of similar alloys varies considerably, there are definite ranges and groups to be recognized. Since the specific stiffness of steel is about 100 million, the values in the following table are also approximately the percentage stiffness, referred to steel.

Material	*Specific stiffness, millions*
Beryllium	650
Silicon carbide	600
Alumina ceramics	400
Mica	350
Titanium carbide cermet	250
Alumina cermet	200
Molybdenum and alloys: silica glass	130
Titanium and alloys; cobalt superalloys; soda-lime glass	110
Carbon and low-alloy steel; wrought iron	105
Stainless steel: nodular cast iron; magnesium and alloys: aluminum and alloys	100
Nickel and alloys; malleable iron	95
Iron silicon alloys (cast); iridium; vanadium	90
Monel alloys: tungsten	80
Gray cast iron: columbium alloys	70
Aluminum bronze: beryllium copper	65
Nickel silver; cupronickel: zirconium	55
Yellow brass; nickel cast iron; bronze; Muntz metal; antimony	50
Copper; red brass; tantalum	45
Silver and alloys: pewter; platinum and alloys: white gold	30
Tin: thorium	25
Gold	20
Tin-lead alloy	10
Lead	5

*Compiled from several sources

Table C.4 Thermal Properties of Pure Metals—Metric Units

	AT ATMOSPHERIC PRESSURE								LIQUID METAL			
				At 100°K		At 25°C (77°F)				Vapor pressure		
Metal	Melting point, °C	Boiling point, °C	Latent heat of fusion, cal/g**	Thermal conductivity, watts/cm °C	Specific heat, cal/g °C**	Specific heat, cal/g °C**	Coeff of linear expansion, (× 10⁶) (°C)⁻¹	Thermal conductivity, watts/cm °C	Specific heat (liquid) at 2000°K, cal/g °C**	10^{-3} atm	10^{-6} atm	10^{-9} atm
										Boiling point temperatures, °K		
Aluminum	660	2441	95	3 00*	115	0 215	25	2 37	26	1,782	1,333	1,063
Antimony	630	1440	38 5	—	040	050	9	185	062	1,007	741	612
Beryllium	1285	2475	324	—	049	436	12	2 18	78	1,793	1,347	1,085
Bismuth	271 4	1660	12 4	—	026	030	13	084	036	1,155	851	677
Cadmium	321	767	13 2	1 03	047	055	30	93	063	655	486	388
Chromium	1860	2670	79	1 58	046	110	6	91	224	1,992	1,530	1,247
Cobalt	1495	2925	66	—	057	10	12	69	164	2,167	1,652	1,345
Copper	1084	2575	49	4 83*	061	092	16 6	3 98	118	1,862	1,391	1,120
Gold	1063	2800	15	3 45*	026	031	14 2	3 15	0355	2,023	1,510	1,211
Iridium	2450	4390	33	—	022	031	6	1 47	0434	3,253	2,515	2,062
Iron	1536	2870	65	1 32*	052	108	12	803	197	2,093	1,594	1,297
Lead	327 5	1750	5 5	0 396	028	031	29	346	033	1,230	889	698
Magnesium	650	1090	88 0	1 69	016	243	25	1 59	32	857	638	509
Manganese	1244	2060	64	—	064	114	22	—	20	1,495	1,131	913
Mercury	− 38 86	356 55	2 7	—	029	033	—	0839	—	393	287	227
Molybdenum	2620	4651	69	1 79	033	060	5	1 4	089	3,344	2,558	2,079
Nickel	1453	2800	71	1 58	055	106	13	899	175	2,156	1,646	1,343
Niobium (Columbium)	2470	4740	68	0 552	045	064	7	52	083	3,523	2,721	2,232
Osmium	3025	4225	34	—	—	031	5	61	039	—	—	—
Platinum	1770	3825	24	0 79*	024	032	9	73	043	2,817	2,155	1,757
Plutonium	640	3230	3	—	019	032	54	08	041	2,200	1,596	1,252
Potassium	63 3	760	14 5	—	150	180	83	99	—	606	430	335
Rhodium	1965	3700	50	—	—	058	8	1 50	092	—	—	—
Selenium	217	700	16	—	—	077	37	005	—	—	—	—
Silicon	1411	3280	430	—	062	17	3	835	217	2,340	1,749	1,427
Silver	961	2212	26 5	4 50*	045	057	19	4 27	.068	1,582	1,179	952
Sodium	97 83	884	27	—	234	293	70	1 34	—	701	504	394
Tantalum	2980	5365	41	0 592	026	034	6 5	54	040	3,959	3,052	2,495
Thorium	1750	4800	17	—	024	03	12	41	047	3,251	2,407	1,919
Tin	232	2600	14 1	0 85	039	054	20	64	058	1,857	1,366	1,080
Titanium	1670	3290	100	0 312	072	125	8 5	2	188	2,405	1,827	1,484
Tungsten	3400	5550	46	2 35*	021	032	4 5	1 78	040	4,139	3,228	2,656
Uranium	1132	4140	12	—	022	028	13 4	25	048	2,861	2,128	1,699
Vanadium	1900	3400	98	—	061	116	8	60	207	2,525	1,948	1,591
Zinc	419 5	910	27	1 32	063	093	35	1 15	—	752	559	449

* Temperatures of maximum thermal conductivity (conductivity values in watts/cm °C): Aluminum 13°K, cond. = 71.5; copper 10°K, cond. = 196; gold 10°K, cond. = 28.2; iron 20°K, cond. = 9.97; platinum 8°K, cond. = 12.9; silver 7°K, cond. = 193; tungsten 8°K, cond. = 85.3.

** To convert to SI units note that 1 cal = 4.186 J

Table C.5 Mechanical Properties of Metals and Alloys:* Typical Composition, Properties, and Uses of Common Materials

For MN/m^2 multiply strength in thousands of psi by 6.895.

FERROUS ALLOYS

Ferrous alloys comprise the largest volume of metal alloys used in engineering The actual range of mechanical properties in any particular grade of alloy steel depends on the particular history and heat treatment. The steels listed in this table are intended to give some idea of the range of properties readily obtainable Many hundreds of steels are available. Cost is frequently an important criterion in the choice of material; in general the greater the percentage of alloying elements present in the alloy, the greater will be the cost

No.	Material	Nominal composition	Form and condition	Typical mechanical properties: Yield strength (0.2% offset), 1000 lb/sq in	Typical mechanical properties: Tensile strength, 1000 lb/sq in	Typical mechanical properties: Elongation, in 2 in., %	Typical mechanical properties: Hardness, Brinell	Comments
	IRON							
1	Ingot iron (Included for comparison)	Fe 99.9	Hot-rolled	29	45	26	90	
			Annealed	19	38	45	67	
	PLAIN CARBON STEELS							
2	AISI-SAE 1020	C 0.20, Mn 0.45, Si 0.25, Fe bal	Hot-rolled	30	55	25	111	Bolts, crankshafts, gears, connecting rods, easily weldable
			Hardened (water-quenched, 1000°F-tempered)	62	90	25	179	
3	AISI 1025	C 0.25, Fe bal., Mn 0.45	Bar stock					
			Hot-rolled	32	58	25	116	
			Cold-drawn	54	64	15	126	
4	AISI-SAE 1035	C 0.35, Mn 0.75	Hot-rolled	39	72	18	143	Medium-strength, engineering steel
			Cold-rolled	67	80	12	163	
5	AISI SAE 1045	C 0.45, Fe bal, Mn 0.75	Bar stock					
			Annealed	73	80	12	170	
			Hot-rolled	45	82	16	163	
			Cold-drawn	77	91	12	179	
6	AISI-SAE 1078	C 0.78, Fe bal., Mn 0.45	Bar stock					
			Hot-rolled, spheroidized	55	100	12	207	
			Annealed	72	94	10	192	
7	AISI-SAE 1095	C 0.95, Fe bal, Mn 0.40						
8	AISI-SAE 1120	C 0.2, Mn 0.8, S 0.1	Cold-drawn	58	69	–	137	Free-cutting, leaded, resulphurized steel: high-speed, automatic machining
	ALLOY STEELS							
9	ASTM A202/56	C 0.17, Mn 1.2, Cr 0.5, Si 0.75	Stress-relieved	45	75	18	—	Low alloy, boilers, pressure vessels

Table C.5 (continued) Mechanical Properties of Metals and Alloys:* Typical Composition, Properties, and Uses of Common Materials

No.	Material	Nominal composition	Form and condition	Typical mechanical properties: Yield strength (0.2% offset), 1000 lb/sq in.	Tensile strength, 1000 lb/sq in.	Elongation, in 2 in., %	Hardness, Brinell	Comments
10	AISI 4140	C 0.40, Si 0.3 Cr 1.0, Mo 0.2 Mn 0.9	Fully-tempered Optimum properties	95 132	108 150	22 18	240 —	High strength; gears, shafts
11	12% Manganese steel	12% Mn, C	Tempered 600°F Rolled and heat-treated stock	200 44	220 160	10 40	— 170	Machine tool parts; wear, abrasion-resistant
12	VASCO 300	Ni 18.5, Ti 0.6 Co 9.0, C 0.03 Mo 4.8	Solution treatment 1500°F, aged 900°F	110	150	18	—	Very high strength, maraging, good machining properties in annealed state
13	T1 (AISI)	W 18.0, V 1.0 Cr 4.0, C 0.7	Quenched, tempered				R(c)	High speed tool steel, cutting tools, punches, etc.
14	M2 (AISI)	W 6.5, Mo 5.0 Cr 4.0, C 0.85 V 2.0	Quenched; tempered				65–66	M-grade, cheaper, tougher
15	Stainless steel type 304	Ni 9.0, C 0.08 max Cr 19.0	Annealed, cold-rolled	35 to 160	85 to 185	60 8	160 to 400	General purpose, weldable, nonmagnetic austenitic steel
16	Stainless steel type 316	Cr 18.0, C 0.10 max Ni 11.0 Mo 2.5, Fe bal.	Annealed	30 to 120	90 to 150	50 8	165 275	For severe corrosive media, under stress; nonmagnetic austenitic steel
17	Stainless steel type 431	Cr 16.0, Si 1.0 Ni 2.0, C 0.20 Mn 1.0, Fe bal.	Annealed Heat-treated	85 150	120 195	25 20	250 400	Heat-treated stainless steel, with good mechanical strength; magnetic
18	Stainless steel 17–4 PH	Cr 17.0, Co 0.35 Ni 4.0, C 0.07 Cu 4.0, Fe bal.	Annealed	110	150	10	363	Precipitation hardening, heat-resisting type; retains strength up to approx. 600°F

Table C.5 (continued) Mechanical Properties of Metals and Alloys:* Typical Composition, Properties, and Uses of Common Materials

CAST IRONS AND CAST STEELS

These alloys are used where large and/or intricate-shaped articles are required or where over-all dimensional tolerances are not critical. Thus the article can be produced with the fabrication and machining costs held to a minimum. Except for a few heat-treatable cast steels, this class of alloys does not demonstrate high-strength qualities.

No.	Material	Nominal composition		Form and condition	Yield strength (0.2% offset), 1000 lb/sq in.	Tensile strength, 1000 lb/sq in.	Elongation, in 2 in., %	Hardness, Brinell	Comments
	CAST IRONS								
19	Cast gray iron ASTM A48–48, Class 25	C 3.4 Mn 0.5	Si 1.8	Cast (as cast)	—	25 min	0.5 max	180	Engine blocks, fly-wheels, gears, machine-tool bases
20	White	C 3.4 Mn 0.6	Si 0.7	Cast	—	25	0	450	
21	Malleable iron ASTM A47	C 2.5 Mn 0.55 max	Si 1.0	Cast (annealed)	33	52	12	130	Automotives, axle bearings, track wheels, crankshafts
22	Ductile or nodular iron (Mg-containing) ASTM A339 ASTM A395	C 3.4 Mn 0.40 Ni 1% Si 2.5	P 0.1 max Mg 0.06 Fe bal.	Cast; Cast (as cast); Cast (quenched, tempered)	53 68 108	70 90 135	18 7 5	170 235 310	Heavy-duty machines, gears, cams, crankshafts
23	Ni-hard type 2	C 2.7 Mn 0.5 Cr 2.0	Si 0.6 Ni 4.5 Fe bal.	Sand-cast; Chill-cast (tempered)	— —	55 75	— —	550 625	Strength, with heat- and corrosion-resistance
24	Ni-resist type 2	C 3.0 Mn 1.0 Cr 2.5	Si 2.0 Ni 20.0 Fe bal.	Cast (as cast)	—	27	2	140	
	CAST STEELS								
25	ASTM A27–62 (60–30)	C 0.3 Si 0.8 Cr 0.4	Mn 0.6 Ni 0.5 Mo 0.2		30	60	24	—	Low alloy, medium strength, general application
26	ASTM A148–60 (105–85)				85	105	17	—	High strength; structural application

Table C.5 (continued) Mechanical Properties of Metals and Alloys:* Typical Composition, Properties, and Uses of Common Materials

No.	Material	Nominal composition	Form and condition	Typical mechanical properties: Yield strength (0.2% offset), 1000 lb/sq in.	Tensile strength, 1000 lb/sq in.	Elongation, in 2 in., %	Hardness, Brinell	Comments
27	Cast 12 Cr alloy (CA-15)	C 0.15 max; Si 1.50 max; Ni 1.00 max; Mn 1.00 max; Cr 11.5-14; Fe bal.	Air-cooled from 1800°F, tempered at 600°F Air-cooled from 1800°F; tempered at 1400°F	150 75	200 100	7 30	390 185	Stainless, corrosion-resistant to mildly corrosive alkalis and acids
28	Cast 29–9 alloy (CE-30) ASTM A296 63T	C 0.30 max; Si 2.00 max; Ni 8–11; Mn 1.50 max; Cr 26–30; Fe bal.	As cast	60	95	15	170	Greater corrosion resistance, especially for oxidizing condition
29	Cast 28–7 alloy (HD) ASTM A297-63T	C 0.50 max; Si 2.00 max; Ni 4–7; Mn 1.50 max; Cr 26–30; Fe bal.	As cast	48	85	16	190	Heat-resistant

SUPER ALLOYS

The advent of engineering applications requiring high temperature and high strength, as in jet engines and rocket motors, has lead to the development of a range of alloys collectively called super alloys. These alloys require excellent resistance to oxidation together with strength at high temperatures, typically 1800°F in existing engines. These alloys are continually being modified to develop better specific properties, and therefore entries in this group of alloys should be considered "fluid". Both wrought and casting-type alloys are represented. As the high temperature properties of cast materials improve, these alloys become more attractive, since great dimensional precision is now attainable in investment castings

No.	Material	Nominal composition	Form and condition	Yield strength	Tensile strength	Elongation	Hardness	Comments
	NICKEL BASE							
30	Hastelloy X	Co 1.5 max; Cr 22.0; W 0.6; C 0.20 max (cast); Fe 18.5; Mo 9.0; C 0.15 max (wrought); Ni bal.	Wrought sheet Mill-annealed As investment cast	52 — 46.5	113.2 67 —	43 17 —	194 172 —	
31	Hastelloy C	Cr 16.0; W 4.0; Mo 17.0; Fe 6.0; C 0.15 max; Ni bal.	Sand-cast (annealed) Rolled (annealed) Investment cast	50 71 50	78 130 80	5 45 10	199 204 215	

Table C.5 (continued) Mechanical Properties of Metals and Alloys:* Typical Composition, Properties, and Uses of Common Materials

No.	Material	Nominal composition	Form and condition	Typical mechanical properties				Comments
				Yield strength (0.2% offset), 1000 lb/sq in.	Tensile strength, 1000 lb/sq in.	Elongation, in 2 in., %	Hardness, Brinell	
	NICKEL BASE (Cont.)							
32	Inconel 713C	Ni (+Co) bal. Mo 4.5 Al 6.0 Cr 13.0 Cb 2.0 Ti 0.6	Investment cast	102	120	6	—	
33	In 100	C 18.0 Mo 3.0 Al 55.0 V 1.0 Cr 10.0 Ti 4.7 Co 15.0	Cast					
34	Taz 8	C 125.0 Mo 4.0 W 4.0 Ta 8.0 Cr 6.0 Al 6.0 Zr 1.0 V 2.5	Cast					
35	Nimonic 90	Ni (+Co) 57.00 Mn 0.50 S 0.007 Cu 0.05 Al 1.65 Co 16.90 C 0.05 Fe 0.45 Si 0.20 Cr 20.55 Ti 2.60	Annealed, wrought	90	155	—	260	General elevated temperature applications
36	Inconel X	Ni (+Co) 72.85 Mn 0.65 S 0.007 Cu 0.05 Al 0.75 Cb (+Ta) 0.85 C 0.04 Fe 6.80 Si 0.30 Cr 15.0 Ti 2.50	Annealed Annealed, age-hardened	50 115	115 175	50 25	150 300	
37	Waspaloy	C 0.08 Mo 4.3 Co 13.5 Cr 19.5 Ti 3.0	Cold-rolled	270	275	8	Rc 51	
38	Rene 41	C 0.09 Mo 10.0 Al 1.5 Cr 19.0 Ti 3.1 Co 11.0	Wrought	100	145	—	—	

Table C.5 (continued) Mechanical Properties of Metals and Alloys:* Typical Composition, Properties, and Uses of Common Materials

No.	Material	Nominal composition	Form and condition	Typical mechanical properties: Yield strength (0.2% offset), 1000 lb/sq in.	Tensile strength, 1000 lb/sq in.	Elongation, in 2 in., %	Hardness, Brinell	Comments
39	Udimet 700	C 0.08, Cr 15.0 Mo 5 0, Ti 3.5 Al 4.3, Co 18.5	Cold-rolled	280	285	6	Rc 53	
40	T D. Nickel	Ni 97.5, ThO_2 2 4	Extended and cold-worked	85	100	13	—	High temperature, jet engine parts
	COBALT BASE							
41	Haynes Stellite alloy 25 (L605)	C 0 15 max, Cr 20 0 Ni 10.0, W 15.0 Mn 1.5, Co bal	Wrought sheet: mill annealed	63	140	60	244	Wrought products
42	Haynes Stellite alloy 21 AMS 5385 (cast)	C 0.25, Mo 5.5 Ni 2.5, Co bal. Cr 28.5	As investment cast	82	103	8	313 max	For castings

ALUMINUM ALLOYS

Although the strength of aluminum alloys is in general less than that attainable in ferrous alloys or copper-base alloys, their major advantage lies in their high strength-to-weight ratio due to the low density of aluminum. Aluminum alloys have good corrosion resistance for most applications except in alkaline solutions.

No.	Material	Nominal composition	Form and condition	Yield strength (0.2% offset), 1000 lb/sq in.	Tensile strength, 1000 lb/sq in.	Elongation, in 2 in., %	Hardness, Brinell	Comments
43	3003 ASTM B221	Cu 0.12, Al bal. Mn 1.2	Annealed-O Cold-rolled-H14 Cold-rolled-H18	6 21 27	16 22 29	40 16 10	28 40 55	Good formability, weldable, medium strength; chemical equipment
44	2017 ASTM B221	Mn 0.5, Mg 0.5 Cu 4 0, Al bal	Annealed-O Heat-treated-T4	10 40	26 62	22 22	45 105	High strength, structural parts, aircraft, heavy forgings
45	2024 ASTM B211	Cu 4.5, Mg 1.5 Mn 0 6, Al bal	Heat-treated-T4	47	68	19	120	
46	5052 ASTM B211	Cr 0.25, Al bal. Mg 2 5	Annealed-O Cold-rolled and stabilized-H34	13 31	28 38	30 14	47 68	Medium strength, good fatigue properties; street-light standards
47	ASTM B209		Cold-rolled and stabilized-H38	37	42	8	77	

Table C.5 (continued) Mechanical Properties of Metals and Alloys:* Typical Composition, Properties, and Uses of Common Materials

No.	Material	Nominal composition	Form and condition	Yield strength (0.2% offset), 1000 lb/sq in.	Tensile strength, 1000 lb/sq in.	Elongation, in 2 in., %	Hardness, Brinell	Comments
48	7075 ASTM B211	Cu 1.6 Cr 0.3 Zn 5.6 Mg 2.5 Al bal.	Annealed-O Heat-treated and artifically aged-T6	15 73	33 83	17 11	60 150	High strength, good corrosion resistance
49	380 ASTM SC84B	Si 9.0 Cu 3.5 Al bal.	Die-cast	24	48	3	—	General purpose die casting
50	195 ASTM C4A	Si 0.8 Cu 4.5 Al bal	Sand-cast; heat-treated-T4 Sand-cast; heat-treated and artificially aged-T6	16 24	32 36	8.5 5	60 75	Structural elements, aircraft. and machines
51	214 ASTM G4A	Mg 3.8 Al bal.	Sand-cast-F	12	25	9	50	Chemical equipment, marine hardware, architectural
52	220 ASTM G10A	Mg 10.0 Al bal.	Sand-cast: heat-treated-T4	26	48	16	75	Strength with shock resistance: aircraft

COPPER ALLOYS

Because of their corrosion resistance and the fact that copper alloys have been used for many thousands of years, the number of copper alloys available is second only to the ferrous alloys. In general copper alloys co not have the high-strength qualities of the ferrous alloys. while their density is comparable. The cost per strength-weight ratio is high: however. they have the advantage of ease of joining by soldering. which is not shared by other metals that have reasonable corrosion resistance.

No.	Material	Nominal composition	Form and condition	Yield strength (0.2% offset), 1000 lb/sq in.	Tensile strength, 1000 lb/sq in.	Elongation, in 2 in., %	Hardness, Brinell	Comments
53	Copper ASTM B152 ASTM B124, B133 ASTM B1, B2, B3	Cu 99.9 plus	Annealed Cold-drawn Cold-rolled	10 40 40	32 45 46	45 15 5	42 90 100	Bus-bars, switches, architectural. roofing. screens
54	Gilding metal ASTM B36	Cu 95.0 Zn 5.0	Cold-rolled	50	56	5	114	Coinage, ammunition
55	Cartridge 70–30 brass ASTM B14 ASTM B19 ASTM B36 ASTM B134 ASTM B135	Cu 70.0 Zn 30.0	Cold-rolled	63	76	8	155	Good cold-working properties; radiator covers. hardware, electrical
56	Phosphor bronze 10% ASTM B103 ASTM B139 ASTM B159	Cu 90.0 P 0.25 Sn 10.0	Spring temper	—	122	4	241	Good spring qualities. high-fatigue strength

Table C.5 (continued) Mechanical Properties of Metals and Alloys:* Typical Composition, Properties, and Uses of Common Materials

No	Material	Nominal composition		Form and condition	Typical mechanical properties				Comments
					Yield strength (0.2% offset), 1000 lb/sq in	*Tensile strength, 1000 lb/sq in.*	*Elongation, in 2 in., %*	*Hardness, Brinell*	
57	Yellow brass (high brass) ASTM B36 ASTM B134 ASTM B135	Cu 65.0	Zn 35.0	Annealed Cold-drawn Cold-rolled (HT)	18 55 60	48 70 74	60 15 10	55 115 180	Good corrosion resistance, plumbing, architectural
58	Manganese bronze ASTM B138	Cu 58 5 Fe 1 0 Mn 0 3	Zn 39 2 Sn 1 0	Annealed Cold-drawn	30 50	60 80	30 20	95 180	Forgings
59	Naval brass ASTM B21	Cu 60 0 Sn 0 75	Zn 39 25	Annealed Cold-drawn	22 40	56 65	40 35	90 150	Condensor tubing: high resistance to salt-water corrosion
60	Muntz metal ASTM B111	Cu 60 0	Zn 40 0	Annealed	20	54	45	80	Condensor tubes; valve stress
61	Aluminum bronze ASTM B169, alloy A ASTM B124 ASTM B150	Cu 92 0	Al 8 0	Annealed Hard	25 65	70 105	60 7	80 210	
62	Beryllium copper 25 ASTM B194 ASTM B197 ASTM B196	Be 1.9 Co or Ni 0 25	Cu bal	Annealed, solution-treated Cold-rolled Cold-rolled	32 104 70	70 110 190	45 5 3	B60 (Rockwell) B81 C40	Bellows, fuse clips, electrical relay parts, valves, pumps
63	Free-cutting brass	Cu 62 0 Pb 2 5	Zn 35 5	Cold-drawn	44	70	18	B80 (Rockwell)	Screws, nuts, gears, keys
64	Nickel silver 18% Alloy A (wrought) ASTM B122, No 2	Cu 65 0 Ni 18.0	Zn 17.0	Annealed Cold-rolled Cold-drawn wire	25 70 —	58 85 105	40 4 —	70 170 —	Hardware, optical goods, camera parts
65	Nickel silver 13% (cast) 10A ASTM B149, No 10A	Ni 12.5 Sn 2 0 Zn 20.0	Pb 9.0 Cu bal	Cast	18	35	15	55	Ornamental castings, plumbing; good machining qualities
66	Cupronickel 10% ASTM B111 ASTM B171	Cu 88 35 Fe 1 25	Ni 10 0 Mn 0 4	Annealed Cold-drawn tube	22 57	44 60	45 15	— —	Condensor, salt-water piping

Table C.5 (continued) Mechanical Properties of Metals and Alloys:* Typical Composition, Properties, and Uses of Common Materials

No	Material	Nominal composition	Form and condition	Typical mechanical properties: Yield strength (0.2% offset), 1000 lb/sq in.	Tensile strength, 1000 lb/sq in.	Elongation, in 2 in., %	Hardness, Brinell	Comments
67	Cupronickel	Cu 70.0 Ni 30.0	Wrought					Heat-exchanger process equipment, valves
68	Red brass (cast) ASTM B30, No 4A	Cu 85.0 Zn 5.0 Pb 5.0 Sn 5.0	As-cast	17	35	25	60	
69	Silicon bronze ASTM B30, alloy 12A	Si 4.0 Fe 2.0 Zn 4.0 Al 1.0 Mn 1.0	Castings					Cheaper substitute for tin bronze
70	Tin bronze ASTM B30, alloy 1B	Sn 8% Zn 4.0	Castings					Bearings, high-pressure bushings, pump impellers
71	Navy bronze		Cast					

TIN AND LEAD-BASE ALLOYS

Major uses for these alloys are as "white"-metal bearing alloys, extruded cable sheathing, and solders. Tin forms the basis of pewter used for culinary applications.

No	Material	Nominal composition	Form and condition	Yield strength (0.2% offset), 1000 lb/sq in.	Tensile strength, 1000 lb/sq in.	Elongation, in 2 in., %	Hardness, Brinell	Comments
72	Lead-base Babbitt ASTM B23, alloy 19	Pb 85.0 Sn 5.0 Sb 10.0 As 0.6 Cu 0.5	Chill cast	—	10	5	19	Bearings, light loads and low speeds
73	Arsenical-lead Babbitt ASTM B23, alloy 15	Pb 83.0 Sn 1.0 Sb 16.0 As 1.1 Cu 0.6	Chill cast	—	10.3	2	20	Bearings, high loads and speeds, diesel engines, steel mills
74	Chemical lead	Pb 99.9 Cu 0.06 Bi 0.005 max	Rolled 95%	1.9	2.5	50	5	
75	Antimonial lead (hard lead)	Pb 94.0 Sb 6.0	Chill cast Rolled 95%	— —	6.8 4.1	22 47	(500 kg) 9	Good corrosion resistance and strength
76	Calcium lead	Pb 99.9 Ca 0.025 Cu 0.10	Extruded and aged	—	4.5	25	—	Cable sheathing, creep-resistant pipe
77	Tin Babbitt alloy ASTM B23-61, grade 1	Sb 4.5 Sn bal Cu 4.5	Chill cast	—	9.3	2	17	General bearings and die casting
78	Tin die-casting alloy ASTM B102-52	Sb 13.0 Sn bal Cu 5.0	Die-cast	—	10	1	29	Die-casting alloy

Table C.5 (continued) Mechanical Properties of Metals and Alloys:* Typical Composition, Properties, and Uses of Common Materials

No.	Material	Nominal composition		Form and condition	Typical mechanical properties: Yield strength (0.2% offset), 1000 lb/sq in.	Tensile strength, 1000 lb/sq in.	Elongation, in 2 in., %	Hardness, Brinell	Comments
79	Pewter	Sn 91.0 Cu 2.0	Sb 7.0	Rolled sheet, annealed	—	8.6	40	9.5	Ornamental and household items
80	Solder 50-50	Sn 50.0	Pb 50.0	Cast	4.8	6.1	60	14	General-purpose solder
81	Solder	Sn 20.0	Pb 80.0	Cast	3.6	5 8	16	11	Coating and joining, filling seams on automobile bodies
MAGNESIUM ALLOYS Because of their low density these alloys are attractive for use where weight is at a premium. The major drawback to the use of these alloys is their ability to ignite in air (this can be a problem in machining); they are also costly. Magnesium alloys are used in both the wrought and die-cast forms, the latter being the most frequently used form.									
82	Magnesium alloy AZ31B	Zn 1.0 Al 3.0	Mn 0.20 min Mg bal.	Rolled-plate (strain-hardened, then partially annealed)	24	37	18	—	Structural applications of medium strength
				Rolled-sheet (strain-hardened, then partially annealed)	32	42	15	73	
				Annealed	22	37	21	56	
				Extruded	28	38	14	—	
83	Magnesium alloy AZ80A	Zn 0.5 Al 8.5	Mn 0.15 min Mg bal.	Extruded	36	49	11	60	General extruded and forged products
				Extruded (age-hardened)	39	53	6	82	
				Forged (age-hardened)	34	50	6	72	
84	Magnesium alloy AZ92A	Zn 2.0 Al 9.0	Mn 0.10 min Mg bal.	Sand-cast (as cast)	14	24	6	50	
				Sand-cast (solution heat-treated)	14	40	12	55	
				Sand-cast (solution heat-treated and aged)	19	40	5	83	
				Sand-cast (age-hardened)	16	30	18	—	
				Sand-cast and tempered	22	40	3	81	Pressure-tight sand and permanent mold castings; high UTS and good yield strength
85	Magnesium alloy ZK60A	Zn 5.7 Zr 0.55	Mg bal.	Extruded	43	52	12	82	

Table C.5 (continued) Mechanical Properties of Metals and Alloys:* Typical Composition, Properties, and Uses of Common Materials

No.	Material	Nominal composition	Form and condition	Typical mechanical properties: Yield strength (0.2% offset), 1000 lb/sq in.	Tensile strength, 1000 lb/sq in.	Elongation, in 2 in., %	Hardness, Brinell	Comments
86	Magnesium alloy AZ91A and AZ91B	Zn 0.6, Al 9.0, Mn 0.13 min, Mg bal.	Die-cast (as cast)	22	33	3	67	General die-casting applications
			BERYLLIUM					
87	Beryllium		 Hot-pressed Cross-rolled	27 38 40 60	33 51 60 90	1-3 10–40	— —	Windows, X-ray tubes Moderator- and reflector-cladding nuclear reactors, heat-shield and structural-member missiles
			NICKEL ALLOYS					
	Nickel and its alloys are expensive and used mainly either for their high-corrosion resistance in many environments or for high-temperature and strength applications. (See Super Alloys, above.)							
88	Nickel (cast)	Ni 95.6, Fe 0.5, Si 1.5, Cu 0.5, Mn 0.8, C 0.8	As cast	25	57	22	110	Good corrosion-resistance applications
89	K Monel	Ni (+Co) 65.25, Mn 0.60, S 0.005, Cu 29.60, Ti 0.45, C 0.15, Fe 1.00, Si 0.15, Al 2.75	Annealed Annealed, age-hardened Spring Spring, age-hardened	45 100 140 160	100 155 150 185	40 25 5 10	155 270 300 335	High strength and corrosion resistance; aircraft parts, valve stems, pumps
90	A nickel ASTM B160 ASTM B161 ASTM B162	Ni (+Co) 99.40, Mn 0.25, S 0.005, Cu 0.05, C 0.06, Fe 0.15, Si 0.05	Annealed Hot-rolled Cold-drawn Cold-rolled	20 25 70 95	70 75 95 105	40 40 25 5	100 110 170 210	Chemical industry for resistance to strong alkalis, plating nickel
91	Duranickel	Ni (+Co) 93.90, Mn 0.25, S 0.005, Cu 0.05, Ti 0.45, C 0.15, Fe 0.15, Si 0.55, Al 4.50	Annealed Annealed, age-hardened Spring Spring, age-hardened	45 125 — —	100 170 175 205	40 25 5 10	160 330 320 370	High strength and corrosion resistance; pump rods, shafts, springs

Table C.5 (continued) Mechanical Properties of Metals and Alloys:* Typical Composition, Properties, and Uses of Common Materials

No.	Material	Nominal composition	Form and condition	Typical mechanical properties: Yield strength (0.2% offset), 1000 lb/sq in.	Tensile strength, 1000 lb/sq in.	Elongation, in 2 in., %	Hardness, Brinell	Comments
92	Cupronickel 55–45 (Constantan)	Cu 55.0 Ni 45.0	Annealed Cold-drawn Cold-rolled	30 50 65	60 65 85	45 30 20	— — —	Electrical-resistance wire; low temperature coefficient, high resistivity
93	Nichrome	Ni 80.0 Cr 20.0						Heating elements for furnaces
94	"S" Monel	Ni 60.0 Cu 29.0 Fe 2.50 max Mn 1.5 max Si 4.0 Al 0.5 max	Sand-casting	80–115	110–145	2	270–350	High-strength casting alloy, good bearing properties for valve seats

TITANIUM ALLOYS

The main application for these alloys is in the aerospace industry. Because of the low density and high strength of titanium alloys, they present excellent strength-to-weight ratios.

No.	Material	Nominal composition	Form and condition	Yield strength (0.2% offset), 1000 lb/sq in.	Tensile strength, 1000 lb/sq in.	Elongation, in 2 in., %	Hardness, Brinell	Comments
95	Commercial titanium ASTM B265–58T	Ti 99.4	Annealed at 1100 to 1350°F (593 to 732°C)	70	80	20	—	Moderate strength, excellent fabricability; chemical industry pipes
96	Titanium alloy ASTM B265–58T-5 Ti-6 Al–4V		Water-quenched from 1750°F (954°C); aged at 1000°F (538°C) for 2 hr	160	170	13	—	High-temperature strength needed in gas-turbine compressor blades
97	Titanium alloy Ti-4 Al–4Mn		Water-quenched from 1450°F (788°C); aged at 900°F (482°C) for 8 hr	170	185	13	—	Aircraft forgings and compressor parts
98	Ti-Mn alloy ASTM B265–58T-7	Fe 0.5 Ti bal. Mn 7.0–8.0	Sheet	140	150	18	—	Good formability, moderate high-temperature strength; aircraft skin

ZINC ALLOYS

A major use for these alloys is for low-cost die-cast products, such as household fixtures, automotive parts, and trim.

No.	Material	Nominal composition	Form and condition	Yield strength (0.2% offset), 1000 lb/sq in.	Tensile strength, 1000 lb/sq in.	Elongation, in 2 in., %	Hardness, Brinell	Comments
99	Zinc ASTM B69	Cd 0.35 Zn bal. Pb 0.08	Hot-rolled	—	19.5	65	38	Battery cans, grommets, lithographer's sheet

Table C.5 (continued) Mechanical Properties of Metals and Alloys:* Typical Composition, Properties, and Uses of Common Materials

No.	Material	Nominal composition	Form and condition	*Typical mechanical properties* Yield strength (0.2% offset), 1000 lb/sq in.	Tensile strength, 1000 lb/sq in.	Elongation, in 2 in., %	Hardness, Brinell	Comments
100	Zilloy-15	Cu 1.00 Zn bal. Mg 0.010	Hot-rolled Cold-rolled	— —	29 36	20 25	61 80	Corrugated roofs, articles with maximum stiffness
101	Zilloy-40	Cu 1.00 Zn bal.	Hot-rolled Cold-rolled	— —	24 31	50 40	52 60	Weatherstrip, spun articles
102	Zamac-5 ASTM 25	Zn (99.99% pure remainder) Al 3.5-4.3 Cu 0.75-1.25 Mg 0.03-0.08	Die-cast	—	47.6	7	91	Die casting for automobile parts, padlocks; used also for die material
	ZIRCONIUM ALLOYS These alloys have good corrosion resistance but are easily oxidized at elevated temperatures in air The major application is for use in nuclear reactors.							
103	Zirconium, commercial	O_2 0.07 C 0.15 Hf 1.90 Zr bal.	Annealed	40	65	27	B80 (Rockwell)	
104	Zircaloy 2	Hf 0.02 Ni 0.05 Fe 0.15 Other 0.25 Sn 1.46 Zr bal.	Annealed	50	75	22	B90 (Rockwell)	Nuclear power-reactor cores at elevated temperatures

*Compiled from various sources

Table C.6 Miscellaneous Properties of Metals and Alloys

Part a. Pure Metals

At Room Temperature

Common name	PROPERTIES (TYPICAL ONLY)						
	Thermal conductivity, Btu/hr ft °F	*Specific gravity*	*Coeff. of linear expansion, μ in / in °F*	*Electrical resistivity, microhm-cm*	*Poisson's ratio*	*Modulus of elasticity, millions of psi*	*Approximate melting point, °F*
Aluminum	137	2.70	14	2 655	0.33	10.0	1220
Antimony	10 7	6 69	5	41 8		11.3	1170
Beryllium	126	1.85	6.7	4 0	0.024–.030	42	2345
Bismuth	4.9	9.75	7 2	115		4 6	521
Cadmium	54	8.65	17	7.4		8	610
Chromium	52	7 2	3 3	13		36	3380
Cobalt	40	8.9	6 7	9		30	2723
Copper	230	8 96	9 2	1 673	0.36	17	1983
Gold	182	19 32	7 9	2 35	0 42	10 8	1945
Iridium	85 0	22.42	3 3	5.3		75	4440
Iron	46.4	7.87	6.7	9 7		28 5	2797
Lead	20 0	11.35	16	20.6	0 40– 45	2 0	621
Magnesium	91.9	1 74	14	4 45	0 35	6 4	1200
Manganese		7 21-7.44	12	185		23	2271
Mercury	4.85	13.546		98 4			−38
Molybdenum	81	10.22	3 0	5 2	0.32	40	4750
Nickel	52 0	8.90	7.4	6 85	0 31	31	2647
Niobium (Columbium)	30	8 57	3 9	13		15	4473
Osmium	35	22 57	2 8	9		80	5477
Platinum	42	21 45	5	10 5	0 39	21.3	3220
Plutonium	4 6	19 84	30	141 4	0 15– 21	14	1180
Potassium	57 8	0 86	46	7 01			146
Rhodium	86 7	12 41	4 4	4 6		42	3569
Selenium	0 3	4 8	21	12 0		8 4	423
Silicon	48 3	2 33	2 8	1×10^5		16	2572
Silver	247	10 50	11	1 59	0 37	10 5	1760
Sodium	77 5	0.97	39	4 2			208
Tantalum	31	16 6	3 6	12 4	0 35	27	5400
Thorium	24	11 7	6 7	18	0 27	8 5	3180
Tin	37	7 31	11	11 0	0 33	6	450
Titanium	12	4.54	4 7	43	0 3	16	3040
Tungsten	103	19.3	2.5	5 65	0 28	50	6150
Uranium	14	18 8	7 4	30	0 21	24	2070
Vanadium	35	6 1	4 4	25		19	3450
Zinc	66 5	7	19	5 92	0 25	12	787

Table C.6 Miscellaneous Properties of Metals and Alloys

Part b. Commercial Metals and Alloys

CLASSIFICATION AND DESIGNATION		*PROPERTIES (TYPICAL ONLY)*					
Material No. (from Table 1-57)	*Common name and classification*	*Thermal conductivity, Btu/hr ft °F*	*Specific gravity*	*Coeff. of linear expansion, μ in./in. °F*	*Electrical resistivity, microhm-cm*	*Modulus of elasticity, millions of psi*	*Approximate melting point, °F*
1	Ingot iron (included for comparison)	42.	7.86	6.8	9.	30	2800
2	Plain carbon steel AISI-SAE 1020	30.	7.86	6.7	10.	30	2760
15	Stainless steel type 304	10.	8.02	9.6	72.	28	2600
19	Cast gray iron ASTM A48-48, Class 25	26.	7.2	6.7	67	13	2150
21	Malleable iron ASTM A47	–	7.32	6.6	30.	25	2250
22	Ductile cast iron ASTM A339, A395	19	7 2	7.5	60.	25	2100
24	Ni-resist cast iron, type 2	23	7.3	9.6	170.	15.6	2250
29	Cast 28-7 alloy (HD) ASTM A297-63T	1.5	7.6	9.2	41.	27	2700
31	Hastelloy C	5	3.94	6.3	139.	30	2350
36	Inconel X, annealed	9	8.25	6.7	122.	31	2550
41	Haynes Stellite alloy 25 (L605)	5.5	9.15	7.61	88.	34	2500
43	Aluminum alloy 3003, rolled ASTM B221	90	2.73	12.9	4	10	1200
44	Aluminum alloy 2017, annealed ASTM B221	95	2.8	12.7	4	10.5	1185
49	Aluminum alloy 380 ASTM SC84B	56	2.7	11.6	7.5	10.3	1050
53	Copper ASTM B152, B124, B133, B1, B2, B3	225	8.91	9.3	1.7	17	1980
57	Yellow brass (high brass) ASTM B36, B134, B135	69	8 47	10.5	7	15	1710
61	Aluminum bronze ASTM B169, alloy A, ASTM B124, B150	41	7 8	9.2	12.	17	1900
62	Beryllium copper 25 ASTM B194	7	8.25	9.3	–	19	1700
64	Nickel silver 18% alloy A (wrought) ASTM B122, No. 2	19	8.8	9.0	29.	18	2030
67	Cupronickel 30%	17	8.95	8.5	35.	22	2240
68	Red brass (cast) ASTM B30, No. 4A	42	8.7	10.	11	13	1825
74	Chemical lead	20	11.35	16 4	21	2	621
75	Antimonial lead (hard lead)	17	10.9	15.1	23	3	554
80	Solder 50-50	26	8.89	13.1	15.	–	420
82	Magnesium alloy AZ31B	45	1 77	14.5	9.	6.5	1160
89	K Monel	11	8.47	7.4	58.	26	2430
90	Nickel ASTM B160, B161, B162	35	8 89	6.6	10.	30	2625
92	Cupronickel 55-45 (Constantan)	13	8.9	8.1	49.	24	2300
95	Commercial titanium	10	5.	4.9	80.	16.5	3300
99	Zinc ASTM B69	62	7.14	18	6.	–	785
103	Zirconium, commercial	10	6.5	2.9	41.	12	3350

*Compiled from several sources.

Table C.7 Composition and Melting Points of Binary Eutectic Alloys:* Binary Alloys and Solid Solutions of Metallic Components

This table represents most of the common binary combinations of metals. For many pairs no eutectic exists; for many others, the informations is uncertain or unavailable. In a fair number of cases, there is complete mutual solubility in all proportions; hence, there is a smooth temperature vs. composition curve, with no point of inflection from the melting point of one constituent to that of the other. For purposes of comparison, all values must be considered approximate in view of the experimental difficulties and the many sources of data.

Those pairs for which the liquidus curve exhibits more than one cusp are designated by a superscript *a*. In a few cases, the cusp selected for this table does not represent the lowest possible melting point for the binary mixture.

Constituents		*Composition*		*Melting point*		*Constituents*		*Composition*		*Melting point*	
A	*B*	*Mol % B*	*Weight % B*	*K*	*deg F*	*A*	*B*	*Mol % B*	*Weight % B*	*K*	*deg F*
Ag	Al	57	25	835	1 044	Au	Bi	86.8	85	514	466
Ag	As	24	18	813	1 004	Au	Cd	70	57.1	773	932
Ag	Ca[a]	37	18	820	1 017	Au	Ce[a]	86	81	793	968
Ag	Ce[a]	80	84	798	977	Au	Ge	27	12	629	673
Ag	Cu	40	28	1 050	1 431	Au	La[a]	83	78	834	1 042
Ag	Ge	25	18	924	1 204	Au	Mg	93	62	848	1 067
Ag	La[a]	72	77	791	964	Au	Mn[a]	32	12	1 233	1 760
Ag	Li	99	89	418	293	Au	Na	17	2.3	1 149	1 609
Ag	Mg[a]	83	52	745	882	Au	Pb	84	85	488	419
Ag	Pb	95.3	97.5	577	579	Au	Sb	34.8	24.8	633	680
Ag	Pd	25.9	25.6	924	1 204	Au	Si	18.6	3.15	636	685
Ag	Sb	41	44	758	905	Au	Sn[a]	29.3	19.9	551	532
Ag	Si	10.5	2.96	1 110	1 539	Au	Te	88	83	689	781
Ag	Sr[a]	77	73	709	817	Au	Tl	72	73	404	268
Ag	Te	65	69	623	662	Au	U	14	16	1 128	1 571
Ag	Th	7.6	15	1 167	1 641	B	Hf	13	71	2 130	3 375
Ag	Zr	97	93	1 100	1 521	B	Ni	57	88	1 263	1 814
Al	Au[a]	59.5	90.0	842	1 056	B	Ti	7	25	1 700	2 601
Al	Ca[a]	65	73	818	1 013	B	Zr	88	98	1 920	2 997
Al	Cd	81	90	1 650	2 511	Ba	Mg	97	87	891	1 144
Al	Ce	69	92	928	1 211	Be	Ni	33	76	1 468	2 183
Al	Cu[a]	17.3	33.0	821	1 018	Be	Pu	97	99	910	1 179
Al	Fe	32	49.34	1 426	2 107	Be	Si	33	61	1 363	1 994
Al	Ge	29	55	700	801	Be	Ti	75	94	1 300	2 061
Al	In	5	18	910	1 179	Be	Y	61	94	1 390	2 043
Al	Mg	70	67.0	710	819	Be	Zr	65	95	1 250	1 791
Al	Ni[a]	76	87	1 658	2 525	Bi	Ca	88	58.5	1 059	1 447
Al	Pt[a]	57	90	1 533	2 300	Bi	Cd	56	40	420	297
Al	Si	13	13	850	1 071	Bi	In[a]	78	66	340	153
Al	Th	80	97	1 510	2 259	Bi	K	50	16	615	648
Al	Zn	88.7	95.0	655	720	Bi	Mg	85	40	820	1 017
As	Co	75	70	1 189	1 681	Bi	Na	22	3.0	500	441
As	Cu[a]	81.6	78.0	958	1 265	Bi	Pb	44	44	397	255
As	Fe	75	69	1 103	2 017	Bi	Sn	57	43	415	288
As	In	13	18	1 004	1 348	Bi	Te	90	84	686	775
As	Mn	57	49	1 143	1 598	Bi	Tl[a]	53	52	465	378
As	Ni[a]	63	57	1 077	1 479	C	Cr	87	96	1 775	2 736
As	Sb	80	87	878	1 121	C	Hf	35	88	3 450	5 751
As	Sn[a]	40	51	852	1 074	C	Mo	17	45	2 480	4 005
As	Zn[a]	20	18	996	1 333	C	Nb	40	84	3 580	5 985

*Compiled from several sources.

Table C.7 (continued) Composition and Melting Points of Binary Eutectic Alloys:* Binary Alloys and Solid Solutions of Metallic Components

Constituents		*Composition*		*Melting point*		*Constituents*		*Composition*		*Melting point*	
A	*B*	*Mol % B*	*Weight % B*	*K*	*deg F*	*A*	*B*	*Mol % B*	*Weight % B*	*K*	*deg F*
C	Ti	36	69	3 050	5 031	Gd	Ni[a]	32	15	943	1 238
C	V	84	96	1 900	2 961	Ge	Mg	38	17	953	1 256
C	W	59	96	2 980	4 905	Ge	Mn[a]	48	41	970	1 287
Ca	Cu	51	62	833	1 040	Hf	Ta	24	24	1 300	1 881
Ca	Mg[a]	32	22	718	833	In	Ni	30	17.97	1 143	1 598
Ca	Na	22	14	983	1 310	In	Sb	68	69	780	945
Ca	Ni	16	22	878	1 121	In	Sn	47	48	390	243
Ca	Sn	19	41	1 010	1 359	Ir	Mo	68	52	2 350	3 771
Cd	Cu	52	38	810	999	Ir	Nb	55	23	2 110	3 339
Cd	In	74	74	400	261	Ir	W	22	12	2 590	4 203
Cd	Pb	71	82	540	513	K	Na	32	21.67	260	− 8.6
Cd	Pu	40	59	1 170	1 647	K	Rb	70	84	307	93
Cd	Sb	7.4	8	563	554	K	Sb[a]	68	84	680	765
Cd	Sn	68	69	450	351	K	Tl	84	96	440	333
Cd	Tl	73	83	475	396	La	Mg[a]	38	9.7	970	1 287
Cd	Zn	27	18	540	513	La	Pb[a]	11	15	1 049	1 429
Ce	Cu[a]	28	15	688	779	La	Sn[a]	10	9	993	1 328
Ce	Ru	33	26	923	1 202	La	Tl	16	22	913	1 184
Co	Gd	65	83	913	1 184	Mg	Ni	11	22.98	780	945
Co	Mo	27	38	1 610	2 439	Mg	Pr	4.9	23	858	1 085
Co	Nb	15	22	1 500	2 241	Mg	Pu	15	63	815	1 008
Co	Si[a]	71	54	1 486	2 215	Mg	Sb[a]	86	97	855	1 080
Co	Sn	21	35	1 380	2 025	Mg	Si	53	57	1 223	1 742
Co	Ti[a]	22	19	1 430	2 115	Mg	Sr[a]	70	89	699	799
Co	V	41	38	1 521	2 278	Mg	Th	7	42	855	1 080
Cr	Mo	14	23	1 973	3 092	Mg	Zn	30	53	615	648
Cr	Ni	46	47	1 610	2 439	Mn	Ni	40	42	1 300	1 881
Cr	Ta	13	34	1 950	3 051	Mn	Pd	26	41	1 398	2 057
Cr	Ti	86	85	950	1 251	Mn	Sb	82	91	843	1 058
Cr	V	33	32	2 050	3 231	Mn	Ti[a]	9	7.9	1 460	2 169
Cs	K	50	23	235	− 36	Mn	U[a]	75	93	988	1 319
Cs	Na	20.9	4.37	241	− 26	Mn	Y[a]	65	75	1 163	1 634
Cs	Rb	50	39	282	48	Mo	Nb	66	65	2 570	4 167
Cu	Ge	34	37	913	1 184	Mo	Ni	64	52	1 590	2 403
Cu	Mg[a]	85.5	69.3	758	905	Mo	Os	21	34	2 650	4 311
Cu	Mn	37	34	1 143	1 598	Mo	Pd	54	57	2 020	3 177
Cu	Pb	15	36	1 230	1 755	Mo	Re	48	64	2 780	4 545
Cu	Pr[a]	69	83	745	882	Mo	Ru	41	42	2 200	3 501
Cu	Sb[a]	63	76	800	981	Mo	Si[a]	17	5.7	2 350	3 771
Cu	Si	30	16	1 075	1 476	Na	Rb	82.1	94.5	269	25
Cu	Te	69	82	617	207	Na	Sb	60	89	678	761
Cu	Ti[a]	27	22	1 133	1 580	Na	Sn	37	75	718	833
Cu	Tl	14.5	35.3	1 357	1 983	Na	Te	55	87	592	606
Cu	U	8.2	25	1 213	1 724	Nb	Ni	58	47	1 450	2 151
Cu	Zr	9.4	13	1 253	1 796	Nb	Pt	54	71	1 970	3 087
Fe	Gd	69	86	1 123	1 562	Nb	Rh	45	31	1 770	2 727
Fe	Mo	21	31	1 725	2 646	Nb	Ru[a]	64	49	2 050	3 231
Fe	Nb	12	18.49	1 643	2 498	Nb	Zr	77	77	2 010	3 159
Fe	Sb	88	94.10	1 021	1 378	Ni	Sb	22	36.90	1 375	2 016
Fe	Si[a]	35	21	1 475	2 196	Ni	Sn	19	32.16	1 403	2 066
Fe	Sn	31	49	1 400	2 061	Ni	Th[a]	35	68	1 303	1 886
Fe	Y	65	75	1 173	1 652	Ni	Ti[a]	39	34	1 390	2 043
Fe	Zr[a]	11	17	1 600	2 421	Ni	V	52	48	1 473	2 192
Ga	Mg[a]	80	58	698	797	Ni	W	20.7	45	1 773	2 732
Ga	Ni	70	66	1 477	2 199	Ni	Zn	69	71	1 148	1 607

Table C.7 (continued) Composition and Melting Points of Binary Eutectic Alloys:* Binary Alloys and Solid Solutions of Metallic Components

Constituents		*Composition*		*Melting point*		*Constituents*		*Composition*		*Melting point*	
A	*B*	*Mol % B*	*Weight % B*	*K*	*deg F*	*A*	*B*	*Mol % B*	*Weight % B*	*K*	*deg F*
Pb	Pr	40	31	1 315	1 908	Si	Th[a]	88	98	1 710	2 619
Pb	Pt	5.3	5.0	563	554	Si	Ti[a]	86	91	1 600	2 421
Pb	Sb	18	11	520	477	Si	Zr[a]	9	24	1 570	2 367
Pb	Sn	73	61	460	369	Sn	Te	84	85	678	761
Pb	Te	85	78	680	765	Sn	Tl	31	44	440	333
Pb	Ti	92	74	998	1 337	Sn	Zn	16	9.5	465	378
Pd	Sb	89	90	868	1 103	Te	Tl	30	41	483	410
Pt	Sn	40	29	1 345	1 962	Th	Ti	40	12	1 463	2 174
Pu	Zn	73	42	1 100	1 521	Th	Zn[a]	49	21	1 220	1 737
Re	W	26	26	3 100	5 121	Ti	U	17	51	933	1 220
Sb	Tl	70	80	468	383	Ti	Y	6.8	12	1 593	2 408
Sb	Zn	33	21	780	945	Ti	Zr	50	66	790	963
Sb	Zr	82	77	1 700	2 601	U	Zr	70	47	879	1 123
Se	Sn	39	49	913	1 184						
Se	Tl	26	48	424	304						

REFERENCES

"Selected Values of Thermodynamic Properties of Metals and Alloys," R. Hultgren, R.L. Orr, P.D. Anderson, K K. Kelley. John Wiley & Sons, inc., 1963; a supplement to this publication has been issued periodically by the University of California, 1964-1971

"Constitution of Binary Alloys," 2nd ed., M. Hansen, McGraw-Hill Book Company, 1958

"Metals Reference Book," 4th ed., C.J. Smithells, Vol. 2, Butterworth & co., London, 1967

"Handbook of Binary Metallic Systems," 2 volumes; translated from Russian, Israel Program for Scientific Translations, JJerusalem. Available from Clearinghouse for Federal Scientific and Technical Information, Springfield, Virginia 22151

See also *Trans. AIME, J. Inst. Metals*, and *Z. Metallkunde*, by indexes.

TABLE C.8 Melting Points of Mixtures of Metals**

Melting Points, °C

Metals	*Percentage of metal in second column* 0%*	10%	20%	30%	40%	50%	60%	70%	80%	90%	100%
Pb. Sn.	326	295	276	262	240	220	190	185	200	216	232
Bi.	322	290			179	145	126	168	205		268
Te.	322	710	790	880	917	760	600	480	410	425	446
Ag.	328	460	545	590	620	650	705	775	840	905	959
Na.	..	360	420	400	370	330	290	250	200	130	96
Cu.	326	870	920	925	945	950	955	985	1005	1020	1084
Sb.	326	250	275	330	395	440	490	525	560	600	632
Al. Sb.	650	750	840	925	945	950	970	1000	1040	1010	632
Cu.	650	630	600	560	540	580	610	755	930	1055	1084
Au.	655	675	740	800	855	915	970	1025	1055	675	1062
Ag.	650	625	615	600	590	580	575	570	650	750	954
Zn.	654	640	620	600	580	560	530	510	475	425	419
Fe.	653	860	1015	1110	1145	1145	1220	1315	1425	1500	1515
Sn.	650	645	635	625	620	605	590	570	560	540	232
Sb. Bi.	632	610	590	575	555	540	520	470	405	330	268
Ag.	630	595	570	545	520	500	505	545	680	850	959
Sn.	622	600	570	525	480	430	395	350	310	255	232
Zn.	632	555	510	540	570	565	540	525	510	470	419

Metals	*Percentage of metal in second column* 0%*	10%	20%	30%	40%	50%	60%	70%	80%	90%	100%
Ni. Sn.	1455	1380	1290	1200	1235	1290	1305	1230	1060	800	232
Na. Bi.	96	425	520	590	645	690	720	730	715	570	268
Cd.	96	125	185	245	285	325	330	340	360	390	322
Cd. Ag.	322	420	520	610	700	760	805	850	895	940	954
Tl.	321	300	285	270	262	258	245	230	210	235	302
Zn.	322	280	270	295	313	327	340	355	370	390	419
Au. Cu.	1063	910	890	895	905	925	975	1000	1025	1060	1084
Ag.	1064	1062	1061	1058	1054	1049	1039	1025	1006	982	963
Pt.	1075	1125	1190	1250	1320	1380	1455	1530	1610	1685	1775
K. Na.	62	17.5	−10	−3 5	5	11	26	41	58	77	97.5
Hg.	.		.	.		90	110	135	162	265	...
Tl.	62 5	133	165	188	205	215	220	240	280	305	301
Cu. Ni.	1080	1180	1240	1290	1320	1355	1380	1410	1430	1440	1455
Ag.	1082	1035	990	945	910	870	830	788	814	875	960
Sn.	1084	1005	890	755	725	680	630	580	530	440	232
Zn.	1084	1040	995	930	900	880	820	780	700	580	419
Ag. Zn.	959	850	755	705	690	660	630	610	570	505	419
Sn.	959	870	750	630	550	495	450	420	375	300	232
Na. Hg.	96 5	90	80	70	60	45	22	55	95	215	...

*The data in this table are compiled from various sources—hence, the variations in the melting point of the metals as shown in this column.

**Based largely on: "Smithsonian Physical Tables," 9th rev. ed., W.E. Forsythe, Ed., The Smithsonian Institution, 1956.

Table C.9 Trade Names, Composition, and Manufacturers of Various Plastics

Trade name	*Composition*	*Manufacturer*
Abson	Acrylonitrile-butadiene, ABS polymers	B F Goodrich Chemical Co
Alathon	Polyethylene resins	E I du Pont de Nemours & Co , Inc
Alkor	Furane resin cement	Atlas Minerals & Chemicals Div The Electric Storage Battery Co
Amres	Phenolics, urea, and melamine resins	American Marietta Co , Pacific Resins & Chemicals Inc
Araldite	Epoxy resins	CIBA Products Co , Div CIBA Corp
Atlac	Polyester resins	Atlas Chemical Industries, Inc
Bakelite	Acrylics epoxies, phenolics, polyethylenes, copolymers	Union Carbide Corp , Chemicals and Plastics Div
Bavick-11	Methyl methacrylate and methylstyrene copolymer	J T Baker Chemical Co
Boltaflex	Supported and unsupported flexible vinyl sheeting	The General Tire & Rubber Co
Boltaron	Rigid polyvinyl chloride sheet	The General Tire & Rubber Co , Chemical & Plastics Div
Butacite	Polyvinyl butyral resins	E I du Pont de Nemours & Co , Inc
Conolite	Polyester resins and laminates	Shellmar-Betner, Div Continental Can Co Woodall Industries, Inc , Conolite Div
Corvel	Epoxies, vinyls Fusion-bond finishes	The Polymer Corp , Export-Polypenco Div
Cumar	Paracoumarone-indene resins	Allied Chemical Corp Plastics Div
Cycolac	ABS polymers acrylonitrile-butadiene-styrene copolymers	Marbon Chemical Div , Borg-Warner Corporation
Dacovin	Polyvinyl chlorides	Diamond Shamrock Corp
Dapon	Diallyl phthalate resins	FMC Corp , Organic Div
Delrin	Acetal resin and pipe	E I du Pont de Nemours & Co Inc
Dylan	Polyethylene	Sinclair-Koppers Co
Dylene	Polystyrene	Sinclair Koppers Co
Dylite	Expandable polystyrene	Sinclair Koppers Co
Epi Rez	Epoxy resins	Celanese Coatings Co Celanese Resin Div
Epolene	Low molecular-weight polyethylene resins	Eastman Chemical Products, Inc , Sub Eastman Kodak Company
Epoxical	Epoxy resins	United States Gypsum Co
Epon	Epoxy resins and curing agents	The Shell Chemical Company, Plastics and Resins Div
Escon	Polypropylene resins	Enjay Chemical Co , Div Humble Oil & Refining Company
Estane	Polyurethane materials	B F Goodrich Chemical Company
Fluorogreen	Teflon with glass and ceramic fibers, fluorocarbons	John L Dore Co
Fluororay	Ceramic-filled fluorocarbons	Raybestos Manhattan, Inc Plastic Products Div
Formica	Melamines	Formica Corp of American Cyanamid
Forticel	Cellulose propionate sheet films, molding powders	Celanese Plastics Co
Fortiflex	Polyethylene resins	Celanese Plastics Co
Fosta-Tuf-Flex	Polystyrene, high-impact	Foster-Grant, Inc
Furnane	Furanes	Atlas Minerals & Chemicals Div , The Electric Storage Battery Co
GenEpoxy	Epoxy resins for adhesives, coatings, etc	General Mills, Inc., Chemical Div
Genetron	Fluorinated hydrocarbons, monomers, and polymers	Allied Chemical Corp , General Chemical Div
Geon	Polyvinyl chloride materials	B F Goodrich Chemical Co
Grex	High-density polyethylenes	Allied Chemical Corp , Plastics Div
Halon	Fluorohalocarbon resins	Allied Chemical Corp
Hetron	Fire-retardant polyester resin	Hooker Chemical Corp Durez Plastics Div
Isothane	Polyurethane foam, ester and ether	Bernel Foam Products Co , Inc
Kel-F	Chlorotrifluoroethylene, molding resins, and dispersions	3M Company
Kralac	High-styrene resins, styrene-butadiene copolymers	Uniroyal Chemical, Div of Uniroyal Inc
Kralastic	ABS polymers, copolymers	Uniroyal Chemical, Div of Uniroyal Inc
Kynar	Polyvinylidene fluoride	Pennsalt Chemical Corp
Lexan	Polycarbonate resin film, and sheet	General Electric Company, Plastics Dept
Lucite	Acrylic resin and syrup	E I du Pont de Nemours & Co , Inc
Lustran	ABS polymers	Monsanto Co
Lustrex	Styrene molding and extrusion resins	Monsanto Co
Lytron	Styrene molding and extrusion resins	Monsanto Co
Madurit	Melamine resins and compounds	Cassella Farbwerke Mainkur, A G
Maraglas	Epoxy-casting resin	The Marblette Corporation, Div of Allied Products
Marlex	Polyethylenes, polypropylenes, copolymers	Phillips Petroleum Co
Marvinol	Vinyl chloride resins and compounds	Uniroyal Chemical, Div of Uniroyal Inc
Merlon	Polycarbonate resins	Mobay Chemical Co
Micarta	Melamines, phenolics, polyesters	Westinghouse Electric Co Industrial Micarta Div
Microthene	Polyethylenes polyolefins	U S Industrial Chemicals Co
Multrathane	Urethane elastomers	Mobay Chemical Company
Nopcofoam	Polyurethane plastics	Nopco Chemical Co , Plastics Div
Novodur	Polyacrylonitrile-butadiene-styrene	Farbenfabriken Bayer, A G
Opalon	Vinyl chloride resins and compounds	Monsanto Co
Paraplex	Polyester resins, acrylic-modified polyester resins	Rohm & Haas Company

Table C.9 (continued) Trade Names, Composition, and Manufacturers of Various Plastics

Trade name	*Composition*	*Manufacturer*
Permelite	Melamines	Melamine Plastics, Inc., Div. of Fiberlite Corp.
Petrothene	Polyethylene resins, polypropylene resins	U.S. Industrial Chemicals Co.
Piccoflex	Styrene-copolymer resins	Pennsylvania Industrial Chemical Corp.
Piccolastic	Styrene-polymer resins	Pennsylvania Industrial Chemical Corp.
Plaskon	Nylons, melamines, phenolics, polyesters	Allied Chemical Corp.
Pleogen	Alkyds, polyesters, copolymers	Mol-Rez Div., American Petrochemical Corp.
Plexiglas	Acrylics	Rohm & Haas Company
Pliovic	Polyvinyl chlorides	The Goodyear Tire & Rubber Co., Chemical Div.
Plyophen	Phenolic resins	Reichhold Chemicals, Inc.
Poly-Eth	Polyethylene resins	Gulf Oil Corp., U.S. Div. of Gulf Oil Corp.
Polylite	Polyester resins	Reichhold Chemicals, Inc.
Polypenco	Acrylics, chlorinated polyethers, fluorocarbons, nylons, polycarbonates	Polymer Corp.
Resimene	Urea and melamine resins	Monsanto Co.
Resinox	Phenolic resins and compounds	Monsanto Co.
Rhonite	Urea resins	Rohm & Haas Company
Roylar	Polyurethanes	Uniroyal Chemical, Div. of Uniroyal Inc.
Ryertex	Laminated phenolics and rigid polyvinyl chloride extrusions	Joseph T. Ryerson & Son, Inc., Industrial Plastics and Bearings Sales Div.
Super Dylan	Polyethylene	Sinclair-Koppers Co.
Supreme	Polyethylenes	Johns-Manville Company
Sylplast	Urea-formaldehyde compounds	FMC Corp., Organic Chemicals Div.
Teflon	Fluorocarbon resins	E. I. du Pont de Nemours & Co., Inc.
Tenite	Cellulose acetate, cellulose-acetate-polyethylene, polypropylenes, urethane elastomers, copolymers	Eastman Chemical Products, Inc., Sub. Eastman Kodak Co.
Tetran	Fluorocarbons	Pennsalt Chemicals Corp.
Texin	Urethane elastomers	Mobay Chemical Company
Thioment	Polyisoprenes	Atlas Minerals & Chemicals Div., The Electric Storage Battery Co.
Ultrapas	Melamine resins	Dynamit Nobel, A.G.
Ultrathene	Ethylene-vinyl acetates	U.S. Industrial Chemicals Co.
Ultron	Polyvinyl chlorides	Monsanto Co.
Vibrathane	Urethane elastomers	Uniroyal Chemical, Div. of Uniroyal Inc.
Vibrin	Polyester resins	Uniroyal Chemical, Div. of Uniroyal Inc.
Vitel	Polyesters	The Goodyear Tire & Rubber Co., Chemical Div.
Viton	Synthetic rubbers	E. I. du Pont de Nemours & Co., Inc.
Vitroplast	Polyester cements	Atlas Minerals & Chemicals Div., The Electric Storage Battery Co.
Vyron	Polyvinyl chlorides	Industrial Vinyls, Inc.

Table C.10 Properties of Commercial Nylon Resins*

Property	Type 6/6	Type 6	Type 6/10	Type 11	Glass-reinforced Type 6/6, 40%	MoS_2-filled, $2\frac{1}{2}$%	Direct polymerized, castable
Mechanical							
Tensile strength, psi	11,800	11,800	8200	8500	30,000	10,000 to 14,000	11,000 to 14,000
Elongation, %	60	200	240	120	1.9	5 to 150	10 to 50
Tensile yield stress, psi	11,800	11,800	8500		30,000		
Flexural modulus, psi	410,000	395,000	280,000	151,000	1,800,000	450,000	
Tensile modulus, psi	420,000	380,000	280,000	178,000		450,000 to 600,000	350,000 to 450,000
Hardness, Rockwell	118R	119R	111R	55A	75E–80E	110R–125R	112R–120R
Impact strength, tensile, ft lb/sq in	76		160			50–180	80–100
Impact strength, Izod ft-lb/in of notch	0.9	1.0	1.2	3.3	3.7**	0.6	0.9
Deformation under load, 2000 psi, 122°F, %	1.4	1.8	4.2	2.02†	0.4§	0.5 to 2.5	0.5 to 1
Thermal							
Heat-deflection temp, °F							
At 66 psi	360	365	300	154	509	400 to 490	400 to 425
At 264 psi	150	152	135	118	502	200 to 470	300 to 425
Coefficient of thermal expansion, per °F	4.5×10^{-5}	4.6×10^{-5}	5×10^{-5}	10×10^{-5}	0.9×10^{-5}	3.5×10^{-5}	5.0×10^{-5}
Coefficient of thermal conductivity, Btu in/hr ft^2 °F	1.7	1.7	1.5				
Specific heat	0.3–0.5	0.4	0.3–0.5	0.58			
Brittleness temp, °F	−112		−166				
Electrical							
Dielectric strength, short time, v/mil	385	420	470	425	480	300 to 400	500 to 600‡
Dielectric constant							
At 60 hz	4.0	3.8	3.9		4.45		3.7
At 10^3 hz	3.9	3.7	3.6	3.3	4.40		3.7
At 10^6 hz	3.6	3.4	3.5		4.10		3.7
Power factor							
At 60 hz	0.014	0.010	0.04	0.03	0.009		0.02
At 10^3 hz	0.02	0.016	0.04	0.03	0.011		0.02
At 10^6 hz	0.04	0.020	0.03	0.02	0.018		0.02
Volume resistivity, ohm-cm	10^{14} to 10^{15}	3×10^{15}	10^{14} to 10^{15}	2×10^{13}	2.6×10^{15}	2.5×10^{13}	
General							
Water absorption, 24 hr, %	1.5	1.6	0.4	0.4	0.6	0.5 to 1.4	0.9
Specific gravity	1.13 to 1.15	1.13	1.07 to 1.09	1.04	1.52	1.14 to 1.18	1.15 to 1.17
Melting point, °F	482 to 500	420 to 435	405 to 430	367	480 to 490	496 ± 9	430 ± 10
Flammability	self ext	self ext	self ext	self ext	self ext	self ext	self ext
Chemical resistance to							
Strong acids	Poor	Poor	Poor	Poor	Poor	Poor	Poor
Strong bases	Good	Good	Good	Fair	Good	Good	Good
Hydrocarbons	Excellent	Excellent	Excellent	Good	Excellent	Excellent	Excellent
Chlorinated hydrocarbons	Good	Good	Good	Fair	Good	Good	Good
Aromatic alcohols	Good	Good	Good	Good	Good	Good	Good
Aliphatic alcohols	Good	Good	Good	Fair	Good	Good	Good

Notes:

Most nylon resins listed in this table are used for injection molding, and test values are determined from standard injection-molded specimens. In these cases, a single typical value is listed. Exceptions are MoS2-filled nylon and direct-polymerized (castable) nylon, which are sold principally in semifinished stock shapes. Ranges of values listed are based on tests on various forms and sizes produced under varying conditions.

Because single values apply only to standard molded specimens, and properties vary in finished parts of different sizes and forms produced by various processes, these values should be used for comparison and preliminary design considerations only. For final design purposes the manufacturer should be consulted for test experience with the form being considered. Limited values should not be used for specification purposes.

† 2000 psi, 73°F.

‡0.040-in. thick.

** 1/2 x 1/4 in. bar

§4000 psi, 122°F

*From: "Nylons," D.D. Carswell, *Machine Design*, 40(29):62, Dec. 12, 1968.

For Conversion factors see Table C.10.

Table C.11 Properties of Silicate Glasses*

Most of the commercially produced glass is for windows, bottles, and inexpensive containers; it is a soda-lime-silica glass of fairly uniform composition, similar to glass No. 0080 in the table below. The following tables on glasses deal largely with that one-tenth of the glass output for which special properties are required. All data are subject to normal manufacturing variations.

Silica glass is inherently high in viscosity and melting point. These are reduced by fluxes such as Na_2O, K2O, and B2O Soda and potash glasses have a high expansion coefficient (column 7), while that of fused silica is very low. Because the borosilicate glasses are intermediate, and their thermal shock resistance is very high (e.g., Corning Code 7740 glasses), they are widely used for laboratory and kitchen glassware. Aluminosilicate glasses are hard, heat-resisting, and of high chemical durability. Glass hardness (indentation) correlates closely with elastic modulus (column 14). Lead oxide is laso used with flux, with a result of reduced softening point and high refractive index: hence, its uses for optical glass and art glass.

Sealing of glass with metal calls for close control of the coefficient of expansion (column 7).

EXPLANATION OF COLUMNS·

Column 5:

B—blown ware
M—multiform
U—panels
P—pressed ware
R—rolled sheet
LC—large castings
S—plate glass
T—tubing and rod

Column 6:

'Since weathering is determined primarily by clouding, which changes transmission, a rating for the opal glasses is omitted.

'These borosilicate glasses may rate differently if subjected to excessive heat treatment

Column 8:

Normal service: No breakage from excessive thermal shock is assumed

Extreme limits· Glass will be very vulnerable to thermal shock Recommendations in this range are based on mechanical stability considerations only Tests should be made before adopting final designs These data are approximate only

Column 9:

Based on plunging sample into cold water after oven heating Resistance of 100°C means no breakage if heated to 110°C and plunged into water at 10°C Tempered samples have over twice the resistance of annealed glass

These data are approximate only

Column 10:

'These data are estimated

Resistance in °C is the temperature differential between the two surfaces of a tube or a constrained plate that will cause a tensile stress of 1000 psi on the cooler surface

Column 11:

Viscosity is given in poises At the strain point the stresses are significantly reduced in a matter of hours, while at the annealing point there is adequate stress reduction in minutes

Column 12:

Data show relative resistance to sandblasting

Column 15:

Data at 25°C are extrapolated from high temperature readings and are approximate only.

* From: "Properties of Selected Commercial Glasses," Publications B-83, Corning Glass Works.

Table C.11 (continued) Properties of Silicate Glasses*

1	2	3	4	5	6			7		8				9		
					Corrosion Resistance			Thermal Expansion 10^{-7} in./in./°C		Upper Working Temperatures (Mechanical Considerations Only)				Thermal Shock Res Plates 6 · 6		
										Annealed		Tempered		Annealed		
Glass Code †	Type	Color	Principal Use	Forms Usually Available	Weathering	Water	Acid	0–300°C (32–572°F)	Room Temp Setting Point	Normal Service °C	Extreme Limit °C	Normal Service °C	Extreme Limit °C	1/8 Thk °C	1/4 Thk °C	1/2 Thk °C
0010	Potash Soda Lead	Clear	Lamp Tubing	T	2	2	2	93	100	110	380			65	50	35
0080	Soda Lime	Clear	Lamp Bulbs	B M T	3	2	2	92	103	110	460	220	250	65	50	35
0120	Potash Soda Lead	Clear	Lamp Tubing	T M	2	2	2	89	98	110	380		–	65	50	35
1720	Aluminosilicate	Clear	Ignition Tube	B T	1	1	3	42	52	200	650	400	450	135	115	75
1723	Aluminosilicate	Clear	Electron Tube	B T	1	1	3	46	54	200	650	400	450	125	100	70
1990	Potash Soda Lead	Clear	Iron Sealing		3	3	4	124	136	100	310		-	45	35	25
2405	Borosilicate	Red	General	B P U				43	51	200	480	–		135	115	75
2475	Soda Zinc	Red	Neon Signs	T	3	2	2	93	-	110	440	–		65	50	35
3320	Borosilicate	Canary	Tungsten Sealing		'1	'1	'2	40	43	200	480		–	145	110	80
6720	Soda Zinc	Opal	General	P	[illegible]	1	2	80	92	110	480	220	275	70	60	40
6750	Soda Barium	Opal	Lighting Ware	B P R	[illegible]	2	2	88	–	110	420	220	220	65	50	35
6810	Soda Zinc	Opal	Lighting Ware	B P R	[illegible]	1	2	69		120	470	240	270	85	70	45
7040	Borosilicate	Clear	Kovar Sealing	B T	'3	'3	'4	48	54	200	430	–	–	–		.
7050	Borosilicate	Clear	Series Sealing	T	'3	'3	'4	46	51	200	440	235	235	125	100	70
7052	Borosilicate	Clear	Kovar Sealing	B M P T	'2	'2	'4	46	53	200	420	210	210	125	100	70
7056	Borosilicate	Clear	Kovar Sealing	B T P	2	2	4	51	57	200	460	–	–	–	-	-
7070	Borosilicate	Clear	Low Loss Electrical	B M P T	'2	'2	'2	32	39	230	430	230	230	180	150	100
7250	Borosilicate	Clear	Seal Beam Lamps	P	'1	'2	'2	36	38	230	460	260	260	160	130	90
7570	High Lead	Clear	Solder Sealing	–	1	1	4	84	92	100	300	-	-			–
7720	Borosilicate	Clear	Tungsten Sealing	B P T	'2	'2	'2	36	43	230	460	260	260	160	130	90
7740	Borosilicate	Clear	General	B P S T U	'1	'1	'1	33	35	230	490	260	290	180	150	100
7760	Borosilicate	Clear	General	B P	2	2	2	34	37	230	450	250	250	160	130	90
7900	96% Silica	Clear	High Temp	B P T U M	1	1	1	8	7	800	1100		–	1250	1000	750
7913	96% Silica	Clear	High Temp	B P R S T	1	1	1	8	7	900	1200	-		-	–	
7940	Fused Silica	Clear	Ultrasonic	U	1	1	1	5.5	7	900	1100	–	-	1250	1000	750
8160	Potash Soda Lead	Clear	Electron Tubes	P T	2	2	3	9[illegible]	100	110	380		-	65	50	35
8161	Potash Lead	Clear	Electron Tubes	P T	2	1	4	90	97	110	390	–	–	–	–	
8363	High Lead	Clear	Radiation Shielding	L C	3	1	4	104	112	100	200			–	-	–
8871	Potash Lead	Clear	Capacitors		2	1	4	102	113	125	300	–	–	55	45	35
9010	Potash Soda Barium	Grey	TV Bulbs	P	2	2	2	89	102	110	380					
9700	Borosilicate	Clear	u.v. Transmission	T U	'1	'1	'2	39	39	220	500		-	150	120	80
9741	Borosilicate	Clear	u.v. Transmission	B U T	'3	'3	'4	39	49	200	390			150	120	80

† Corning Glass Works code numbers are used in this table.

Table C.11 (continued) Properties of Silicate Glasses*

10	11				12	13	14			15			16			17	18
Thermal Stress Resistance °C	Viscosity Data†				Impact Abrasion Resistance	Density grams per C C	Young's Modulus		Poisson's Ratio	Log$_{10}$ of Volume Resistivity			Dielectric Properties at 1 Mc and 20°C			Refractive Index Sod D Line (5893 Microns)	Glass Code
	Strain Point °C	Annealing Point °C	Softening Point °C	Working Point °C			(10^6lb / sq in)	(10^6kg/cm^2)		25°C 77°F	250°C 482°F	350°C 662°F	Power Factor	Dielectric Const	Loss Factor		
19	395	435	625	985	0 8	2 86	8 9	0 63	21	17 +	8 9	7 0	16	6 7	1	1 539	0010
17	470	510	695	1005	1 2	2 47	10 0	0 70	24	12 4	6 4	5 1	9	7 2	6 5	1 512	0080
20	395	435	630	980	0 8	3 05	8 6	0 60	22	17 +	10 1	8 0	12	6 7	8	1 560	0120
28	670	715	915	1190	2 0	2 52	12 7	0 89	0 25		11 4	9 5	38	7 2	2 7	1 530	1720
25	670	710	910	1175	2 0	2 64	12 5	0 88	0 25		13 5	11 3	16	6 3	1 0	1 547	1723
14	330	360	500	755		3 47	8 4	0 59	25		10 1	7 7	04	8 3	33		1990
37	500	530	770	1085	–	2 50	9 9	0 70	0 21	·						1 507	2405
17	440	480	690	1040		2 59	10 0	0 70	·	–	7 8	6 2		·		1 511	2475
40	500	540	780	1155		2 27	9 4	0 66	0 19	–	8 6	7 1	30	4 9	1 5	1 481	3320
19	510	550	775	1010	-	2 58	10 2	0 72	21	-	·	·	–		–	1 507	6720
18	445	485	670	1040		2 59	–	·				-				1 513	6750
23	490	530	770	1010	-	2 65	–		–	–		·				1 508	6810
37	450	490	700	1080	-	2 24	8 6	0 60	23	–	9 6	7 8	20	4 8	1 0	1 480	7040
39	460	500	705	1025	-	2 24	8 7	0 61	22	16	8 8	7 2	33	4 9	1 6	1 479	7050
41	435	480	710	1115	-	2 28	8 2	0 58	22	17	9 2	7 4	26	4 9	1 3	1 484	7052
34	470	510	720	1045		2 29	9 2	0 65	21		10 2	8 3	27	5 7	1 5	1 487	7056
66	455	495		1070	4 1	2 13	7 4	0 52	22	17 +	11 2	9 1	06	4 1	25	1 469	7070
48	490	540	780	1190	3 2	2 24	9 2	0 65	20	15	8 2	6 7	27	4 7	1 3	1 475	7250
21	340	365	440	560	·	5 42	8 0	0 56	28		10 6	8 7	22	15	3 3	–	7570
49	485	525	755	1140	3 2	2 35	9 1	0 64	20	16	8 8	7 2	27	4 7	1 3	1 487	7720
53	515	565	820	1245	3 1	2 23	9 1	0 64	20	15	8 1	6 6	50	4 6	2 6	1 474	7740
52	480	525	780	1210	·	2 23	9 1	0 64	-	17	9 4	7 7	18	4 5	79	1 473	7760
202	820	910	1500	–	3 5	2 18	10 0	0 70	19	17	9 7	8 1	05	3 8	19	1 458	7900'
211	820	910	1500	–	3 5	2 18	9 6	0 67	19	–	9 7	8 1	04	3 8	0 15	1 458	7913
290	990	1050	1580	·	3 6	2 20	10 5	0 74	16	·	11 8	10 2	001	3 8	0038	1 459	7940
18	395	435	630	975	-	2 98		–	–	·	10 6	8 4	09	7 0	63	1 553	8160
22	400	435	600	860		4 00	7 8	0 55	24	·	12 0	9 9	06	8 3	0 50	1 659	8161
19	300	315	380	460		6 22	7 4	0 52	27	-	9 2	7 5	19	17 0	3 2	1 97	8363
17	350	385	525	785		3 84	8 4	0 59	26		11 1	8 8	05	8 4	42		8871
18	405	445	650	1010		2 64	9 8	0 69	21	·	8 9	7 0	17	6 3	1 1	1 507	9010
45	520	565	805	1200		2 26	9 6	0 67	20	15	8 0	6 5		-		1 478	9700
55	410	450	705			2 16	7 2	0 51	23	17 +	9 4	7 6				1 468	9741

† Viscosities at these four temperatures are approximately as follows $10^{14.5}$ poises at the strain point. 10^{13} poises at the annealing point. $10^{7.8}$ poises at the softening point. at 10^4 poises at the working point

Table C.12 Properties of Window Glass*: Transmittance of Sheet and Plate Glass

Type or tint	Nominal thickness, in.	Weight, lb/ft²	Transmittance	
			Total visible daylight, %	Direct 90° solar energy, %
Sheet	$\frac{1}{16}$	0.81	91	89
Sheet	$\frac{5}{64}$	1.00	91	88
Sheet	$\frac{3}{32}$	1.22	90	87
Sheet	$\frac{1}{8}$	1.64	90	86
Sheet	$\frac{3}{16}$	2.47	89	84
Sheet	$\frac{7}{32}$	2.85	89	82
Plate or float	$\frac{1}{8}$	1.64	90	86
Plate or float	$\frac{1}{4}$	3.28	88	79
Plate or float	$\frac{5}{16}$	4.09	88	77
Plate or float	$\frac{3}{8}$	4.91	87	74
Plate or float	$\frac{1}{2}$	6.55	86	70
Plate or float	$\frac{5}{8}$	8.18	85	65
Plate or float	$\frac{3}{4}$	9.83	83	60
Plate or float	$\frac{7}{8}$	11.45	81	55
Plate or float	1	13.13	79	49
Gray[a]	$\frac{1}{4}$	3.28	43	46
Bronze[a]	$\frac{1}{4}$	3.28	49	45
Green[a]	$\frac{1}{4}$	3.28	75	46
Double[b]	$\frac{1}{4}$ each	6.56	78	–

Note Many types of glass are available, including tempered heat-strengthened glass, laminated shatter-proof glass, conductive-coated glass, and reflective-coated glass. Several double-pane combinations are offered.

Direct 90° transmittance of solar ultraviolet radiation through non-tinted window glass is about 85 percent az high as the values for toal solar energy transmittance Ultraviolet transmittance of gray or bronze glass is lower

Infrared transmittance is considerably lower than visual transmittance. This is significant in view of the large percentage of infrared radiation from most sources.

Approximate shading coefficients, ASHRAE, 1/4-in. glass only: clear, 0.93; gray, 0.67, bronze, 0.65; green, 0.67

Overall heat transfer coefficient of window area (air to air) is usually assumed to be 1.0 Btu/ft² hr, but it is lower if there is no wind.

[a] Transmittance of tinted glass depends on depth of tint..

[b] Two 1/2-in panes with 1/2-in air space, sealed.

* Tables compiled from several sources.

Table C.13 Properties and Uses of American Woods*

Species	Specific gravity		Characteristics	Uses	Weight		
	Green	Dry			lb/cu ft, green	lb/cu ft, air-dry 12%	lb/1000 board ft, air-dry 12%
Alder, red	0 37	0 41	Low shrinkage, moderate in strength, shock resistance, hardness, and weight†	Furniture, sash, doors, millwork	46	28	2330
Ash, black	0 45	0 49	Light in weight†	Cabinets, veneer, cooperage, containers	52	34	2830
Ash, Oregon	0 50	0 55	Similar to but lighter than white ash†	Similar to white ash	46	38	3160
Ash, white	0 54	0 58	Heavy; hard; stiff, strong, high shock resistance†	Handles, ladder rungs; baseball bats, farm implements, car parts	48	41	3420
Bald cypress (Southern cypress)			Moderate in strength, weight, hardness, and shrinkage**	Building construction, beams; posts, ties, tanks; ships, paneling	51	32	2670
Beech, American	0 56	0 64	Heavy, high strength, shock resistance, and shrinkage, uniform texture†	Flooring; furniture, handles, kitchenwear, ties (treated)	54	45	3750
Birch	0 57	0 63	Heavy, high strength, shock resistance, and shrinkage; uniform texture†	Interior finish, dowels, ties (treated), veneer, musical instruments	57	44	3670
Cottonwood	0 37	0 40	Uniform texture, does not split readily; moderate in weight, strength, hardness, and shrinkage	Crates, trunks; car parts, farm implements	49	28	2330
Douglas fir	0 41	0 44	Moderate in strength, weight, shock resistance, and shrinkage‡	Building and construction, poles, veneer, plywood; ships, furniture; boxes	38	34	2830
Elm	0 57	0 63	Moderate in strength, weight, and hardness, high in shock resistance and shrinkage, good in bending†	Cooperage, baskets, crates, veneer, vehicle parts	54	34	2920
Hemlock, Eastern	0 38	0 40	Moderate in weight, strength, and hardness†	Building and construction, boxes	50	28	2330
Hemlock, Western	0 38	0 42	Moderate in weight, strength, and hardness†	Sash; doors, posts, piles, building and construction	41	29	2420
Hickory, true	0.65	0 73	High toughness, hardness, shock resistance, strength, and shrinkage†	Dowels, spokes; poles, shafts, gymnasium equipment	63	51	4250
Incense cedar	0.35		Uniform texture, easy to season, low shrinkage; shock resistance, weight, and stiffness**	Lumber; fence posts, ties, poles, shingles	45		
Larch, Western	0 48	0 52	Moderate in strength, weight, shock resistance, hardness, and shrinkage‡	Doors; sash; posts; pilings, building and construction	48	36	3000

Table C.13 (continued) Properties and Uses of American Woods*

Species	*Specific gravity*		*Characteristics*	*Uses*	*Weight*		
	Green	*Dry*			*lb/cu ft, green*	*lb/cu ft, air-dry 12%*	*lb/1000 board ft, air-dry 12%*
Locust, black	0.66	0.69	High in shock resistance, weight, and hardness, very high strength, moderate shrinkage**	Mine timbers, posts, poles; ties	58	48	4000
Maple	0.44	0.48	High in hardness, weight, strength, shock resistance, and shrinkage; uniform texture†	Flooring; furniture, trim, spools, farm implements	54	40	3330
Oak, red and white	0.57	0.63	High in hardness, weight, strength, shock resistance, and shrinkage; red†, white‡	Trim, ships, flooring; ties, furniture, cooperage piles	64	44	3670
Pine, jack			Coarse texture, low strength, stiffness, shock resistance, and shrinkage	Box lumber, fuel, mine timber, ties, poles, posts			
Pine, lodgepole	0.38	0.41	Moderate in weight, hardness, strength, shock resistance, and shrinkage, easy to work‡	Poles, mine timber ties, construction	39	29	2420
Pine			High shrinkage, moderate strength, stiffness, hardness, and shock resistance	General construction, ties, poles, posts			
Pine, Ponderosa	0.38	0.40	Moderate in weight, shock resistance, shrinkage, and hardness, easy to work†	Building, paneling, sash, frames	45	28	2330
Pine, S yellow	0.47	0.51	Moderate in shock resistance, shrinkage, and hardness, high in strength‡	Building and construction poles, pilings boxes	55	41	3420
Pine sugar	0.35	0.36	Low shock resistance, easy to work moderate strength†	Sash, counters blinds, patterns	52	25	2080
Pine, Western white	0.36	0.38	Moderate in strength, shock resistance, shrinkage, and hardness, easy to work†	Building and construction patterns, boxes	35	27	2250
Red cedar, Eastern and Western	0.44	0.47	High shock resistance, low stiffness and shrinkage, moderate in strength and hardness**	Fence posts, closet liners, chests, flooring	37	37	2750
Redwood	0.38	0.40	Low shrinkage, medium in weight, strength, hardness, and shock resistance**	Posts, doors, interiors, cooling towers	50	28	2330
Spruce, Eastern	0.38	0.40	Moderate in hardness, shock resistance, weight, shrinkage, and strength†	Building, millwork, boxes, ladders	34	28	2330

Table C.13 (continued) Properties and Uses of American Woods*

Species	Specific gravity		Characteristics	Uses	Weight		
	Green	Dry			lb/cu ft, green	lb/cu ft, air-dry 12%	lb/1000 board ft, air-dry 12%
Spruce, Engelmann	0 31	0 33	Generally straight grained, light in weight; low strength as a beam or post, low shock resistance, moderate shrinkage	Mine timber, ties, poles, flooring, studding, paper	39	23	1920
Spruce, Sitka	0 37	0 40	Moderate in weight, hardness, strength, shock resistance, and shrinkage†	Important in boat and plane construction; sash; doors, boxes; siding	33	28	2330
Sycamore	0 46	0 49	High shrinkage; moderate in weight, strength, hardness, and shock resistance†	Boxes, ties, posts, veneer, flooring, butcher blocks	52	34	2830
Tamarack	0 49	0 53	Coarse texture, moderate in strength, hardness, shrinkage, and shock resistance	Ties, mine timber; posts, poles; tanks; scaffolding	47	37	3080
Tupelo			Uniform texture, moderate in strength, hardness, shock resistance, high shrinkage, interlocked grain makes splitting difficult‡	Flooring; planking; crates, furniture			
Walnut, black	0 51	0 55	Moderate shrinkage; high weight, strength, hardness, and shock resistance, easily worked and glued**	Gun stocks, cabinets, plywood, furniture, veneer	58	38	3170
White cedar	0.31	0 32	Low shrinkage, weight, shock resistance, and strength, soft, easily worked**	Poles, posts; ties, tanks, ships	24	23	1920
Willow, black			High strength and shock resistance, low beam strength and weight, interlocked grain	Lumber, veneer, charcoal, furniture, sub-flooring, studding			

†Decay resistance low
‡Decay resistance medium
**Decay resistance high
*From "Materials Data Book", E R Parker, McGraw Hill Book Company, 1967, pp 252 255

Note: For weight-density in kg/m^3, multiply value in lb/ft^3 by 16.02.

Table C.14 Properties of Natural Fibers*

Because there are great variations within a given fiber class, average properties may be misleading. The following typical values are only a rough comparative guide.

Name	*Specific gravity*	*Tenacity, g/denier*	*Tensile strength, 10^3 psi*	*Elongation at break (dry), %*	*Standard regain, % of dry*[b]	*Fiber diameter, microns*	*Fiber length, in*	*Fiber shape and kind*	*Resistant to*
ANIMAL ORIGIN									
Wool	1.32	1.0-1.7	17-29	23-35	15-18	17-40	1.5-5	Oval, crimped, scales	Age, weak acids, solvents
Silk	1.25	3.5-5	90	20-25	10	10-13		Flexible, soft, smooth	Heat, solvents, weak acids, wear
Cashmere						15-16	1-4	Round, scales, soft	
Mohair	1.32	1.2-1.5		30	13	24-50	6-12	Round, silky	Wear, age, solvents, weak acids
Camel hair	1.32	1.8		40	13	10-40	1-6	Oval, striated	Age, solvents
VEGETABLE ORIGIN									
Cotton	1.54	2-5	30-120	5-11	7.5-8.5	10-20	0.5-2	Flat, convoluted, ribbon	Age, heat, washing, wear, solvents, alkalies, insects
Jute (bast)	1.5		50	1-1.5	14	15-20		Woody, rough, polygon	
Sisal (leaf)	1.49	2.2	75	2-2.5	13	10-30	Strand 30–40	Stiff, straight	
Flax (bast)	1.52	4-7		2-3	12	15-18	Strand 40-50	Soft, fine	Age, solvents, washing, insects, weak acids, and alkalies
Kenaf (bast)			45			15-30		Polygon or oval	
Hemp (bast)	1.48			2		18-25	Strand 30-70	Polygon or oval, irregular	
Henequen (leaf)			60				Strand 30-60	Finer than sisal	
Abaca (leaf) (Manila)	1.48	2.3-2.9	100	2-3	13		Strand 30-120		
MINERAL ORIGIN									
Asbestos	2.5		40-200			Various	0.5-10	Smooth, straight	Heat to 400 deg C, acids, chemicals, organisms
Glass[a]	2.5	7-12	200-500	3-4.5	0	Various		Circular, smooth	Chemicals, insects
Silicate[a] (Ca, Al, Mg)	2.85				0				Heat to 900 deg C, most chemicals, insects, rot

Note Wide variations may be expected, especially for different grades of cotton. Wet strength is lower (for rayon, very much lower), but it depends on the duration of soaking. The strength of yarn is only a fraction of the cumulative strength of all individual fibers.

Most fibers exhibit relaxation of stress at constant strain and also increase in elongation at constant load (creep). The stress-strain curve is greatly affected by the rate of extension. When the stress is removed, there is a quick elastic recovery, a delayed recovery, and a permanent set. Hence the elastic behavior of any fiber depends on its stress-strain history. The elastic recoveries of nylon and wool are high, those of cotton, flax, and rayon are much lower.

The heat capacity (specific heat) of most fibers is about one-third that of water.

Other fibers Fur hair is slightly coarser than silk fibers. Camel and llama hairs are almost as coarse as wool but only about one-third the size of human hair. Horse hair is over 100 microns, hog bristles, over 200 microns. Jute, sisal, and hemp are intermediate between cotton and wool. These are rough average sizes, and many natural fibers range 50% above or below such averages.

[a] Here classified as natural fibers for convenience, although they are man-made by processing.

[b] Expected equilibrium moisture regain of dry fiber, in percent of dry weight, when exposed in air at 70 deg F, 65% relative humidity.

*Compiled from several sources.

Table C.15 Properties of Manufactured Fibers*

Chemical class, common name (sources)	*Specific gravity*	*Tenacity, g/denier*	*Tensile strength, 10^3 psi*	*Elongation at break, %*	*Regain (standard)*	*Softening point, deg C*	*Melting point, deg C*	*Flammability*	*Brittleness temp, deg C*
CELLULOSE FIBERS (NATURAL)									
Acetate	1 30	1 -1 3	18-25	20-30	6 5	140	230	Melts and burns	
Triacetate	1 32	1 2-1 4	20-28	25-30	3-4 5	225	300	Melts and burns	
Viscose rayon	1.51	2-2 6	30-46	17-25	13.		200[a]	Burns readily	
High-tenacity viscose	1 53	3-5	60-80	10-12	10		200[a]	Burns readily	< −114
Polynosic viscose	1 53	3-5	60-80	8-20	7		200[a]	Burns readily	
Cuprammonium rayon (cupro)	11 52	1 7-2.3	30-45	10-17	12 5		250[a]	Burns readily	
PROTEIN FIBERS (NATURAL)									
Animal casein (milk)	1.3	1.0	15	60-70	14	100	150	Slow	
Vegetable – seed soybeans, peanuts, corn	1.3	0 7-0 9	11-14	40-60	11-15	150	250	Slow	
Vegetable – latex rubber (vulcanized)	1.0	0 4-0 6	4-7	700-900	0	300		Burns	−60
SYNTHETIC FIBERS									
Polyacrylonitrile (acrylic)	1.17	2-5	50-75	25-40	2	190	260	Burns	
Polyamide (nylon)	1.14	4-9	70-120	20-40	4	200	215-250	Slow	< −100
Polyester (PET dacron)	1.38	4-8	70-120	10-50	0.4	225	250-290	Low	
Polyethylene (olefin, low density)	0.92	3-6	40-70	25-40	0.15	90-120	120	Slow	−114
Polyethylene (olefin, high density)	0 95	5 7	60-80	10-20	0 01	120-130	140	Slow	−114
Polypropylene (olefin)	0 91	4 5-8	45-80	15-30	0-0 5	145	160-170	Self-ext low	−70
Polyurethane (spandex)	1 1	0 5-1 0	7-16	500-700	1 0	190	250	Burns	
Polyvinyl chloride (PVC)	1 38	0 7-2	12 17	100 125	0 1	70	140[a]	No; chars	< −100
Polyvinyl alcohol (PVA)	1 3	3-7	60-90	15-28	5	230	240	Slow	
Polyvinylidene chloride (saran)	1 7	2	40	20-30	0 1	115-135	170	No	
Polytetrafluoroethylene (PTFE)	2 1	1 2-1 4	33	15-30	0	225	300[a]	No	

Note Mechanical properties are for room temperature and humidity and based on unstressed cross section.

[a]Decomposition, does not melt

*Compiled from several sources

Table C.16 Properties of Rubbers and Elastomers*

Elastomers cannot be classified in any brief and simple manner, nor are they well characterized by the usual mechanical tests. The terms *rubber* and *synthetic rubber* are loosely applied to a great variety of elastic materials, from pure gum natural rubber and pure synthetics to cured, compounded, filled, and even reinforced products.

ASTM designations (D1418) by chemical polymer description are used in the following table; yet within each class the properties can vary widely, depending on the exact composition, heat treatment service temperature, and application. Typical uses, such as rubber springs and cushioning, permit an almost unlimited number of combinations of design variables.

Mechanically, rubbers may be expected to lose strength rapidly with increase in temperature, to show a large hysteresis in stress-strain behavior, to exhibit marked creep and set, and to be greatly affected by rates of load application or frequency of repeated stress. "Heat build-up", i.e., increase in temperature in service, as well as deterioration from environment (sunlight, oils, ozone, etc.) will reduce the valuable properties of many rubbers, both natural and synthetic.

The following data apply to typical samples of commercial elastomers for common uses.

KEY

A – Acetone	J – Alkalies	S – Salts
B – Benzene	K – Ketones	T – Heat or high temperature
C – Carbon tetrachloride	L – Alcohols	U – Ultraviolet
D – Carbon disulfide	M – Ammonia	V – Vegetable oils
E – Phenol	N – Turpentine	W – Weathering
F – Sulfur compounds	O – Coal derivatives, bitumens	X – Oxidation
G – Glycerol or glycol	P – Petroleum products	Y – Aging
H – Hexane	R – Aromatics	Z – Ozone
I – Acids		

Chemical name	*Polyisoprene*	*Butadiene*	*Styrene-butadiene*	*Acrylonitrile butadiene*
Other names	*Natural (or synthetic) rubber NR (IR)*	*BR Cis 4*	*Buna S Styrene SBR, GR-S*	*Nitrile, Buna N Hycar NBR, GR-A*
CHEMICAL AND PHYSICAL				
Specific gravity	0.93	1.0	1.0	1.0
Specific heat	0.40	0.45	0.40	0.47
Thermal conductivity				
W/cm·K	0.001 7	0.002 5	0.002 6	0.002 5
Btu/hr ft deg F	0.10	0.14	0.15	0.14
Service temperature, deg C				
min	−25	−40	−20	−20
max	90	90	75	110
Solvents, softeners	D.K.P.V	D.H.N.P	K.P.R.V	C.K.O.R
Resistant to	A.I.J.L	G.I.J.W.Y	G.I.L.S.X	G.I.K.L.P.S. T.V.W
Swelled by	D.P.V	A.P.V	P.V	A.E.N
MECHANICAL AND ELECTRICAL				
Tensile strength				
kg/cm^2 (max)	300	210	210	295.
kpsi (max)	4.3	3.0	3.0	4.2
Elongation at break, %	600	700	600.	600.
Vol. resistivity, ohm-cm	10^{15}	10^{15}	10^{14}	10^{10}
Dielectric strength				
kV/cm	235		235	185
V/mil	600.		600.	475.
Dielectric constant	3.0	2.3	2.8	3.0
Power factor (50-100 Hz)	0.003	0.005	0.005	0.007
Rebound	Good	Good	Fair	Good
COMPARATIVE RATINGS – RESISTANCE TO				
Abrasion	Good	Excellent	Good	Excellent
Cold flow (set)	Excellent		Good	Good
Tearing	Good		Poor	Fair
Air permeability	Fair	Good	Fair	Excellent
Oxidation	Fair	Fair	Fair	Fair
Flame	Poor		Poor	Poor

*Compiled from several sources.

Table C.16 (continued) Properties of Rubbers and Elastomers

Chemical name	*Polychloro-prene*	*Isobutylene-isoprene*	*Polysulfide*	*Polymethane*
Other names	*Neoprene*[a] *CR, GR-M*	*Butyl IIR, GR-I*	*Thiokol*[a] *PS, GR-P*	*Adiprene*[a] *PU*
CHEMICAL AND PHYSICAL				
Specific gravity	1.25	0.95	1.4	1 2
Specific heat	0 5	0.45	0.31	0 45
Thermal conductivity				
W/cm·K	0 002 1	0.001 3	0.003	0.001 3
Btu/hr·ft deg F	0.12	0.075	0.17	0.075
Service temperature, deg C				
min	−20	−40	−15	−35
max	100	120	90	120
Solvents, softeners	A,B,C,D,I,N,R	D,P	C	
Resistant to	G,L,P,S,T,U,V, W,Y,Z	E,G,J,S,U,V, W,X,Y,Z	L,P,U,Z	P,V,X,Z
Swelled by	C,D,N,R	D,H,P	C,R	B,C,K,R
MECHANICAL AND ELECTRICAL				
Tensile strength				
kg/cm^2 (max)	240.	175.	90	350.
kpsi (max)	3.5	2.5	1 3	5 0
Elongation at break, %	800.	700.	500	550.
Vol. resistivity, ohm-cm	10^{11}	10^{17}	10^{8}	10^{11}
Dielectric strength				
kV/cm	195	295	125	195
V/mil	500	750	325	500
Dielectric constant	7	2.4	8.	7
Power factor (50–100 Hz)	04	0.004	0 02	0 04
Rebound	Good	Poor	Poor	
COMPARATIVE RATINGS RESISTANCE TO				
Abrasion	Excellent	Fair	Poor	Excellent
Cold flow (set)	Excellent	Fair	Poor	Poor
Tearing	Good	Good	Poor	Excellent
Air permeability	Good	Excellent	Good	Excellent
Oxidation	Good	Good	Good	Good
Flame	Excellent	Poor	Poor	Poor

[a]Proprietary

Appendix D. Gases and Vapors

Table D.1 SI Units — Definitions, Abbreviations and Prefixes

BASIC UNITS—MKS

Length	meter	m	Electric current	ampere	A
Mass	kilogram	kg	Thermodynamic temperature	kelvin	K
Time	second	s	Luminous intensity	candela	cd

DERIVED UNITS

Property	*Units*†	*Abbreviations and dimensions*	
Acceleration	meter per second squared	m/s^2	
Activity (of radioactive source)	1 per second	s^{-1}	
Angular acceleration	radian per second squared	rad/s^{-1}	
Angular velocity	radian per second	rad/s	
Area	square meter	m^2	
Density	kilogram per cubic meter	kg/m^3	
Dynamic viscosity	newton-second per sq meter	$N{\cdot}s/m^2$	
Electric capacitance	farad	F	$(A \cdot s/V)$
Electric charge	coulomb	C	$(A \cdot s)$
Electric field strength	volt per meter	V/m	
Electric resistance	ohm		(V/A)
Entropy	joule per kelvin	J/K	
Force	newton	N	$(kg \cdot m/s^2)$
Frequency	hertz	hz	(s^{-1})
Illumination	lux	lx	(lm/m^2)
Inductance	henry	H	$(V \cdot s/A)$
Kinematic viscosity	sq meter per second	m^2/s	
Luminance	candela per sq meter	cd/m^2	
Luminous flux	lumen	lm	$(cd \cdot sr)$
Magnetomotive force	ampere	A	
Magnetic field strength	ampere per meter	A/m	
Magnetic flux	weber	Wb	(V s)
Magnetic flux density	tesla	T	(Wb/m^2)
Power	watt	W	(J/s)
Pressure	newton per square meter	N/m^2	
Radiant intensity	watt per steradian	W/sr	
Specific heat	joule per kilogram kelvin	J/kg K	
Thermal conductivity	watt per meter kelvin	W/m K	
Velocity	meter per second	m/s	
Volume	cubic meter	m^3	
Voltage, potential difference, electromotive force	volt	V	(W/A)
Wave number	1 per meter	m^{-1}	
Work, energy, quantity of heat	joule	J	$(N \cdot m)$

PREFIX NAMES OF MULTIPLES AND SUBMULTIPLES OF UNITS

Decimal equivalent	*Prefix*	*Pronunciation*	*Symbol*	*Exponential expression*
1,000,000,000,000	tera	tĕr'ȧ	T	10^{-12}
1,000,000,000	giga	jĭ'gȧ	G	$10^{\cdot 9}$
1,000,000	mega	mĕg'ȧ	M	$10^{\cdot 6}$
1,000	kilo	kĭl'ō	k	$10^{\cdot 3}$
100	hecto	hĕk'tō	h	$10^{\cdot 2}$
10	deka	dĕk'ȧ	da	10
0.1	deci	dĕs'ĭ	d	10^{-1}
0.01	centi	sĕn'tĭ	c	10^{-2}
0.001	milli	mĭl'ĭ	m	10^{-3}
0.000 001	micro	mī'krō	μ	10^{-6}
0.000 000 001	nano	năn'ō	n	10^{-9}
0.000 000 000 001	pico	pē'kō	p	10^{-12}
0.000 000 000 000 001	femto	fĕm'tō	f	10^{-15}
0 000 000 000 000 000 001	atto	ăt'tō	a	10^{-18}

Appendix E. Miscellaneous

Table E.1 Sizes and Allowable Unit Stresses for Softwood Lumber

American Softwood Lumber Standard. A voluntary standard for softwood lumber has been developing since 1922. Five editions of Simplified Practice Recommendation R16 were issued from 1924–53 by the Department of Commerce; the present NBS voluntary Product Standard PS 20-70, "American Softwood Lumber Standard," was issued in 1970. It was supported by the American Lumber Standards Committee, which functions through a widely representative National Grading Rule Committee.

Part a. Nominal and Minimum-Dressed Sizes of Lumber[*]

Item	Thicknesses			Face widths		
	Nominal	Minimum-dressed		Nominal	Minimum-dressed	
		Dry,[a] inches	Green, inches		Dry,[a] inches	Green, inches
Boards[b]				2	1 1/2	1 9/16
				3	2 1/2	2 9/16
				4	3 1/2	3 9/16
				5	4 1/2	4 5/8
	1	3/4	25/32	6	5 1/2	5 5/8
				7	6 1/2	6 5/8
	1 1/4	1	1 1/32	8	7 1/4	7 1/2
				9	8 1/4	8 1/2
	1 1/2	1 1/4	1 9/32	10	9 1/4	9 1/2
				11	10 1/4	10 1/2
				12	11 1/4	11 1/2
				14	13 1/4	13 1/2
				16	15 1/4	15 1/2
Dimension				2	1 1/2	1 9/16
				3	2 1/2	2 9/16
				4	3 1/2	3 9/16
	2	1 1/2	1 9/16	5	4 1/2	4 5/8
	2 1/2	2	2 1/16	6	5 1/2	5 5/8
	3	2 1/2	2 9/16	8	7 1/4	7 1/2
	3 1/2	3	3 1/16	10	9 1/4	9 1/2
				12	11 1/4	11 1/2
				14	13 1/4	13 1/2
				16	15 1/4	15 1/2
Dimension				2	1 1/2	1 9/16
				3	2 1/2	2 9/16
				4	3 1/2	3 9/16
				5	4 1/2	4 5/8
	4	3 1/2	3 9/16	6	5 1/2	5 5/8
	4 1/2	4	4 1/16	8	7 1/4	7 1/2
				10	9 1/4	9 1/2
				12	11 1/4	11 1/2
				14		13 1/2
				16		15 1/2
Timbers	5 and thicker		1/2 off	5 and wider		1/2 off

[a]Maximum moisture content of 19% or less.

[b]Boards less than the minimum thickness for 1 in. nominal but 5/8-in. or greater thickness dry (11/16-in. green) may be regarded as American Standard Lumber, but such boards shall be marked to show the size and condition of seasoning at the time of dressing. They shall also be distinguished from 1-in. boards on invoices and certificates.

[*]Reprinted from: "american Softwood Lumber Standard," NBS PS 20-70, National Bureau of Standards, 1970; available from Superintendent of documents.

Note: This table applies to boards, dimensional lumber, and timbers. The thicknesses apply to all widths and all widths to all thicknesses.

Table E.1 (continued) Sizes and Allowable Unit Stresses for Softwood Lumber

The "American Softwood Lumber Standard", PS 20-70, gives the size and grade provisions for American Standard lumber and describes the organization and procedures for compliance enforcement and review. It lists commercial name classifications and complete definitions of terms and abbreviations.

Eleven softwood species are listed in PS 20-70, viz., cedar, cypress, fir, hemlock, juniper, larch, pine, redwood, spruce, tamarack, and yew. Five dimensional tables show the standard dressed (surface planed) sizes for almost all types of lumber, including matched tongue-and-grooved and shiplapped flooring, decking, siding, etc. Dry or seasoned lumber must have 19% or less moisture content, with an allowance for shrinkage of 0.7-1.0% for each four points of moisture content below the maximum. Green lumber has more than 19% moisture. Table A illustrates the relation between nominal size and dressed or green sizes.

National Design Specification. Part b is condensed from the 1971 edition of "National Design Specification for Stress-Grade Lumber and Its Fastenings," as recommended and published by the National Forest Products Association, Washington, D.C. This specification was first issued by the National Lumber Manufacturers Association in 1944; subsequent editions have been issued as recommended by the Technical Advisory Committee. The 1971 edition is a 65-page bulletin with a 20-page supplement giving "Allowable Unit Stresses, Structural Lumber," from which Part b has been condensed. The data on working stresses in this Supplement have been determined in accordance with the corresponding ASTM Standards, D245-70 and D2555-70.

Part b. Species, Sizes, Allowable Stresses, and Modulus of Elasticity of Lumber

Normal loading conditions: Moisture content not over 19%, No. 1 grade, visual grading. To convert psi to N/m², multiply by 6 895.

Species[a]	*Sizes, nominal*	*Typical grading agency, 1971*[b]	*Allowable unit stresses, psi*[d]				*Modulus of elasticity, psi*
			Extreme fiber in bending[c]	*Tension parallel to grain*	*Compression perpendicular*	*Compression parallel*	
CEDAR							
Northern white	2 × 4	NL, NH	1 100	600	205	675	800 000
	2 or 4 × 6+	NL, NH	1 000	575	205	675	800 000
Western	2 × 4	NC	1 450	725	285	975	1 100 000
	2 or 4 × 6+	NC, WW	1 250	725	285	975	1 100 000
FIR							
Balsam	2 × 4	NL, NH	1 300	675	170	825	1 200 000
	2 or 4 × 6+	NL, NH	1 150	650	170	825	1 200 000
Douglas (larch)	2 × 4	WC, NC	2 400	1 200	385	1 250	1 800 000
	2 or 4 × 6+	WC, NC	1 750	1 000	385	1 250	1 800 000
HEMLOCK							
Eastern (tamarack)	2 × 4	NL, NH	1 750	900	365	1 050	1 300 000
	2 or 4 × 6+	NL, NH	1 500	875	365	1 050	1 300 000
Hem-fir	2 × 4	WC, NC	1 600	825	245	1 000	1 500 000
	2 or 4 × 6+	WC, NC	1 400	800	245	1 000	1 500 000
Mountain	2 × 4	WC, WW	1 700	850	370	1 000	1 300 000
	2 or 4 × 6+	WC, WW	1 450	850	370	1 000	1 300 000
PINE							
Idaho white	2 × 4	WW	1 400	725	240	925	1 400 000
	2 or 4 × 6+	WW	1 200	700	240	925	1 400 000
Lodgepole	2 × 4	WW	1 500	750	250	900	1 300 000
	2 or 4 × 6+	WW	1 300	750	250	900	1 300 000

Table E.1 (continued) Sizes and Allowable Unit Stresses for Softwood Lumber

Species[a]	Sizes, nominal	Typical grading agency, 1971[b]	Allowable unit stresses, psi[d]				Modulus of elasticity, psi
			Extreme fiber in bending[c]	Tension parallel to grain	Compression perpendicular	Compression parallel	
PINE (continued)							
Northern	2 × 4	NL, NH	1 600	825	280	975	1 400 000
	2 or 4 × 6 +	NL, NH	1 400	800	280	975	1 400 000
Ponderosa (sugar)	2 × 4	WW, NC	1 400	700	250	850	1 200 000
	2 or 4 × 6 +	WW, NC	1 200	700	250	850	1 200 000
Red	2 × 4	NC	1 350	700	280	825	1 300 000
	2 or 4 × 6 +	NC	1 150	675	280	825	1 300 000
Southern	2 × 4	SP	2 000	1 000	405	1 250	1 800 000
	2 or 4 × 6 +	SP	1 750	1 000	405	1 250	1 800 000
REDWOOD							
California	2 or 4 × 2 or 4	RI	1 950	1 000	425	1 250	1 400 000
	2 or 4 × 6 to 12	RI	1 700	1 000	425	1 250	1 400 000
SPRUCE							
Eastern	2 × 4	NL, NH	1 500	750	255	900	1 400 000
	2 or 4 × 6 +	NL, NH	1 250	750	255	900	1 400 000
Engelmann	2 × 4	WW	1 300	675	195	725	1 200 000
	2 or 4 × 6 +	WW	1 150	650	195	725	1 200 000
Sitka	2 × 4	WC	1 550	775	280	925	1 500 000
	2 or 4 × 6 +	WC	1 300	775	280	925	1 500 000

Note Allowable unit stresses in horizontal shear are in the range of 60–100 psi for No. 1 grade.

[a]Grade designations are not entirely uniform. Values in the table apply approximately to "No. 1" There is seldom more than one better grade than No 1, and this may be designated as select, select structural, dense, or heavy. In addition to lower grades 2 and 3, there may be other lower grades, designated as construction, standard, stud, and utility In bending and tension the allowable unit stresses in the lowest recognized grade (utility) are of the order of $\frac{1}{8}$ to $\frac{1}{6}$ of the allowable stresses for grade No 1. The tabular values for allowable bending stress are for the extreme fiber in "repetitive member uses," and edgewise use. The original tables give correction factors, which are less than unity for moist locations and for short-time loading; they are greater than unity if the moisture content of the wood in service is 15% or less In general, all data apply to uses within covered structures. From the extensive tables, only the No. 1 grade in nominal 2 × 4 size and 2-in. or 4-in planks, 6 in., and wider have been selected for illustration.

In a few cases the allowable stresses specified for the Canadian products will vary slightly from those given here for the same species by the U.S. agencies.

[b]Grading agencies represented by letters in this column are as follows.

NC = National Lumber Grades Authority (a Canadian agency)
NH = Northern Hardwood and Pine Manufacturers Association
NL = Northern Lumber Manufacturers Association
RI = Redwood Inspection Service
SP = Southern Pine Inspection Bureau
WC = West Coast Lumber Inspection Bureau
WW = Western Wood Products Association

[c]It is assumed that all members are so framed, anchored, tied, and braced that they have the necessary rigidity.

[d]For short term loads, these values may be increased add 15% for 2-month snow load; add 33% for wind or earthquake; add 100% for impact load.

REFERENCES

"Wood Handbook." Handbook No. 72, U.S. Department of Agriculture, 1955.

"Timber Construction Manual." American Institute of Timber Construction, John Wiley & Sons, Inc., 1966.

"National Design Specification for Stress-Grade Lumber and its Fastenings." National Forest Products Association, Washington D.C., 1971.

Table E.2 Standard Grades of Bolts

Part a: SAE Grades for Steel Bolts

SAE grade no.	Size range incl.	Proof strength,† kpsi	Tensile strength,† kpsi	Material	Head marking
1	$\frac{1}{4}$–$1\frac{1}{2}$			Low- or medium-carbon steel	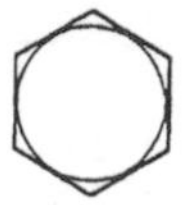
2	$\frac{1}{4}$–$\frac{3}{4}$	55	74		
	$\frac{7}{8}$–$1\frac{1}{2}$	33	60		
5	$\frac{1}{4}$–1	85	120	Medium-carbon steel, Q & T	
	$1\frac{1}{8}$–$1\frac{1}{2}$	74	105		
5.2	$\frac{1}{4}$–1	85	120	Low-carbon martensite steel, Q & T	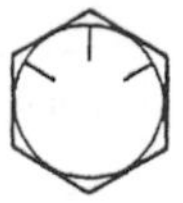
7	$\frac{1}{4}$–$1\frac{1}{2}$	105	133	Medium-carbon alloy steel, Q & T‡	
8	$\frac{1}{4}$–$1\frac{1}{2}$	120	150	Medium-carbon alloy steel, Q & T	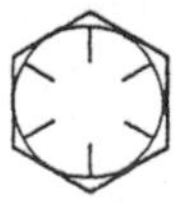
8.2	$\frac{1}{4}$–1	120	150	Low-carbon martensite steel, Q & T	

†Minimum values.
‡Roll threaded after heat treatment.
SOURCES. See "Helpful Hints," by Russell, Burdsall & Ward Corp., Mentor, Ohio 44060; and Chap. 23.

Table E.2 (continued) Standard Grades of Bolts

Part b: ASTM Grades for Steel Bolts

ASTM designation	Size range incl.	Proof strength,† kpsi	Tensile strength,† kpsi	Material	Head marking
A307	$\frac{1}{4}$ to 4			Low-carbon steel	
A325 type 1	$\frac{1}{2}$ to 1 $1\frac{1}{8}$ to $1\frac{1}{2}$	85 74	120 105	Medium-carbon steel, Q & T	
A325 type 2	$\frac{1}{2}$ to 1 $1\frac{1}{8}$ to $1\frac{1}{2}$	85 74	120 105	Low-carbon martensite steel, Q & T	
A325 type 3	$\frac{1}{2}$ to 1 $1\frac{1}{8}$ to $1\frac{1}{2}$	85 74	120 105	Weathering steel, Q & T	
A354 grade BC				Alloy steel, Q & T	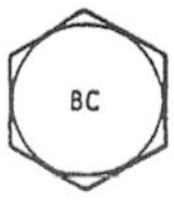
A354 grade BD	$\frac{1}{4}$ to 4	120	150	Alloy steel, Q & T	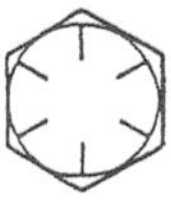
A449	$\frac{1}{4}$ to 1 $1\frac{1}{8}$ to $1\frac{1}{2}$ $1\frac{3}{4}$ to 3	85 74 55	120 105 90	Medium-carbon steel, Q & T	
A490type	$\frac{1}{2}$ to $1\frac{1}{2}$	120	150	Alloy steel, Q & T	
A490type 3				Weathering steel, Q & T	

†Minimum value.

Sources: See "Helpful Hints," by Russell, Burdsall & Ward Corp., Mentor, Ohio 44060; and Chapter 23.

Table E.2 (continued) Standard Grades of Bolts

Part c: Metric Mechanical Property Classes for Steel Bolts, Screws, and Studs

Property class	Size range incl.	Proof strength, MPa	Tensile strength, MPa	Material	Head marking
4.6	M5–M36	225	400	Low- or medium-carbon steel	
4.8	M1.6–M16	310	420	Low- or medium-carbon steel	
5.8	M5–M24	380	520	Low- or medium-carbon steel	
8.8	M16–M36	600	830	Medium-carbon steel, Q & T	
9.8	M1.6–M16	650	900	Medium-carbon steel, Q & T	
10.9	M5–M36	830	1040	Low-carbon martensite steel, Q & T	
12.9	M1.6–M36	970	1220	Alloy steel, Q & T	

Sources: "Helpful Hints," by Russell, Burdsall & Ward Corp., Mentor, Ohio 44060; see also Chapter 23 and SAE standard J1199, and ASTM standard F569.

Table E.3 Steel Pipe Sizes

					Surface Area		*Areas and Weights Cross-sectional*		*Weight*
Nominal Pipe Size, in.	*Outside Diameter, in.*	*Schedule Number or Weight*	*Wall Thickness, in.*	*Inside Diameter, in.*	*Outside, ft^2/ft*	*Inside, ft^2/ft*	*Metal Area, $in.^2$*	*Flow Area, $in.^2$*	*Pipe lb/ft*
¾	1.05	40	0 113	0 824	0 275	0 216	0 333	0 533	1 131
		80	0.154	0 742	0 275	0 194	0 434	0 432	1 474
1	1.315	40	0 133	1 049	0 344	0 275	0 494	0 864	1 679
		80	0 179	0.957	0 344	0 250	0 639	0 719	2 172
1¼	1.660	40	0 140	1.38	0 434	0 361	0 668	1 496	2 273
		80	0 191	1.278	0 434	0 334	0 881	1 283	2 997
1½	1 900	40	0 145	1 61	0 497	0 421	0 799	2 036	2 718
		80	0 200	1 50	0.497	0 393	1 068	1 767	3 632
2	2.375	40	0 154	2 067	0 622	0 541	1 074	3 356	3.653
		80	0 218	1 939	0 622	0 508	1 477	2 953	5 022
2½	2 875	40	0 203	2 469	0 753	0 646	1 704	4 79	5 794
		80	0 276	2.323	0 753	0 608	2 254	4 24	7 662
3	3 5	40	0 216	3 068	0 916	0 803	2 228	7 30	7 58
		80	0 300	2 900	0 916	0 759	3 016	6 60	10 25
3½	4 0	40	0 226	3 548	1 047	0 929	2 680	9 89	9 11
		80	0 318	3 364	1 047	0 881	3 678	8 89	12 51
4	4 5	40	0 237	4 026	1 178	1 054	3 17	12 73	10 79
		80	0 337	3 826	1 178	1 002	4 41	11 50	14 99
5	5.563	10 S	0 134	5 295	1 456	1 386	2 29	22 02	7 77
		40	0 258	5 047	1 456	1 321	4 30	20 01	14 62
		80	0 375	4 813	1 456	1 260	6 11	18 19	20 78
6	6.625	10 S	0 134	6 357	1 734	1 664	2 73	31 7	9 29
		40	0 280	6 065	1 734	1 588	5 58	28 9	18 98
		80	0 432	5 761	1 734	1 508	8 40	26 1	28 58
8	8.625	10 S	0 148	8 329	2 258	2 180	3 94	54 5	13 40
		30	0 277	8 071	2 258	2 113	7 26	51 2	24 7
		80	0 500	7 625	2 258	1 996	12 76	45 7	43 4
10	10 75	10 S	0 165	10 420	2 81	2 73	5 49	85 3	18 7
		30	0 279	10 192	2 81	2 67	9 18	81 6	31 2
		Extra heavy	0 500	9 750	2 81	2 55	16 10	74 7	54 7
12	12.75	10 S	0 180	12 390	3 34	3 24	7 11	120 6	24 2
		30	0 330	12 09	3 34	3 17	12 88	114 8	43 8
		Extra heavy	0 500	11 75	3 34	3 08	19 24	108 4	65 4
14	14 0	10	0 250	13 5	3 67	3 53	10 80	143 1	36 7
		Standard	0 375	13 25	3 67	3 47	16 05	137 9	54 6
		extra heavy	0 500	13 00	3 67	3 40	21 21	132 7	72 1
16	16.0	10	0 250	15 50	4 19	4 06	12 37	188 7	42 1
		Standard	0 375	15 25	4 19	3 99	18 41	182 7	62 6
		extra heavy	0 500	15 00	4 19	3 93	24 35	176 7	82 8
18	18.0	10 S	0 188	17 624	4 71	4 61	10 52	243 9	35 8
		Standard	0 375	17 25	4 71	4 52	20 76	233 7	70 6
		extra heavy	0 500	17 00	4 71	4 45	27 49	227 0	93 5
20	20.0	10 S	0 218	19 564	5 24	5 12	13 55	300 6	46 1
		Standard	0 375	19 25	5 24	5 04	23 12	291	78 6
		extra heavy	0 500	19 00	5 24	4 97	30 6	283 5	104 1
22	22 0	10	0 250	21 50	5 76	5 63	17 1	363	58 1
		Standard	0 375	21 25	5 76	5 56	25 5	355	86 6
		extra heavy	0 500	21 00	5 76	5 50	33 8	346	114 8
24	24 0	10	0 250	23 50	6 28	6 15	18 7	434	63 4
		Standard	0.375	23 25	6 28	6 09	27 8	425	94.6
		extra heavy	0 500	23 00	6 28	6 02	36 9	415	125 5
26	26 0	Standard	0 375	25 25	6 81	6 61	30 2	501	102 6
		extra heavy	0.500	25 00	6 81	6 54	40 1	491	136 2
30	30.0	10	0 312	29 376	7 85	7 69	29 1	678	98 9
		Standard	0 375	29 250	7 85	7 66	34 9	672	118 7
		extra heavy	0 500	29 00	7 85	7 59	46 3	661	157 6
34	34.0	Standard	0 375	33 250	8 90	8 70	39 6	868	134 7
		extra heavy	0 500	33.00	8 90	8 64	52 6	855	178.9
36	36.0	Standard	0 375	35.25	9.42	9 23	42 0	976	142 7
		extra heavy	0 500	35.00	9 42	9 16	55 8	962	189 6
42	42.0	Standard	0 375	41 25	11.0	10 8	49 0	1336	166 7
		extra heavy	0.500	41 00	11.0	10 73	65 2	1320	221.6

*Reprinted with permission, from: "Design Properties of Pipe," ©1958, Chemetron Corporation.

Table E.4 Commercial Copper Tubing*

The following table gives dimensional data and weights of copper tubing used for automotive, plumbing, refrigeration, and heat exchanger services. For additional data see the standards handbooks of the Copper Development Association, Inc., the ASTM standards, and the "SAE Handbook."

Dimensions in this table are actual specified measurements, subject to accepted tolerances. Trade size designations are usually by actual OD, except for water and drainage tube (plumbing), which measures 1/8-in. larger OD. A 1/2-in. plumbing tube, for example, measures 5/8-in. OD, and 2-in. plumbing tube measures 2 1/8-in. OD.

KEY TO GAGE SIZES

Standard-gage wall thicknesses are listed by numerical designation (14 to 21), BWG or Stubs gage. These gage sizes are standard for tubular heat exchangers. The letter *A* designates SAE tubing sizes for automotive service. Letter designations *K* and *L* are the common sizes for plumbing services, soft or hard temper.

OTHER MATERIALS

These same dimensional sizes are also common for much of the commercial tubing available in aluminum, mild steel, brass, bronze, and other alloys. Tube weights in this table are based on copper at 0.323 lb/in^3. For other materials the weights should be multiplied by the following approximate factors.

aluminum	0.30	monel	0.96
mild steel	0.87	stainless steel	0.89
brass	0.95		

Size, OD		*Wall Thickness*			*Flow Area*		*Metal Area,*	*Surface Area*		*Weight,*
in.	*mm*	*in*	*mm*	*gage*	*in.*^{2}	*mm*^{2}	*in*^{2}	*Inside, ft*^{2}*/ft*	*Outside, ft*^{2}*/ft*	*lb/ft*
1/8	3.2	.030	0.76	A	0.003	1.9	0.012	0.017	0.033	0.035
3/16	4.76	.030	0.76	A	0.013	8.4	0.017	0.034	0.049	0.058
1/4	6.4	.030	0.76	A	0.028	18.1	0.021	0.050	0.066	0.080
1/4	6.4	.049	1.24	18	0.018	11.6	0.031	0.038	0.066	0.120
5/16	7.94	.032	0.81	21A	0.048	31.0	0.028	0.065	0.082	0.109
3/8	9.53	.032	0.81	21A	0.076	49.0	0.033	0.081	0.098	0.134
3/8	9.53	.049	1.24	18	0.060	38.7	0.050	0.072	0.098	0.195
1/2	12.7	.032	0.81	21A	0.149	96.1	0.047	0.114	0.131	0.182
1/2	12.7	.035	0.89	20L	0.145	93.6	0.051	0.113	0.131	0.198
1/2	12.7	.049	1.24	18K	0.127	81.9	0.069	0.105	0.131	0.269
1/2	12.7	.065	1.65	16	0.108	69.7	0.089	0.97	0.131	0.344
5/8	15.9	.035	0.89	20A	0.242	156	0.065	0.145	0.164	0.251
5/8	15.9	.040	1.02	L	0.233	150	0.074	0.143	0.164	0.285
5/8	15.9	.049	1.24	18K	0.215	139	0.089	0.138	0.164	0.344
3/4	19.1	.035	0.89	20A	0.363	234	0.079	0.178	0.196	0.305
3/4	19.1	.042	1.07	L	0.348	224	0.103	0.174	0.196	0.362
3/4	19.1	.049	1.24	18K	0.334	215	0.108	0.171	0.196	0.418
3/4	19.1	.065	1.65	16	0.302	195	0.140	0.162	0.196	0.542
3/4	19.1	.083	2.11	14	0.268	173	0.174	0.151	0.196	0.674
7/8	22.2	.045	1.14	L	0.484	312	0.117	0.206	0.229	0.455
7/8	22.2	.065	1.65	16K	0.436	281	0.165	0.195	0.229	0.641
7/8	22.2	.083	2.11	14	0.395	255	0.206	0.186	0.229	0.800
1	25.4	.065	1.65	16	0.594	383	0.181	0.228	0.262	0.740
1	25.4	.083	2.11	14	0.546	352	0.239	0.218	0.262	0.927
1 1/8	28.6	.050	1.27	L	0.825	532	0.176	0.268	0.294	0.655

*Compiled and computed.

Table E.4 (continued) Commercial Copper Tubing*

Size, OD		*Wall Thickness*			*Flow Area*		*Metal Area,*	*Surface Area*		*Weight,*
in.	*mm*	*in.*	*mm*	*gage*	*in*^{2}	*mm*^{2}	*in*^{2}	*Inside, ft*^{2}*/ft*	*Outside, ft*^{2}*/ft*	*lb/ft*
1 1/8	28.6	.065	1.65	16K	0.778	502	0.216	0.261	0.294	0.839
1 1/4	31.8	.065	1.65	16	0.985	636	0.242	0.293	0.327	0.938
1 1/4	31.8	.083	2.11	14	0.923	596	0.304	0.284	0.327	1.18
1 3/8	34.9	.055	1.40	L	1.257	811	0.228	0.331	0.360	0.884
1 3/8	34.9	.065	1.65	16K	1.217	785	0.267	0.326	0.360	1.04
1 1/2	38.1	.065	1.65	16	1.474	951	0.294	0.359	0.393	1.14
1 1/2	38.1	.083	2.11	14	1.398	902	0.370	0.349	0.393	1.43
1 5/8	41.3	.060	1.52	L	1.779	1148	0.295	0.394	0.425	1.14
1 5/8	41.3	.072	1.83	K	1.722	1111	0.351	0.388	0.425	1.36
2	50.8	.083	2.11	14	2.642	1705	0.500	0.480	0.628	1.94
2	50.8	.109	2.76	12	2.494	1609	0.620	0.466	0.628	2.51
2 1/8	54.0	.070	1.78	L	3.095	1997	0.449	0.520	0.556	1.75
2 1/8	54.0	.083	2.11	14K	3.016	1946	0.529	0.513	0.556	2.06
2 5/8	66.7	.080	2.03	L	4.77	3078	0.645	0.645	0.687	2.48
2 5/8	66.7	.095	2.41	13K	4.66	3007	0.760	0.637	0.687	2.93
3 1/8	79.4	.090	2.29	L	6.81	4394	0.950	0.771	0.818	3.33
3 1/8	79.4	.109	2.77	12K	6.64	4284	1.034	0.761	0.818	4.00
3 5/8	92.1	.100	2.54	L	9.21	5942	1.154	0.897	0.949	4.29
3 5/8	92.1	.120	3.05	11K	9.00	5807	1.341	0.886	0.949	5.12
4 1/8	104.8	.110	2.79	L	11.92	7691	1.387	1.022	1.080	5.38
4 1/8	104.8	.134	3.40	10K	11.61	7491	1.682	1.009	1.080	6.51

Table E.5 Standard Gages for Wire, Sheet, and Twist Drills

Gage	(1) Mfrs steel sheet	(2) USS steel sheet (old)	(3) Birmingham or Stub	(4) W & M or Roebling steel wire	(5) AWG or B & S non-ferrous wire or sheet	Numbered twist drills	Copper wire (AWG) Circular mils	Ohms/1000 ft, 77°F	Lb/1000 ft	Sheet steel Lb/sq ft
0000000		0 500		0 4900						20 00
000000		0 469		0 4615	0 580					18 75
00000		0 438		0 4305	0 516					17 50
0000		0 406	454	0 3938	0 460		212,000	0 0500	641 0	16 25
000		0 375	425	0 3625	0 410		168,000	0 0630	508 0	15
00		0 344	380	0 3310	0 365		133,000	0 0795	403 0	13 75
0		0 313	340	0 3065	0 325		106,000	0 100	319 0	12 50
1		0 281	300	0 2830	0 289	0 2280	83,700	0 126	253 0	11 25
2		0 266	284	0 2625	0 258	0 2210	66,400	0 159	201 0	10 625
3	2391	0 250	259	0 2437	0 229	0 2130	52,600	0 201	159 0	10
4	2242	0 234	238	0 2253	0 204	0 2090	41,700	0 253	126 0	9 375
5	2092	0 219	220	0 2070	0 182	0 2055	33,100	0 319	100 0	8 75
6	1943	0 203	203	0 1920	0 162	0 2040	26,300	0 403	79 5	8 125
7	1793	0 188	180	0 1770	0 144	0 2010	20,800	0 508	63 0	7 5
8	1644	0 172	165	0 1620	0 128	0 1990	16,500	0 641	50 0	6 875
9	1495	0 156	148	0 1483	0 114	0 1960	13,100	0 808	39 6	6 25
10	1345	0 141	134	0 1350	0 102	0 1935	10,400	1 02	31 4	5 625
11	1196	0 125	120	0 1205	0 0907	0 1910	8,230	1 28	24 9	5
12	1046	0 109	109	0 1055	0 0808	0 1890	6,530	1 62	19 8	4 375
13	0897	0 0937	095	0 0915	0 0720	0 1850	5,180	2 04	15 7	3 75
14	0747	0 0781	083	0 0800	0 0641	0 1820	4,110	2 58	12 4	3 125
15	0673	0 0703	072	0 0720	0 0571	0 1800	3,260	3 25	9 86	2 813
16	0598	0 0625	065	0 0625	0 0508	0 1770	2,580	4 09	7 82	2 5
17	0538	0 0562	058	0 0540	0 0453	0 1730	2,050	5 16	6 20	2 25
18	0478	0 0500	049	0 0475	0 0403	0 1695	1,620	6 51	4 92	2
19	0418	0 0437	042	0 0410	0 0359	0 1660	1,290	8 21	3 90	1 75
20	0359	0 0375	035	0 0348	0 0320	0 1610	1,020	10 4	3 09	1 50
21	0329	0 0344	032	0 0318	0 0285	0 1590	810	13 1	2 45	1 375
22	0299	0 0312	028	0 0286	0 0253	0 1570	642	16 5	1 94	1 25
23	0269	0 0281	025	0 0258	0 0226	0 1540	509	20 8	1 54	1 125
24	0239	0 0250	022	0 0230	0 0201	0 1520	404	26 2	1 22	1
25	0209	0 0219	020	0 0204	0 0179	0 1495	320	33 0	0 970	0 875
26	0179	0 0187	018	0 0181	0 0159	0 1470	254	41 6	0 769	0 75
27	0164	0 0172	016	0 0173	0 0142	0 1440	202	52 5	0 610	0 6875
28	0149	0 0156	014	0 0162	0 0126	0 1405	160	66 2	0 484	0 625
29	0135	0 0141	013	0 0150	0 0113	0 1360	127	83 4	0 384	0 5625
30	0120	0 0125	012	0 0140	0 0100	0 1285	101	105	0 304	0 5
31	0105	0 0109	010	0 0132	0 0089	0 1200	79 7	133	0 241	0 4375
32	0097	0 0102	009	0 0128	0 0080	0 1160	63 2	167	0 191	0 4063
33	0090	0 0094	008	0 0118	0 0071	0 1130	50 1	211	0 152	0 375
34	0082	0 0086	007	0 0104	0 0063	0 1110	39 8	266	0 120	0 3438
35	0075	0 0078	005	0 0095	0 0056	0 1100	31 5	335	0 0954	0 3125
36	0067	0 0070	004	0 0090	0 0050	0 1065	25 0	423	0 0757	0 2813
37	0064	0 0066		0 0085	0 0045	0 1040	19 8	533	0 0600	0 2656
38	0060	0 0062		0 0080	0 0040	0 1015	15 7	673	0 0476	0 25
39				0 0075	0 0035	0 0995	12 5	848	0 0377	
40				0 0070	0 0031	0 0980	9 9	1070	0 0200	
41				0 0066	0 0028	0 0960				
42				0 0062	0 0025	0 0935				
43				0 0060	0 0022	0 0890				
44				0 0058	0020	0 0860				
45				0 0055	0018	0 0820				
46				0 0052	0016	0 0810				
47				0 0050	0014	0 0785				
48				0 0048	0012	0 0760				
49				0 0046	0011	0 0730				
50				0 0044	0010	0 0700				

Note The present trend, especially for sheet and strip, is to quote thickness as decimal or fraction of an inch rather than gage number ANSI Standard preferred thicknesses have been adopted These preferred sizes for thickness of uncoated sheet, strip, and plate under 0 25 in are as follows 224, 220, 180, 160, 140, 125, .112, 100, 090, 080, 071, 063, 056, 050, 045, 040, 036, 032, 028, .025, .022, .020, .018, 016, .014, 012, 011, 010, 009, 008, 007, 006, .005, 004

KEY: (1) Manufacturer's standard for hot- and cold-rolled uncoated carbon steel sheet and most alloy steel sheet
(2) U S Standard for cold-rolled steel strip and stainless and nickel alloy sheet
(3) Birmingham or Stub for hot-rolled carbon and alloy steel strip and tubing
(4) Washburn and Moen, Roebling, or U S Steel for steel wire
(5) American wire gage or Brown and Sharpe for non-ferrous wire, sheet, and strip

Dimensions in approximate decimals of an inch.

Table E.6 Properties of Typical Gaseous and Liquid Commercial Fuels*

Gaseous fuels	*Composition, percent by volume*								*Mol wt of fuel*	*Theor air/fuel ratio by wt*	*Higher heating value, Btu/lb$_m$*	*Density, lb$_m$/ft^3*
	H_2	N_2	O_2	CH_4	CO	CO_2	C_2H_4	C_6H_6				
Blast furnace gas	1.0	60.0	—	—	27.5	11.5	—	—	29.6	0.667	1,170	.075 5[a]
Blue water gas	47.3	8.3	0.7	1.3	37.0	5.4	—	—	16.4	3.759	6,550	.042 2[a]
Carb. water gas	40.5	2.9	0.5	10.2	34.0	3.0	6.1	2.8	18.3	7.299	11,350	.046 6[a]
Coal gas	54.5	4.4	0.2	24.2	10.9	3.0	1.5	1.3	12.1	10.87	16,500	.031 1[a]
Coke-oven gas	46.5	8.1	0.8	32.1	6.3	2.2	3.5	0.5	13.7	17.24	17,000	.032 6[a]
Natural gas (15.8% C_2H_6)	—	0.8	—	83.4	—	—	—	—	18.3	17.24	24,100	.045 1[a]
Producer gas	14.0	50.9	0.6	3.0	27.0	4.5	—	—	24.7	14.29	2,470	.063 6[a]

Liquid commercial fuels	*Vapor*		*Gravity, API, 60°F*	*Distillation*			*Flash point, °F*	*Viscosity, centi-stokes, 100°F*	*Mol wt of fuel*	*Theor air/fuel ratio by wt*	*Higher heating value, Btu/lb$_m$*	*Density, lb$_m$/ft^3*
	c_p, 60°F	*c_p/c_v, 60°F*		*10%, °F*	*90%, °F*	*End point, °F*						
	(approximately)											
Gasoline	0.4	1.05	63	121	320	397	0	—	113	14.93	20,460	43.8[b]
Gasoline	0.4	1.05	63	118	330	410	0	—	126[c]	14.97	20,260	46.1[b]
Kerosene	0.4	1.05	41.9	370	510	546	130	—	154[c]	14.99	19,750	51.5[b]
Diesel oil (1-D)	0.4	1.05	42	—	550	—	100	1.4–2.5	170	15.02	19,240	54.6[b]
Diesel oil (2-D)	0.4	1.05	36	—	540–576	—	125	2.0–5.8	184	15.06	19,110	57.4[b]
Diesel oil (4-D)	0.4	1.05	—	—	—	—	130	5.8–26.4	198	14.93	18,830	59.9[b]

[a] Based on dry air at 25°C and 760 mm Hg
[b] Based on H_2O at 60°F, 1 atm (ρ = 62.367 lb$_m$/ft^3)
[c] Estimated
* Abridged from: "Engineering Experimentation", G L Tuve and L C Domholdt, McGraw-Hill Book Company, 1966, and "The Internal Combustion Engine", 2nd ed., C F Taylor and E S. Taylor, International Textbook Co., 1961

Note. For heating value in J/kg, multiply the value in Btu/lb$_m$ by 2324. For density in kg/m^3, multiply the value in lb/ft^3 by 16.02.

Table E.7 Combustion Data for Hydrocarbons*

Hydrocarbon	Formula	Higher heating value (vapor), Btu/lb_m	Theor air/fuel ratio, by mass	Max flame speed, ft/sec	Adiabatic flame temp (in air), °F	Ignition temp (in air), °F	Flash point, °F	Flammability limits (in air), % by volume	
				PARAFFINS OR ALKANES					
Methane	CH_4	23875	17.195	1.1	3484	1301	gas	5.0	15.0
Ethane	C_2H_6	22323	15.899	1.3	3540	968-1166	gas	3.0	12.5
Propane	C_3H_8	21669	15.246	1.3	3573	871	gas	2.1	10.1
n-Butane	C_4H_{10}	21321	14.984	1.2	3583	761	−76	1.86	8.41
iso-Butane	C_4H_{10}	21271	14.984	1.2	3583	864	−117	1.80	8.44
n-Pentane	C_5H_{12}	21095	15.323	1.3	4050	588	< −40	1.40	7.80
iso-Pentane	C_5H_{12}	21047	15.323	1.2	4055	788	< −60	1.32	9.16
Neopentane	C_5H_{12}	20978	15.323	1.1	4060	842	gas	1.38	7.22
n-Hexane	C_6H_{14}	20966	15.238	1.3	4030	478	−7	1.25	7.0
Neohexane	C_6H_{14}	20931	15.238	1.2	4055	797	−54	1.19	7.58
n-Heptane	C_7H_{16}	20854	15.141	1.3	3985	433	25	1.00	6.00
Triptane	C_7H_{16}	20824	15.141	1.2	4035	849	–	1.08	6.69
n-Octane	C_8H_{18}	20796	15.093	–	–	428	56	0.95	3.20
iso-Octane	C_8H_{18}	20770	15.093	1.1	–	837	10	0.79	5.94
				OLEFINS OR ALKENES					
Ethylene	C_2H_4	21636	14.807	2.2	4250	914	gas	2.75	28.6
Propylene	C_3H_6	21048	14.807	1.4	4090	856	gas	2.00	11.1
Butylene	C_4H_8	20854	14.807	1.4	4030	829	gas	1.98	9.65
iso-Butene	C_4H_8	20737	14.807	1.2	–	869	gas	1.8	9.0
n-Pentene	C_5H_{10}	20720	14.807	1.4	4165	569	–	1.65	7.70
				AROMATICS					
Benzene	C_6H_6	18184	13.297	1.3	4110	1044	12	1.35	6.65
Toluene	C_7H_8	18501	13.503	1.2	4050	997	40	1.27	6.75
p-Xylene	C_8H_{10}	18663	13.663	–	4010	867	63	1.00	6.00
				OTHER HYDROCARBONS					
Acetylene	C_2H_2	21502	13.297	4.6	4770	763-824	gas	2.50	81
Naphthalene	$C_{10}H_8$	17303	12.932	–	4100	959	174	0.90	5.9

*Based largely on "Gas Engineers Handbook" American Gas Association, Inc Industrial Press 1967

REFERENCES

"American Institute of Physics Handbook", 2nd ed. D.E. Gray, Ed. McGraw-Hill Book Company, 1963.

"Chemical Engineers' Handbook", 4th ed., R H Perry, C.H. Chilton, and S.D Kirkpatrick, Eds. McGraw-Hill Book Company, 1963

"Handbook of Chemistry and Physics", 53rd ed. R.C. Weast, Ed., The Chemical Rubber Company, 1972, gives the heat of combustion of 500 organic compounds.

"Handbook of Laboratory Safety", 2nd ed., N V. Steere, Ed., The Chemical Rubber Company, 1971

"Physical Measurements in Gas Dynamics and Combustion", Princeton University Press, 1954

Note For heating value in J/kg, multiply the value in Btu/lb_m by 2324. For flame speed in m/s, multiply the value in ft/s by 0.3048.

Index

K

L

M

N

O

P

R

S

T

U

V

Milton Keynes UK
Ingram Content Group UK Ltd.
UKHW051950071024
449327UK00026B/2247

9 780367 399160